The Immunoglobulin Receptors and their Physiological and Pathological Roles in Immunity

The authors would like to acknowledge Drs. Peter Artymiuk, Dennis Barton, Geoff Ford and Jenny Wood for their kind contribution to the cover and Dr. Ian McKenzie for helpful suggestions on the format of the book.

Immunology and Medicine Series

VOLUME 26

Series Editors:

Dr. Graham Bird, *Churchill Hospital, Oxford, UK*
Professor Keith Whaley, *University of Leicester, Leicester, UK*

A list of titles in the series can be found at the end of this volume.

The Immunoglobulin Receptors and their Physiological and Pathological Roles in Immunity

Edited by

Jan G.J. van de Winkel
Utrecht University, Utrecht, The Netherlands

and

P. Mark Hogarth
Austin Research Institute, Melbourne, Australia

SPRINGER-SCIENCE+BUSINESS MEDIA, B.V.

Library of Congress Cataloging-in-Publication Data is available.

ISBN 978-94-010-6106-3 ISBN 978-94-011-5018-7 (eBook)
DOI 10.1007/978-94-011-5018-7

Printed on acid-free paper

Contents

Preface

Antibodies are crucial to the fine specificity of the immune system. Effective functioning of these molecules requires interaction with immune cells, mediated by antibody (Fc) receptors. These receptors thus provide a critical link between the humoral and cellular branches of the immune system. Although the existence of Fc receptors was postulated more than 20 years ago, our insight into their molecular and functional complexity has increased greatly over the past decade. Engagement of Fc receptors triggers a plethora of biological functions as diverse as antibody transport, regulation of serum antibody levels, triggering of phagocytosis and cytolysis, induction of inflammatory cascades, and down-modulation of immune responses. In addition, Fc receptor heterogeneity between individuals, or polymorphism, has been identified as a crucial diagnostic and a prognostic factor in a number of human diseases. Our present understanding of their physiological and pathological roles, furthermore, has stimulated the development of novel types of immunotherapy for cancer, autoimmune and infectious diseases.

This book provides a state-of-the-art overview of the crucial coordinating role of Fc receptors in immunity. These recent developments are discussed by a panel of internationally recognized experts, each of whom has contributed significantly to this rapidly growing field.

J.G.J. van de Winkel and P.M. Hogarth

Series Editor's Note

Over the last 30 years there has been a continuing and exponential increase in the understanding of basic immunology mechanisms and their application to the practice of medicine. The aim of this series is to provide a comprehensive and up-to-date review of the scientific basis and clinical relevance of a variety of immunologically-based diseases. The series, producing three new volumes a year, will cover the majority of specialist areas of medicine with diseases of immunological importance and over a four-year cycle will aim continuously to update knowledge and experience in the individual specialist areas. The series is designed for the general or specialist physician or the clinical scientist with an interest in clinical or related problems. We intend to provide in single volumes a synthesis of information which is otherwise difficult to assemble from original papers which are often produced in a variety of specialist medical and immunological journals.

G. Bird and K. Whaley

List of Contributors

A. ASTIER
Dana Farber Cancer Institute
Boston
USA

R. BAKER
Haematology Department
Royal Perth Hospital
University of Western Australia

C. BOUCHARD
IMT Phillipps-Universität Marburg
Germany

W. BOYLE
University of Melbourne
Australia

E. J. BROWN
Department of Medicine
Division of Infectious Diseases
Washington University School of Medicine
St. Louis, MO 63110
USA

J. C. CAMBIER
Department of Pediatrics
National Jewish Medical & Research Center
1400 Jackson Street
Denver, Colorado
USA

O. H. CHOI
Laboratory of Molecular Immunology
National Heart, Lung and Blood Institute
NIH, Building 10, Room 8N109
9000 Rockville Pike
Bethesda, MD 20892
USA

D. H. CONRAD
Virginia Commonwealth University
Department of Microbiology and Immunology
Box 980678
Richmond, VA
USA

M. DAËRON
Unit Inserm 255
Immunologie Cellulaire et Clinique
Institut Curie
26 rue d'Ulm
75005 Paris
France

B. DALE
Coagulation Unit
Royal Perth Hospital
University of Western Autralia

M. De HAAS
Central Laboratory of the Netherlands
Red Cross Blood Transfusion Service and
Laboratory for Clinical and Experimental Immunology
Academic Medical Center
University of Amsterdam
The Netherlands

M. W. FANGER
Department of Microbiology
Dartmouth Medical School
1 Medical Center Drive
Lebanon, NH 03756
USA

W. H. FRIDMAN
Unit Inserm 255
Immunologie Cellulaire et Clinique
Institut Curie
26 rue d'Ulm
75005 Paris
France

J. GALON
National Institute of Helath
Bethesda, Maryland
USA

A. GAVIN
Austin Research Institute
Studley Road
Heidelberg, Victoria 3084
Australia

A. GREINACHER
Institute for Immunology and Transfusion Medicine
Ernst-Moritz-Arndt University
Sauerbruchstraβe
Greifswald
Germany

C. A. GUYRE
Department of Physiology
Dartmouth Medical School
1 Medical Center Drive
Lebanon, NH 03756
USA

P. M. GUYRE
Department of Physiology
Dartmouth Medical School
1 Medical Center Drive
Lebanon, NH 03756
USA

B. HEYMAN
Department of Genetics and Pathology
Uppsala University
Uppsala
Sweden

P. M. HOGARTH
Austin Research Institute
Kronheimer Building
Studley Road
Heidelberg
Victoria, 3084 Australia

P. G. HOLBROOK
Laboratory of Molecular Immunology
National Heart, Lung and Blood Institute
National Institutes of Health
Building 10, Room 8N109
9000 Rockville Pike
Bethesda, MD 20892
USA

Z. K. INDIK
Department of Medicine
University of Pennsylvania School of Medicine
Philadelphia, PA 19104
USA

R. P. JUNGHANS
Harvard Institute of Human Genetics
Harvard Medical School
Division of Hematology-Oncology
Beth Israel Medical Center
Boston, MA
USA

C. S. KAETZEL
Department of Pathology and Laboratory Medicine
University of Kentucky
MS117 Chandler Medical Center
Lexington, KY 40536-0084
USA

R. P. KIMBERLY
University of Alabama at Birmingham
1900 University Boulevard
Birmingham, Alabama 35294-006
USA

H. R. KOENE
Central Laboratory of the Netherlands
Red Cross Blood Transfusion Service and Laboratory for Clinical and Experimental Immunology
Academic Medical Center
University of Amsterdam
The Netherlands

R. G. LYNCH
Department of Pathology and Microbiology
College of Medicine
University of Iowa
Iowa City, Iowa 52242
USA

S. E. McKENZIE
DuPont Hospital for Children
Wilmington, DE 19899
USA

H. METZGER
National Institutes of Health
10 Center Drive MSC 1820
Bethesda
MD 20892-1820
USA

H. C. MORTON
Laboratory for Immunohistochemistry and Immunopathology (LIIPAT)
Institute of Pathology
University of Oslo
The National Hospital
Rikshospitalet
N-0027 Oslo
Norway

LIST OF CONTRIBUTORS

K. MOSTOV
Department of Anatomy
University of California San Francisco
513 Parnassus Avenue
San Francisco, CA 94143-0452
USA

J. O'SHEA
Department of Physiology
Dartmouth Medical School
1 Medical Center Drive
Lebanon, NH 03756
USA

M. S. POWELL
Austin Research Institute
Studley Road
Heidelberg, Victoria 3084
Australia

J. V. RAVETCH
Laboratory of Molecular Genetics and Immunology
Rockefeller University, 1230 York Avenue
New York, NY 10021
USA

R. REPP
Department of Medicine III
University of Erlangen-Nuremberg
Germany

D. ROOS
Central Laboratory of the Netherlands
Red Cross Blood Transfusion Service and Laboratory for Clinical and
Experimental Immunology
AMC, University of Amsterdam
The Netherlands

J. E. SALMON
Department of Medicine
Cornell University College of Medicine and Hospital for Special Surgery
541 E 71st Street
New York, NY
USA

M. SANDOR
Department of Pathology
University of Wisconsin
Madison, WI 53706
USA

C. SAUTÈS
Unit Inserm 255
Immunologie Cellulaire et Clinique
Institut Curie
26 rue d'Ulm, 75005 Paris
France

A. D. SCHREIBER
Department of Medicine
University of Pennsylvania School of Medicine
Philadelphia, PA 19104
USA

N. E. SIMISTER
Rosenstiel Center for Basic Biomedical Sciences
W.M. Keck Institute for Cellular Visualization and Biology Department
Brandeis University
Waltham, MA 02254-9110
USA

T. TAKAI
Department of Molecular Embryology
Institute of Development, Aging and Cancer
Tohoku University, Seiryo 4-1
Sendai 980-8575
Japan

J.-L. TEILLAUD
Unit Inserm 255
Immunologie Cellulaire et Clinique
Institut Curie
26 rue d'Ulm
75005 Paris
France

J. G. J. van de WINKEL
Department of Immunology and Medarex Europe
University Hospital Utrecht
Heidelberglaan 100
3584 CX Utrecht
The Netherlands

M. van EGMOND
Department of Immunology
University Hospital Utrecht
Heidelberglaan 100
3584 CX Utrecht
The Netherlands

A. E. C. Kr. von dem BORNE
Department of Clinical Hematology
Academic Medical Center
Amsterdam
The Netherlands

P. K. WALLACE
Department of Microbiology
Dartmouth Medical School
1 Medical Center Drive
Lebanon, NH 03756
USA

1
Introduction to the field

W. H. FRIDMAN

Since the elucidation of the structure of antibodies, it has been recognized that immunoglobulins (Ig) are bifunctional molecules which assemble the products of several genes, allowing the almost infinite diversity of antigen recognition and a large array of effector and regulatory functions. Families of V genes encode the variable regions of Ig heavy (H) and light (L) chains whereas a set of C genes encodes the H and L constant regions which define Ig isotypes. Associations of the variable regions form the antigen-binding site while the C terminal constant regions form the Fc part of the Ig molecule and bear the sites for functional activities. The use of proteolytic enzymes has allowed separation of the Fc region from the rest of the molecule. An Fc-less Ig is a pure antigen-binding unit (and has therefore be called F(ab) when monovalent and $F(ab)'_2$ when divalent) whereas intact molecules, after binding to antigen, exert multiple effector and regulatory functions [1,2].

The biological effects of antibodies depend on their interaction with effector systems. These interactions, or the triggering of effector mechanisms, are initiated by their binding to antigen. Thus, with the exception of Ig transport through epithelial barriers such as intestine or placenta, antigen-free antibodies have no activity. When they encounter antigen, and bind to it, antibodies may trigger an array of functions, some of them being independent of host cells, most requiring the participation of cells of the immune system as effectors or targets (Table 1.1).

The cell-independent activities of antibodies are the result of the triggering of a cascade of plasmatic enzymes, the complement (C) system which is initiated by the binding of the first component of complement (C1) on the Fc portion of certain Ig isotype when fixed on antigens. Complement activation then results in the destruction of invading agents such as Gram-positive bacteria, in the lysis of

J.G.J. van de Winkel and P.M. Hogarth (eds.), The Immunoglobulin Receptors and their Physiological and Pathological Roles in Immunity. 1–7.

Table 1.1 Biological functions of antibodies

Cell independent
Lysis of bacteria
Cytotoxicity
Activation of inflammatory mediators
Cell dependent
Transport
Antigen internalization
Endocytosis of immune complexes
Phagocytosis of opsonized particles
Exocytosis of preformed granules
Release of inflammatory mediators
Antibody dependent cell-mediated cytotoxicity
Immunoregulation
Inhibition of activating receptors
Induction of cytokines
Induction of soluble receptors

virus-infected cells, in the release of inflammatory mediators and in the production of chemotactic agents. It represents an effective defence against many infections.

The cell-dependent activities of antibodies are the consequence of their binding, or the binding of immune complexes, to specific receptors on cells of the immune system. Since, as for complement, Ig interact with these receptors through their Fc portion, the latter have been called Fc receptors (FcR) [3]. With the exception of the antigen-independent Ig transport across epithelial barriers, all the cell-dependent activities of Ig require FcR aggregation by antigen–antibody complexes. This aggregation may be one or two steps. Thus, low affinity receptors only bind immune complexes and binding and aggregation are made in one – or two very close – step(s), while high affinity receptors bind monomeric Ig. This binding does not trigger any detectable event. Only when antigen aggregates FcR-bound Ig does it result in biological activity. In any case, these cell-dependent biological effects can be classified into three categories [4]:

(1) the internalization of antigen in soluble (endocytosis) or particulate (phagocytosis) form. Antigen internalization may result in antigen presentation, initiating immune responses and, in the case of phagocytosis of microorganisms, in the destruction of the latter;

(2) the exocytosis of preformed granules may result in the release of inflammatory mediators by mast cells and basophils or of cytotoxic agents by natural killer cells;

(3) the interaction of antibodies with immune cells also regulates immune reactions by exerting a negative feed-back effect on cell activation, by inducing the production of immunoregulatory cytokines or by enhancing the release of soluble receptors which may buffer immune reactions.

The types of biological activity which are triggered by antibodies depend on their capacity to interact with the various effector molecules such as complement and FcR. They also depend on the expression of FcR on different cells and on the machinery of the latter. It has been known for a long time that the different Ig isotypes exert different functions, which reflect their differential ability to bind to complement and FcR. Table 1.2 summarizes a few characteristics of human Ig isotypes and their major functions. The complement-binding isotypes (IgM, IgG_1, IgG_3) are potent in defence against bacteria, the mast cell- and basophil-binding IgE trigger inflammatory and allergic responses, NK cells are activated by IgG while IgA are the main antibodies in mucosal defences. These observations clearly lead to the concept that there are different FcR for the different IgG isotypes expressed on different cells [5].

The existence of such receptors has, for a long time, be disputed. It is know accepted, although all FcR have not yet been molecularly characterized. Rather than presenting a careful analysis of the historical dates of the field, I shall underline what, in my view, have been the major contributions of the FcR scientists to the comprehension of the immune system.

LESSONS FROM THE FcR FIELD

Specificity

Mirroring the set of genes encoding Ig heavy chains and determining Ig isotypes is a set of genes coding for the corresponding FcR. Grossly, the specificity of FcR matches that of Ig isotypes. However, several different FcR bind to the same isotype and a given FcR may react with several isotypes. Thus IgE binds to two structurally different Fcε receptors (FcεR), a high affinity receptor (FcεRI) member of the Ig superfamily [6] and a low affinity receptor (CD23 or FcεRII) member of the family of animal lectins [7]. Several different Fcγ receptors (FcγR) bind with different affinities to several IgG isotypes, and, under some conditions even IgE [5]. Thus, the ligand recognition by FcR in both highly specific and, as often in the immune system, has a degree of degenerescence.

Structure

Most FcR belong to the Ig supergene family [5]. It is striking, however that several Ig-binding proteins, with affinity and specificity for their ligand belong to different families. Low affinity receptors for IgE or for IgD are lectins [7,8]. The diversity of cell-associated Ig-binding structures reflects that of the bacterial FcR-like molecules (Protein A, Protein G, etc.) or the parasite-associated Ig binding sites [9].

Table 1.2 Physiochemical properties of human immunoglobulin isotypes

Property	*IgG*	*IgA*	*IgM*	*IgD*	*IgE*
Usual molecular form	Monomer	Monomer, dimer	Pentamer, hexamer	Monomer	Monomer
Sub-classes	G1, G2, G3, G4	A1, A2	None	None	None
Molecular weight (kDa)	150	160, 400	950, 1150	175	190
Serum level (mg/ml)	9.5–12.5	1.5–2.6	0.7–1.7	<0.04	0.0003
Complement activation	+	–	++	–	–
Biological properties	Placental transfer secondary Ab for most anti-pathogen responses	Secretory immunoglobulin	Primary Ab responses to microorganisms	Marker for mature B cells	Allergy, anti-parasite responses

Expression

Not only are the FcR encoded by genes from different families, but they also have, in general, a broad cellular distribution. FcR are more or less ubiquitously expressed, with a rather restricted pattern for FcεRI and a wide pattern for FcγRII, but any FcR, even the more restricted, is physiologically found in several cell types [10]. In certain conditions, such as tumour cells, or virally infected cells, an ectopic expression of FcR has been reported which may have functional implications [11]. Another striking feature of FcR expression is that, with the possible exception of NK cells, a single cell bears several different FcR, for different Ig isotypes, allowing its interaction at the same time with antibodies from different classes [5,10]. This property opens the possibility of a fine tuning of the activities of these cells at different times of an immune response, depending on the isotype composition of the antibodies produced.

Triggering

There are high affinity receptors for IgE, IgG and IgA which bind monomeric Ig and low affinity receptors for IgE, IgG, IgD, IgM and probably IgA which bind immune complexes [5]. The pioneering experiments of Metzger and his colleagues, and their comparison with that of other groups have established a law for FcR, and other receptors: whatever the affinity for ligand is, receptor-mediated cell triggering requires receptor aggregation [6]. For FcR, this is achieved by antigen under the form of immune complexes (one step signalling) or when binding to cell-associated FcR (two step signalling).

Diversity of FcR functions

The diversity of the cell-dependent antibody functions (Table 1.1) reflects that of their corresponding FcR, which in turn reflects FcR structure. Specific sequences in the intracellular region, or associated chains, of FcR allow receptor-mediated phagocytosis, endocytosis or cellular activation [12]. Here again, different strategies are used by different FcR to exert these functions. A subfamily of FcR, specially those associated with a common γ chain, endocytose and initiate cell activation through the binding of protein tyrosine kinases to a phosphorylated tyrosine [13,14], part of a consensus immunoreceptor tyrosine-based activation motif (ITAM), found in the T cell and B cell receptors. Other FcR, such as CD23, utilize completely different motifs to endocytose and possibly trigger [7]. Finally, extracellular interactions of FcR with other signalling molecules such as complement receptors (CR3 and CR4) may initiate different pathways [15], explaining, in part, the triggering capacities of a tail-less, GPI-linked receptor, such as the human FcγRIII (CD16) found on neutrophils.

Negative regulation of cell activation

The understanding of the mechanisms underlining the negative IgG feed-back for antibody production or for immunotherapy of allergic patients has led to the identification, in the intracellular region of FcγRIIb, of a motif required for the negative regulation of ITAM-containing receptors, such as the T and B cell receptors or the activating FcR [16–18]. This inhibitory motif also contains a tyrosine which needs to be phosphorylated, and then binds a phosphatase [19, 20]. It has thus been called 'immunoreceptor tyrosine-based inhibitory motif (ITIM)' [18]. A recent interesting series of findings have led to the concept that ITIMs are not restricted to FcγR, but are present on many different regulatory molecules, on NK cells, mast cells, T cells or B cells. They all function in a similar general may, binding SH2-containing phosphatases, thus expanding a finding from the FcR field to a general concept for immunoregulation [21].

Soluble receptors

Another general concept originated from the FcR field, that of the production and circulation in biological fluids of soluble receptors. First, demonstrated for FcγR, soluble forms were reported for all FcR isotypes [22]. They are usually generated by proteolytic cleavage at the cell membrane but can also originate from alternative splincing of mRNA, deleting the transmembrane exon [23]. An interesting property of soluble FcR is that, in addition to their possible role in inhibiting Ig-mediated functions, some of them (CD23, CD16) have been reported to bind, at least *in vitro*, to non-Ig ligands and eventually exert Ig-independent biological activities [24,25]. Again, these observations initially made in the study of FcR, can be extended to many receptors, including cytokine receptors, which soluble forms may exert agonistic or antagonistic effects.

References

1. Nezlin R. Immunoglobulin Structure and Function. In: Van Oss CJ, Van Regenmortel MHV, eds, Immunochemistry. New York: Marcel Dekker Inc; 1994:3–45.
2. Fridman WH. Structure and function of immunoglobulins. In: Fridman WH, Sautès C, eds, Cell Mediated Effects of Immunoglobulins. Heidelberg: RG Landes Company, Springer; 1997:1–28.
3. Paraskevas F, Lee ST, Orr KB, Israels G. A Receptor for Fc on Mouse B-Lymphocytes. J Immunol. 1972;108:1319–27.
4. Fridman WH, Sautès C. Cell Mediated Effects of Immunoglobulins. Heidelberg: RG Landes Company, Springer; 1997:1–201.
5. Hulett MD, Hogarth MP. Molecular basis of the Fc receptor function. Adv Immunol. 1994;57:1–127.
6. Metzger H, Alcaraz G, Hohman R, Kinet JP, Pribluda V, Quarto R. The receptor with high affinity for immunoglobulin E. Annu Rev Immunol. 1986;4:419–70.
7. Capron M, Grangette C, Torpier G, Capron A. The second receptor for IgE in eosinophil effector function. Chem Immunol. 1989;47:128–78.
8. Coico RF, Siskind GW, Thorbecke J. Role of IgD and T cells in the regulation of humoral immune response. Immunol Rev. 1988;105:45–68.

9. Fridman WH. Fc Receptors and immunoglobulin-binding factors. FASEB J. 1991;5:2684–90.
10. Ravetch JV, Kinet JP. Fc receptors. Annu Rev Immunol. 1991;9:457–92.
11. Ran M, Katz B, Kimchi N et al. *In vivo* acquisition of Fcγ RII expression on polyoma virus transformed cells derived from tumors of long latency. Cancer Res. 1991;51:612–8.
12. Daëron M. Fc receptor biology. Annu Rev Immunol. 1997;15:203–34.
13. Bonnerot C, Amigorena S, Choquet D, Pavlovich R, Choukroun V, Fridman WH. Role of associated γ chain in tyrosine kinase activation via murine FcγRIII. EMBO J. 1992;11:2747–57.
14. Amigorena S, Salamero J, Davoust J, Fridman WH, Bonnerot C. Tyrosine-containing motif that transduces cell activation signals also determines internalization and antigen presentation via type III receptors for IgG. Nature. 1992;358:337–41.
15. Zhou MJ, Poo H, Todd III RF, Petty HR. Surface-bound immune complexes trigger transmembrane proximity between complement receptor type 3 and the neutrophil's cortical microfilaments. J Immunol. 1992;148:3550–3.
16. Amigorena S, Bonnerot C, Drake J et al. Cytoplasmic domain heterogeneity and functions of IgG Fc receptors in B lymphocytes. Science. 1992;256:1808–12.
17. Muta T, Kurosaki T, Misulovin Z, Sanchez M, Mussenzweig MC, Ravetch JV. A13-aminoacid motif in the cytoplasmic domain of FcγRIIB modulates B cell receptor signalling. Nature. 1994;368:70–3.
18. Daëron M, Latour S, Malbec O et al. The same tyrosine-based inhibition motif in the intracytoplasmic domain of FcγRIIB, regulates negatively BCR-, TCR- and FcR-dependent cell activation. Immunity. 1995;3:635–46.
19. D'Ambrosio D, Hippen KH, Minskoff SA et al. Recruitment and activation of PTP1C in negative regulation of antigen receptor signaling by FcγRIIB1. Science. 1995;268:293–6.
20. Ono M, Bolland S, Tempst P, Ravetch JV. Role of the inositol phosphatase SHIP in negative regulation of the immune system by the receptor FcγRIIB. Nature. 1996;383:263–6.
21. Scharenberg AM, Kinet JP. The emerging field of receptor-mediated inhibitory signaling: SHP or SHIP? Cell. 1996;87:961–4.
22. Fridman WH, Rabourdin-Combe C, Neauport-Sautès C, Gisler RH. Characterization and function of T cell Fcγ receptor. Immunological Rev. 1981;56:51–88.
23. Sautès C. Soluble Fc receptor. In: Fridman WH, Sautès C, eds, Cell Mediated Effects of Immunoglobulins. Heidelberg: RG Landes Company, Springer; 1997:139–63.
24. Lecoanet-Henchoz S, Gauchat J, Aubry J et al. CD23 regulates monocyte activation through a novel interaction with the adhesion molecules CD11b-CD18 and CD11c-CD18. Immunity. 1995;3:119–25.
25. Galon J, Gauchat JF, Mazières N et al. Soluble Fcγ receptor type III (FcγRIII, CD16) triggers cell activation through interaction with complement receptors. J Immunol. 1996;157:1184–92.

Part A

PHYSIOLOGICAL ROLE OF Fc RECEPTORS

2
Molecular basis for the interaction of Fc receptors with immunoglobulins

A. GAVIN, M. HULETT and P. M. HOGARTH

INTRODUCTION

Fc receptors form one of the front lines of membrane receptors in immune defence mechanisms by providing humoral immunity with powerful cell-mediated effector mechanisms. These are activated by the binding of antigen–antibody complexes to specific Fc receptors. The interaction of immunoglobulins with Fc receptors has profound biological effects, including: activation of cell mediated killing, induction of mediator release, uptake, removal and destruction of antibody-coated particles, including pathogens, transport of immunoglobulins, and the regulation of immunity. Since the molecular cloning of Fc receptor genes and the development of sophisticated molecular analytical techniques, there have been major advances in understanding the structure of Fc receptors and defining the molecular basis of the interaction between these receptors and their ligands. In this chapter we will attempt to provide an overview on what is known about this interaction and we will especially review the work relating to the high affinity IgG receptor FcγRI. In recent years there have been several extensive reviews of Fc receptor structure, especially in relation to the IgG receptors and FcεRI and we would draw the reader's attention to these [1–5].

Fc RECEPTOR STRUCTURE

General comments

Molecular cloning of Fc receptors has indicated that for the most part the

J.G.J. van de Winkel and P.M. Hogarth (eds.), The Immunoglobulin Receptors and their Physiological and Pathological Roles in Immunity. 11–35.

leukocyte Fc receptors are composed of extracellular globular domains, virtually all of which belong to the Ig superfamily. The leukocyte FcR shows considerable amino acid identity including the IgG receptors, FcγRI, FcγRII and FcγRIII, the IgA receptor FcαRI, the IgE receptor FcεRI. With the exception of FcγRI the members of these classes contain two extracellular domains – FcγRI contains three. Extensive analysis of the IgG receptors and FcεRI indicates that similar regions in these molecules are involved in the interaction with Ig (see below). This is perhaps not surprising given that their ligands, i.e. immunoglobulins, also show conservative evolutionary history. The exception to this is FcεRII which is a type 2 membrane glycoprotein with lectin-like activity. This receptor bears no homology with the other leukocyte receptors.

Several other Fc receptors not found on leukocytes have been defined. Such receptors being involved principally in the transport of immunoglobulin (i.e. FcRn and the poly Ig receptor). Although these receptors are also part of the Ig superfamily, they are not related to the leukocyte Fc receptors. The poly Ig receptor is composed of typical Ig-like domains, but these are a subset distinct from those of the leukocyte receptors. The FcRn is very closely related to the major histocompatibility complex Class I molecules. Thus in the development of this diverse array of receptors, the primordial Ig fold has acted as a platform on which has been built a series of diverse receptor structures which have been fine tuned to have unique specificities and affinities for the various immunoglobulins.

The receptors which have been most extensively characterized are FcγRII and FcγRIII as well as FcεRI [6–10]. These analyses have indicated that several regions of the receptors are involved in the interaction with Ig and that these regions are located in similar positions in the different receptors. Most of the data used in these analyses has been generated on the basis of:

(1) analyses of genetic polymorphisms that influence the Fc receptor Ig interaction;

(2) structure function studies by traditional mutagenesis or the use of chimeric receptors;

(3) epitope mapping and antibody blocking studies.

A definitive analysis of this interaction by x-ray crystallography or NMR has not been reported for any of the leukocyte receptors. Although several models have been proposed [1–4,9,10].

The non-leukocyte FcRn (which is MHC related) has been crystallized and sites of interaction have been defined precisely. It is clear that the interaction between FcRn and Ig is quite distinct from that of the leukocyte Fcγ receptors and IgG. Thus a definitive answer by X-ray crystallography on the interaction with Ig and leukocyte Fc receptors will be extremely interesting.

LIGAND BINDING INTERACTIONS OF LEUKOCYTE FcR

FcγRI (CD64)

Human FcγRI (CD64) is a 70 kDa glycoprotein with a predicted core size of 45 kDa [11–15]. Mouse FcγRI is also 70 kDa (similar to that of the human FcγRI) and has a protein core of 49 kDa after removal of N-linked glycosylation [16] (Table 2.1). Both human and mouse FcγRI consists of three Ig-like extracellular domains [17,18]. The first two extracellular domains are homologous with the two domains of FcγRII and FcγRIII, but the third extracellular domain is distinct and confers the unique high affinity IgG binding characteristics of FcγRI. Both human and mouse FcγRI are non-covalently associated with the homo-dimer of the γ subunit initially described for FcεRI [19–23]. Unlike FcεRI α chain and FcγRIIIa, FcγRI does not require the presence of the γ subunit for assembly and transport to the cell surface for efficient surface expression *in vitro* [17,18,24]. However *in vivo*, in the absence of the γ subunit, expression is reduced (80% decrease) [25]. Moreover, a role for the γ subunit in enhancing ligand binding affinity has been recently demonstrated for human and mouse FcγRI [23].

Ligand binding affinity and specificity

Setting FcγRI apart from the other FcγR is its capacity to bind monomeric IgG with high affinity(10^7–10^9 M^{-1}) [11,17,23,26,27]. Human FcγRI exhibits varying affinities for the different IgG subclasses, binding human $IgG_1 \geqslant IgG_3 >> IgG_4$ but not human IgG_2 [11,28,29]. Human FcγRI also binds the mouse IgG_{2a}, IgG_3 and rat IgG_{2b} [27,28,34].

Interestingly, the pentameric, calcium-binding acute phase serum protein C-reactive protein (CRP) is also able to bind to human FcγRI on U937 cells (monocyte cell line) [35] and to COS-7 cells transfected with human FcγRI cDNA [36]. CRP also binds C-polysaccharide, phosphatidylcholine (PC) and through the PC binding site is also able to interact directly with chromatin [37], fibronectin [38], histones [39], laminin [40]. Although CRP binds to monocytes via a distinct receptor (CRP-R) [41,42], CRP is also able to interact with hFcγRI. The binding of CRP to FcγRI on monocytes is inhibited by IgG, at high concentrations, CRP can also inhibit IgG binding to hFcγRI, indicating that FcγRI has a relatively low affinity for CRP and due to the direct competition, the interaction is probably at or near the IgG binding site.

It is also interesting to note that human FcγRI on monocytes and transfected cells is able to interact with a fluorescent dye used in flow cytometry, PE-CY5 [43]. This binding was not observed of other dyes, could be blocked by incubation with 20% human serum, aggregated IgG, a monoclonal antibody that blocks ligand binding, implicating the interaction of FcγRI with PE-CY5 is at or near the ligand binding site [43].

Mouse FcγRI displays high affinity binding for monomeric IgG, although the measured affinity constant is slightly lower (10^7–10^8 M^{-1}) than that of human

Table 2.1 Characterization of FcγRI[a]

	Human			Mouse
Genes	*FcγRIA*	*FcγRIB*	*FcγRIC*	*FcγRI*
Transcripts	hFcγRIa	hFcγRIb1[b], hFcγRIb2[b]	hFcγRIc[b]	mFcγRI
Chromosome location	1q21.1	1q21.1	1q21.1	3
Extracellular domains	3 (d1,d2,d3)	2[c] (d1,d2)	2[c] (d1,d2)	3 (d1,d2,d3)
Receptor form	Transmembrane	Transmembrane, soluble	Soluble	Transmembrane
Molecular mass (kDa)	70	ND[d]	ND	70
Protein core size (kDa)	45	22[e]	24[e]	49
Affinity for IgG (Ka)	10^7–10^9 M^{-1}	ND, (< 10^7 M^{-1})[e]	ND	10^7–10^8 M^{-1}
Specificity				
Human IgG	1 ≥ 3 > 4 > > > 2	ND	ND	3 > 1 > 4 > > > 2
Mouse IgG	2a=3 > > > 1,2b			2a > > > 3,1,2b
Regions of FcγRI involved in binding IgG	Domain 3 provides high affinity binding	ND	ND	Domain 3 provides high affinity and specificity of binding. Domains 1 and 2 can bind IgG with low affinity
Main regions of IgG[f] involved in receptor binding	Hinge proximal Cγ2 domain (Leu^{234}–Ser^{239}) of human IgG_3, supported by Cγ2 Gly^{316}–Lys^{338}	ND	ND	ND

[a]See text for details

[b]Transcripts by PCR only

[c]No known protein product, structures predicted from cDNA sequence

[d]ND, not determined

[e]Protein tested after transcript transfected into COS cells

[f]Numbering based on Reference 189

FcγRI [18,22,23,44,45]. Mouse IgG_{2a} appears to be the only subclass that mFcγRI binds well, binds human $IgG_1 \geqslant IgG_3 >> IgG_4$ and *not* IgG_2) [18,45–47]. A polymorphic mouse FcγRI isolated from non-obese diabetic mice is able to bind $mIgG_{2a}$ and $hIgG_1$ with a 10-fold increase in affinity [22].

Regions of FcγRI involved in IgG binding

The first two domains of mFcγRI can bind IgG, but the third domain appears to confer both specificity and high affinity binding. Mutant FcγRI and naturally occurring splice variants lacking the third domain only bind complexes and in the mouse have radically altered specificity [45,48]. Interaction with the γ subunit also influences ligand binding affinity 2–5-fold. This effect appears to be mediated through the transmembrane domain [23].

Regions of IgG involved in FcγRI binding

Human FcγRI binds only one IgG molecule [49] and studies using hybrid monoclonal antibodies have demonstrated that only one heavy chain is required for the binding [33,50]. In IgG the sequence Leu^{234}, Leu^{235}, Gly^{236}, Gly^{237}, Pro^{238}, Ser^{239} of Cγ2 is involved in the interaction with hFcγRI, with the Cγ3 domain contributing a stabilizing role [51–57]. Human IgG_1 and IgG_3, mouse IgG_{2a}, rat IgG_{2b} and rabbit IgG all contain these residues, and all bind FcγRI. The segment Gly^{316} to Ala^{339} plays a role in the IgG:FcγRI interaction [28,54, 58,59].

Residues in contact with the oligosaccharides of Cγ2 also influence FcγRI binding: e.g. Asp^{265} makes contact with the primary GlcNac sugar attached to Asn^{297}, and replacement of this residue with alanine results in loss of FcγRI binding [60]. Further studies demonstrated that even single changes of amino acid residues within the oligosaccharide interaction site can influence the galactosylation and sialylation of the IgG saccharide chains, consequently influencing interaction with FcγRI [61].

Analysis of the CRP sequence reveals a region that has similarity with the 'binding motif' of IgG. CRP contains the sequence Tyr Leu Gly Gly Pro, very similar to the IgG sequence Leu Leu Gly Gly Pro involved in the FcγRI interaction. Mutation of the CRP Tyr Leu Gly Gly Pro sequence to that of Glu Leu Gly Gly Pro ablated the FcγRI binding, implying this region was indeed involved [36]. Both CRP and IgG may stimulate inflammatory cells via FcγRI. Mice do not have a direct equivalent of CRP [62] and the homologous mouse acute phase protein, serum amyloid P component (SAP) does not contain the Tyr Leu Gly Gly Pro sequence [63,64]. Mouse SAP is unable to inhibit human CRP binding to Fc receptors on mouse macrophages [65], so the interaction of human acute phase proteins and human FcγRs appears to be different to the mouse acute phase proteins and mouse FcγRs.

FcγRII (CD32)

Human and mouse FcγRII are members of 40 kDa glycoprotein family [66–77]. Consisting of two Ig-like domains, FcγRII (CD32) exhibits high avidity binding of complexed IgG but does not bind IgG monomer in conventional whole cell assays. Numerous isoforms within this class of receptor have been identified, all of which contain two Ig-like domains. The extracellular and transmembrane regions are highly conserved between isoforms, although the cytoplasmic domains demonstrate heterogeneity, suggesting distinct signalling and functional capacities of each isoform; these have been reviewed [2,78]. Characteristics of FcγRII are summarized in Table 2.2.

Ligand binding affinity and specificity

Human and mouse FcγRII bind monomeric IgG poorly ($Ka < 10^7\ M^{-1}$) [68,73,79,80], but bind complexed IgG avidly. The affinity of FcγRII for IgG complexes can be modified by protease treatment and by GM-CSF (on eosinophils), suggesting that FcγRII may act as a 'standby' receptor that is activated by these processes at sites of inflammation [76,81,82].

Complexes of the human IgG isotypes IgG_3 and IgG_1 are bound well by hFcγRIIa, which binds IgG_4 poorly [79,83,84] (Table 2.2). Only one allelic form of FcγRIIa1 will bind IgG_2;FcγRII-H^{131} [85]. The binding of mouse IgG_1 and rat IgG_{2b} is also dependent upon the allelic form of FcγRIIa1: as only the FcγRII-R^{131} form can bind mIgG_1, whereas for rat IgG_{2b}, only the FcγRII-H^{131} binds strongly [33]. Both allelic forms can bind mouse IgG_{2a} and IgG_{2b} [86,87]. The binding specificity of the hFcγRIIb1 and FcγRIIb2 isoforms has been assayed and these isoforms bind: human $IgG_3 >$ h$IgG_4 \geqslant$ h$IgG_1 >>>$ hIgG_2 [78,88]. Mouse isotype binding to hFcγRIIb1 has also been assessed with the following binding pattern: $IgG_{2a} = IgG_{2b} >> IgG_1$ [78].

Like human FcγRII, mouse FcγRII binds mouse IgG_1, IgG_{2a}, IgG_{2b}, is unable to bind mIgG_3 [45,47,89–95]. A distinct receptor for mouse IgG_3 has been described on macrophages but has not yet been isolated [96]. Human IgG_1 and IgG_3 bind preferentially to mouse FcγRII over IgG_2 and IgG_4 [46]. Low affinity binding of mouse IgE to mouse FcγRII but not human FcγRII has also been demonstrated [97].

Regions of FcγRII involved in IgG binding

Using IgG binding polymorphisms, the mapping of the epitopes of ligand blocking monoclonal antibodies, the use of chimeric receptors and site-directed mutagenesis, the binding site of human FcγRII has been localized to the extracellular domain 2 [1,6–8,77,86,87,98].

Genetic studies indicate that residue 131 (either Arg or His) is involved in Ig binding (see above) [85–87]. The *Ly-17* alloantigenic system of the mouse depends on residues 116 and 161 and antibodies detecting this *Ly-17* alloantigen

Table 2.2 Characterization of FcγRII[a]

	Human			Mouse
Genes	*FcγRIIA*	*FcγRIIB*	*FcγRIIC*	*FcγRII*
Transcripts	hFcγRIIa1, hFcγRIIa2	hFcγRIIb1, hFcγRIIb2, hFcγRIIb3	hFcγRIIc	mFcγRIIb1, FcγRIIb2, FcγRIIb3
Alleles	HR, LR[b]	–	–	Ly17.1, Ly17.2
Chromosome location	1q23-24	1q23-24	1q23-24	1
Extracellular domains	2	2	2	2
Receptor f orms	Transmembrane, soluble	Transmembrane	Transmembrane	Transmembrane, soluble
Molecular mass (kDa)	40	40	40	40-60
Protein core size (kDa)	31	29,27	31	33,29
Affinity for IgG (Ka)	$<10^7$ M^{-1}	$<10^7$ M^{-1}	ND[c]	$<10^7$ M^{-1}
Specificity:				
Human IgG	LR 3≥1=2>>>4 HR 3≥1>>2>4	3≥1>4>2	ND	3≥1>2>>4
Mouse IgG	LR 2a=2b>>1 HR 2a=2b=1	2a=2b≥1	ND	1=2a=2b>>>>3 mouse IgE also
Regions of FcγRII involved in binding IgG	Domain 2 F-G loop Asn^{154}–Ser^{161} B-C loop Ser^{109}–Val^{116} C-E′ loop Phe^{129}–Thr^{135}	ND	ND	Anti-Ly17.2 blocking antibody recognizes Leu^{116} and Leu^{161} in domain 2
Main regions of IgG involved in receptor binding	Hinge proximal Cγ2 domain (Leu^{234}–Gly^{237}) of human IgG$_3$	ND	ND	ND

[a]See text for details; [b]High responder and low responder alleles of hFcγRIIa; [c]NC, not determined

block Fc binding implying these residues are close to the Ig binding site [71,72,77,100–102].

Segments of FcγRII involved in the interaction with IgG have been identified by the use of chimeric Fc receptors, wherein similar regions of functionally distinct but structurally homologous receptors were exchanged. These demonstrated that domain 2 of FcγRII and FcεRI are the principal Ig interactive domains. In FcγRII the segment containing residues Asn^{154}–Ser^{161} plays a major role as well as residues Ser^{109}–Val^{116} and Phe^{129}–Thr^{135}. Domain 1 probably acts by either contributing secondary binding sites or stabilising the receptor:Ig complex [6–8]. By molecular modelling analysis of FcγRII (based on CD4 structure), the binding sites within domain 2 appear to form a hydrophobic pocket on loops within or near the interface between domains 1 and 2 (Figure 2.1) [8].

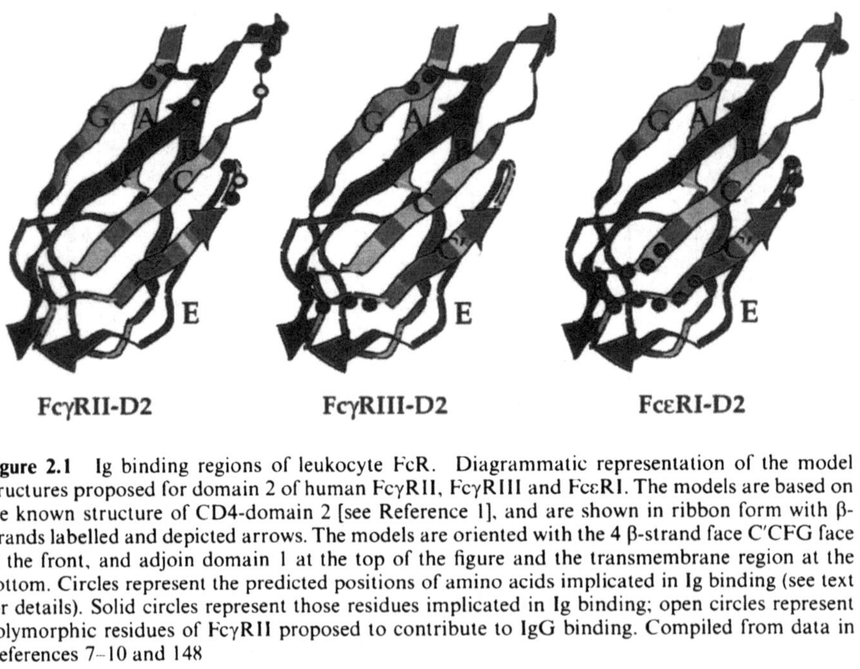

Figure 2.1 Ig binding regions of leukocyte FcR. Diagrammatic representation of the model structures proposed for domain 2 of human FcγRII, FcγRIII and FcεRI. The models are based on the known structure of CD4-domain 2 [see Reference 1], and are shown in ribbon form with β-strands labelled and depicted arrows. The models are oriented with the 4 β-strand face C'CFG face at the front, and adjoin domain 1 at the top of the figure and the transmembrane region at the bottom. Circles represent the predicted positions of amino acids implicated in Ig binding (see text for details). Solid circles represent those residues implicated in Ig binding; open circles represent polymorphic residues of FcγRII proposed to contribute to IgG binding. Compiled from data in References 7–10 and 148

Regions of ligand involved in FcγRII binding

Mapping of the regions of the Fc portion of IgG involved in the binding to FcγRII has been difficult due to the nature of the low affinity interaction with IgG. Aglycosylation of the Cγ2 domain of the human isotypes IgG_1, IgG_3 leads to loss of hFcγRII binding, implicating this region of the Fc portion [83].

Experiments using chimeric mIgE/hIgG$_1$ antibodies indicated that both Cγ2 and Cγ3 domains from hIgG$_1$ were required to maintain binding of sensitised erythrocytes by hFcγRII on K562 cells (FcγRI⁻) [53]. As is the case for hFcγRI, the region Leu ^{234}–Leu^{234}–Gly^{236} –Gly^{237} in Cγ2 of IgG is also involved in the binding interaction with hFcγRII although the region may interact with FcγRII differently. Other regions of Ig also have to interact, as isotypes such as mouse IgG$_1$, IgG$_{2b}$ and human IgG$_2$ which do not contain intact Leu Leu Gly Gly consensus sequences can still bind to hFcγRII.

Carbohydrate moieties and non-covalent interactions with protein are also important in FcγRII recognition. The replacement of Asp^{265} in human IgG$_3$ or mouse IgG$_{2b}$ which makes contact with the primary GlcNac sugar attached to Asn^{297} resulted in loss of recognition of hFcγRII and mouse FcγRII, respectively [60].

FcγRIII (CD16)

Two isoforms of membrane bound human FcγRIII that exhibit different cellular distribution have been defined. These isoforms each contain two Ig-like domains and exhibit low to intermediate binding affinity to monomeric IgG and can bind IgG complexes avidly. Signalling pathways involving subunit association with FcγRIIIa have been elucidated, this class of FcγR is the only one expressed on, involved in, NK cell functions (see Table 2.3 for a summary of characteristics).

Biochemical analyses of human FcγRIII (CD16) revealed a heterogeneous protein ranging in size from 50–80 kDa [103–107]. The heterogeneity was caused by differences in the size of the polypeptide of different isoforms and also differential glycosylation of two allelic forms of human FcγRIIIb, namely hFcγRIIIb ^{NA1} and hFcγRIIIb ^{NA2} [108,109]. Mouse FcγRIII also demonstrated heterogeneous size due to glycosylation, ranging from 40–60 kDa with a protein core of 33 kDa [90].

Two genes, FcγRIIIA and FcγRIIIB, encode the two forms of hFcγRIII (FcγRIIIa and FcγRIIIb respectively) that are selectively expressed on different cell populations [108,110–113] and differ by only six amino acids. Both receptors have two Ig-like extracellular domains. The hFcγRIIIA gene encodes a transmembrane receptor while the hFcγRIIIB gene encodes a receptor anchored with a glycophosphatidylinositol linkage [108,114,115]. At position 203, a serine specifies the formation of a glycophosphatidylinositol anchor (GPI) in FcγRIIIb, while hFcγRIIIa contains a phenylalanine at this position, leading to the retention of the transmembrane spanning and cytoplasmic tail domains in hFcγRIIIa [110,114,116,117].

Both human and mouse transmembrane forms of FcγRIII require subunit association for efficient surface expression and signalling. In macrophages, human FcγRIIIa is associated with disulphide linked homodimers of the γ subunit for FcεRI and in NK cells is associated with homo- and or heterodimers of the γ subunit and the ζ subunit of the CD3-TCR complex [24,118,119].

Table 2.3 Characterization of FcγRIII[a]

	Human		*Mouse*
Genes	*hFcγRIIIA*	*hFcγRIIIB*	*mFcγRIII*
Transcripts	hFcγRIIIa	hFcγRIIIb	mFcγRIII
Alleles	–	FcγRIIIb$^{NA1^b}$, FcγRIIIb^{NA2}	–
Chromosome location	1q23-24	1q23-24	1
Extracellular domains	2	2	2
Receptor form	Transmembrane	GPI-anchored	Transmembrane
Molecular mass (kDa)	50-80	50-80	40-60
Protein core size (kDa)	33	29 (NA1) 33 (NA2)	33
Associated subunits	γ-chain, FcεRI ζ-chain, TCR/CD3	–	γ-chain FcεRI β-chain FcεRI
Affinity for IgG (Ka)	2×10^7 M^{-1}	$<10^7$ M^{-1}	$<10^7$ M^{-1}
Specificity:			
Human IgG	1=3>>>2=4	1=3>>>2=4	3=1>2>>4
Mouse IgG	3>2a>2b>>1	3>2a>2b>>1	1=2a=2b>>>3
Regions of FcγRIII involved in binding IgG	ND[c]	Domain 2 F-G loop Lys^{162}-Val^{164} F β strand Arg^{156} C-C′ loop Trp^{124}-Tyr^{131}	ND
Main regions of IgG involved in receptor binding	Not clearly defined but Cγ2 and/or Cγ3 play a role	ND	ND

[a]See text for details

[b]Neutrophil antigen system comprising two allelic forms of hFcγRIIIb

[c]Not determined

In the mouse, however, FcγRIII can only associate with the γ subunit and the mouse ζ subunit of the CD3-TCR complex does not rescue surface expression of mouse FcγRIII in fibroblasts [118–120].

Ligand binding affinity and specificity

Both isoforms of human FcγRIII display similar affinities for monomeric IgG with hFcγRIIIa exhibiting an affinity of $Ka \approx 2 \times 10^7$ M^{-1} [121] and hFcγRIIIb a slightly lower affinity of $Ka \approx 10^7$ M^{-1} [103,108,122]. The higher affinity of FcγRIIIa for IgG has been attributed to the association of this receptor form with the γ subunit [23]. The specificity of human FcγRIII isoforms is similar to that of hFcγRII, bind human IgG_1=IgG_3 preferentially over IgG_2 and IgG_4 [107,123]. The specificity for mouse IgG reveals preference for IgG_3 over IgG_{2a} over IgG_{2b} with poor binding measured for IgG_1 complexes [124–126]. The stoichiometry of the FcγRIII:IgG interaction is 1:1 [127].

Regions of FcγRIII involved in IgG binding

While there are many similarities between FcγRII and FcγRIII, there are also several key differences. Like FcγRII, epitopes detected by ligand blocking monoclonal antibodies map within domain 2 [9,128]. Further mutagenesis studies demonstrate that residues Lys^{162} and Val^{164} in the putative FG loop are also important in binding (Figure 2.1) [10]. However, unlike FcγRII, residues 124–131 within a postulated C–C′ loop are a major binding site [9]. This region would probably be juxtaposed to the membrane and is distinct from the regions of FcγRII – B–C loop, C′–E and F–G loop which may lie at the d1:d2 interface [8–10]. These observations raise the possibility that FcγRII and FcγRIII interact with IgG in a slightly different way.

Regions of ligand involved in FcγRIII binding

The region of IgG involved in the interaction with FcγRIII has not been clearly identified, however there is some evidence that this may be distinct from the region involved in binding to FcγRII. Conflicting reports indicate that either the interface between Cγ2 and Cγ3 of IgG was required for binding as protein A, which binds this region, was able to block the binding of IgG to FcγRIII [129], or a region in Cγ3 was involved with a second region in Cγ2 important for ADCC function of the receptor [130]. Carbohydrate on the Fc portion also influences interactions with FcγRIII as a deglycosylated human IgG_3 was unable to bind FcγRIII on NK cells (as assessed by ADCC) [131].

FcεRI

FcεRI, a multi-subunit tetrameric complex comprised of α, β, γ dimer subunits, is intimately involved in the triggering of type 1 hypersensitivity reactions upon cross-linking of the FcεRI complex with IgE and multivalent antigen. The α subunit of FcεRI is homologous to the other FcRs, containing two Ig-like domains, yet the α subunit of FcεRI binds IgE with extremely high affinity [reviewed in Reference 132]. The β and γ subunits have been shown to associate with other FcR and are necessary for FcεRIα chain expression. A summary of FcεRI characteristics can be found in Table 2.4. The FcεRIα subunit requires the presence of either the γ subunit (human FcεRIα) or both the β and γ subunits (rat and mouse FcεRIα) for efficient surface expression [133–135].

Ligand binding affinity and specificity

Human, rat and mouse FcεRI bind monomeric IgE of the same species with high affinity ($Ka \approx 10^{10}\ M^{-1}$) [133,136,137]. Although human FcεRI can bind IgE from all three species, mouse and rat FcεRI are able to bind only rodent IgE and not human IgE [138]. Hydrodynamic studies of human IgE with a soluble form of FcεRIα suggests the stoichiometry of IgE binding to receptor is 1:1 [139].

Table 2.4 Characterization of FcεRI[a]

Genes	*Human FcεRI*	*Rat FcεRI*	*Mouse FcεRI*
Associated subunits[b]	α, β, γ	α, β, γ	α, β, γ
Chromosome location	1q23 (α), 11q13 (β), 1q23 (γ)	ND	1 (α), 19 (β), 1 (γ)
Extracellular domains	2 (α)	2 (α)	2 (α)
Receptor form	Transmembrane	Transmembrane	Transmembrane
Molecular mass (kDa)	45-65 (α), 32 (β),7-8 (γ)	45-65 (α), 32 (β), 7-9 (γ)	45-65 (α), 32 (β), 7-9 (γ)
Protein core size (kDa)	26.4 (α), 25.9 (β), 7-8 (γ)	25.2 (α), 27 (β), 7.8 (γ)	25.8 (α), 25.9 (β), 7.8 (γ)
Affinity for IgE (Ka)[c]	$>10^{10}$ M^{-1}	$>10^{10}$ M^{-1}	$>10^{10}$ M^{-1}
Specificity IgE	Human IgE Rat IgE Mouse IgE	Rat IgE and mouse IgE only	Rat IgE and mouse IgE only
Regions of FcεRI involved in binding mouse IgE	Domain 2 B-C Trp^{87}–Lys^{128} C′-E Tyr^{129}–Asp^{145} F-G Lys^{154}–Glu^{161}	ND[d]	ND
Regions of FcεRI involved in binding human IgE	Domain 2 Ser^{93}–Phe^{104} Arg^{111}–Glu^{125} Tyr^{129}–Ile^{134} Lys^{154}–Glu^{161}		
Main regions of IgE involved in receptor binding	Cε3 Asp^{330}–Leu^{363} with support from Cε2 and Cε4	Cε3 with support from Cε2. Cε4 not rerquired for mouse IgE binding	ND

[a]See text for details

[b]α-chain is the IgE binding subunit

[c]Affinity of receptors for their species specific ligand

[d]Not determined

Regions of FcεRIα involved in IgE binding

Chimeric receptors comprising extracellular domains of FcεRIα, transmembrane and cytoplasmic tails of other receptors, indicated only the extracellular domains of the alpha subunit were required for IgE binding, that the β and γ subunits did not appear to play a role in direct binding [6,140]. Soluble receptors comprising only the two extracellular domains are also able to bind IgE with high affinity [141–144]. As mentioned above, mapping of the epitopes of monoclonal antibodies that were able to block IgE binding by the FcεRIα subunit indicated the second Ig-like extracellular domain played a role in IgE binding [145]. Chimeric receptors wherein extracellular domains between structurally homologous receptors (FcγRII, FcγRIII and FcεRI) were exchanged, clearly demonstrated that multiple regions within domain two were involved in IgE interactions by human FcεRI [6,7,146].

The role of domain 1 in the high affinity interaction with IgE was demonstrated by exchanging extracellular domains between hFcεRIa and hFcγRIIa. Replacement of domain 1 of FcεRIa with that of domain 1 of FcγRIIa generated a receptor that could only bind complexes of IgE, indicating the principle binding site was in domain 2 FcεRI, but that domain 1 FcεRI must contribute to the high affinity [6,147]. In this regard, the rat and human FcεRI appear to interact slightly differently with ligand, however, as receptors with the domain 1 of human FcεRI replaced with that of human FcγRII exhibited only low affinity to binding to monomeric IgE [6]. This was in contrast to rat FcεRI where the exchange of the first domain with that of human FcγRIII did not appear to alter affinity for IgE [146].

At least three independent subregions within FcεRI domain 2 have been identified that are directly capable of binding IgE. These regions encompass Trp^{87}–Lys^{128}, $Tyr^{129}Asp^{445}$, and Lys^{154}–Glu^{161}, when inserted into FcγRIIa are able to impart mIgE binding to hFcγRIIa [6]. Similar regions were also identified in a study using human FcεRI and FcγRIII, where complete loss of IgE binding was observed when regions of FcεRI, namely Ser^{93}–Phe^{104}, Arg^{111}–Glu^{125} and Asp^{123}–Ser^{137} were substituted by equivalent sequence of FcγRIII [146]. A peptide derived from FcεRI has recently been identified that is able to inhibit IgE binding and is derived from the C′loop, C′strand, E loop sequences [148].

Molecular modelling based on the CD4 domain 2 structure reveals these interactive regions are likely to be situated in loop regions juxtaposed at the interface with domain 1, specifically the F–G, C′–E and B–C loops, with contributions also from the B and C strands [6,146]. The influence of domain 1 on the Ig binding by domain 2 is consistent with the interdomain interface comprising the IgE binding site of human FcεRIα, is homologous to the regions proposed as the binding site for FcγRII and FcγRIII [8,10].

Regions of ligand involved in FcεRIα binding

The isolation of the region on the Fc portion of IgE that is involved in binding to FcεRI has received much attention, with the aim of developing therapeutic drugs to block the type 1 hypersensitivity reaction triggered after IgE:FcεRI crosslinking. Many laboratories have addressed these issues using a variety of technologies and this is comprehensively covered in Chapter 3 of this volume and readers are also directed to references [148–157].

FcαRI

Like most of the leukocyte Fc receptors, the FcαRI is a membrane spanning glycoprotein with two extra-cellular domains each of which belongs to the C-2 set of Ig-like domains [see Reference 158 for review]. These domains are related to the Fcγ receptors and FcεRI and contain six possible sites of N-linked glycosylation [159] but some O-glycosylation has also been suggested [160].

Like other FcRs, FcαRI also co-associates with the FcεRI γ subunit in monocytes [161].

Both IgA_1 and IgA_2 bind to FcαRI as does dimeric IgA [162–164] and the affinity of this interaction has been calculated to be approximately $5 \times 10^7\ M^{-1}$ and has been determined both on normal leukocytes and transfected cells (Table 2.5). In addition there is no evidence for the binding of other Ig classes to this receptor.

The regions of FcαRI responsible for interaction with its ligand are largely unknown. However, a number of different isoforms of this receptor have been cloned and there has been some analysis of the effect of sequence differences within these isoforms on the interaction of this receptor with IgA [165,166]. Analysis of an unusual splice variant which contains only the second Ig-like domain indicates this domain is able to bind secretory IgA but not serum IgA, suggesting that there is a decrease in the overall affinity of the interaction [165]. This report indicates there is clearly a role for the first domain in the interaction with IgA. Such a role has not been reported for the FcγR but has been implied in the rat FcεRI [146]. However a more precise localisation of the binding site is presently not available.

The region of IgA known to interact with FcαRI has been recently defined. Recent evidence has also emerged that the interface between Cα2 and Cα3 of IgA is involved in the interaction with FcαRI [167]. This is also quite distinct from the interaction between FcγR and IgG wherein the hinge proximal region of Cγ2 is important in that interaction and not the Cγ2/Cγ3 interface which is the site for binding of staphylococcal protein A. Clearly then, there may be distinct differences between the IgG receptors and the IgA receptors in the way they interact with their respective ligands.

Table 2.5 Characteristics of human FcαRI (CD89)[a]

Transcripts	FcαRIa	FcαRIb
Receptor form	Transmembrane	Cell associated and soluble (lacks TM but not GPI-linked)
Extracellular domains	2	2
Molecular mass (kDa)	55-72[b]	?
Associated subunits	FcεRI γ subunit	
Affinity for IgA (Ka)	$2 \times 10^7\ M^{-1}$	?
Specificity	IgA monomers or multimers. Human not mouse IgA	IgA monomer and multimers
Regions of FcαRI involved in interaction with IgA	? Domain 1	? Domain 1[c]
Regions of IgA involved in receptor binding	Cα2-Cα3 interface	Cα2-Cα3 interface[c]

[a]See text for details

[b]Cell type dependent

[c]Presumed by analogy with FcαRIa

NON-LEUKOCYTE Fc RECEPTORS

FcRn

The transepithelial transport of IgG in neonatal suckling rodents is performed by the MHC related Fc receptor, FcRn. This is a 45 kDa membrane glycoprotein which is structurally related to MHC class 1 but bears little relationship to the leukocyte IgG receptors – FcγR. The binding of IgG to FcRn is pH dependent with binding occurring at acid pH and subsequent dissociation occurring at neutral pH [168–171]. This receptor has been identified both in rodents and also in man [172] and has been shown also to be responsible for the regulation of a catabolism of plasma IgG and is likely to be the previously defined 'protection receptor' defined by Brambell [173,174]. The rat FcRn shows greater affinity for IgG_{2a} ($K_a=10^8$ M^{-1}) than do IgG_{2b}, IgG_1 and IgG_{2c}, which are bound with decreasing affinity (Table 2.6 [175].

X-ray crystallographic studies combined with mutagenesis studies have provided much information on the nature of the FcRn:IgG interaction [176–178]. IgG Fc interacts with FcRn in areas quite distinct from the equivalent of the Class I peptide groove or the CD8 interactive site. The α2 domain of FcRn is the principal contact area and residues are contributed by α1 and probably α3 as well as β_2-microglobulin [178]. It is clear, however, from mutagenesis studies that residues within FcRn which lie outside the Ig binding site affect interaction with Ig, probably affecting the dimerization of this receptor which is part of its natural biology [179,182]. Moreover, the pH dependence of the FcRn:Ig interaction is likely to be due to the involvement of histidine residues both in FcRn and Ig which influence FcRn dimerisation as well as the receptor:ligand interaction [179,182]. The Cγ2 and Cγ3 interface have also been shown to be involved in the interaction with FcRn and in this region several key residues have been identified [172–180]. This is distinct from the area of Ig that is involved in interaction with the leukocyte Fcγ receptors, i.e. the hinge proximal region, but is similar to the region involved in the interaction with staphylococcal protein A.

Polymeric immunoglobulin receptor (Poly Ig receptor; pIgR)

The transepithelial transport of polymeric immunoglobulins (i.e. dimeric IgA or IgM) is performed by a specialised receptor known as the poly Ig receptor (pIgR; Table 2.6). This receptor is a membrane spanning molecule consisting of five immunoglobulin-like domains, domains 1–4 being related to Ig-variable regions and domain 5 being related to Ig-constant regions [181] and much of the biology has been reviewed [182]. The interaction of the poly Ig receptor with IgA has been well defined as enzymatic cleavage of domains 1 and 2 once bound to IgA forms a secretory component of the IgA complex. This interaction of receptor/secretory component with IgA has been localized to domain 1 by mutagenesis and the use of synthetic peptides [183–186] with the principle region involving a highly conserved 23 amino acid sequence that is found in

Table 2.6 Characteristics of non-leukocyte FcR

	Poly Ig receptor	*FcRn*
Receptor form	Transmembrane (soluble cleavage products secretory component)	Transmembrane
Extracellular domains	5 (I–V)	3 ($\alpha1$, $\alpha2$, $\alpha3$)
Associated subunits	None	β_2-microglobulin
Affinity	High	2×10^8 M^{-1} (pH6)
Specificity	Polymeric Ig i.e. dimeric IgA IgM	Rat $IgG_{2a} > IgG_{2b}$, IgG_1, IgG_{2c}[a]
Regions involved in interaction with Ig	DI (15–37)	$\alpha1$, $\alpha2$, $\alpha3$, β_2[b]
Regions of Ig interacting with FcR	IgH hinge	$C\gamma2$–$C\gamma3$ interface[b]

[a]Rat FcRn and rat Ig clones

the mammalian receptors analysed [187]. It is interesting to note that the 23 amino acid binding site is located in a possible loop region of domain 1, i.e. the equivalent of the complementarity determining region-1 of the immunoglobulin variable region, thus, domain 1 is related to the variable regions of Ig. It is also interesting that in the case of FcγRII and FcεRI, the binding regions thereto seem to be composed of multiple loop regions.

The role of the other domains is not clear however it is apparent that the binding of polymeric Ig to the pIgR does induce conformational changes (measured by loss of antibody epitopes) which involve domains 2 and 3 [188].

CONCLUSIONS

As the study of the Fc receptor:Ig interaction progresses, it is becoming increasingly clear that Fc receptors that belong to the Ig superfamily have a great many differences yet a great many similarities in the structural components which enable this interaction to take place e.g. it appears that the loop regions of the Ig fold in FcγRII, FcεRI and the poly Ig receptor are all responsible or play a great role in the interaction between these receptors and their ligands. This is clearly depicted in a comparison of the regions of FcγRII, III and FcεRI involved in binding (Figure 2.1). The leukocyte Fc receptors are major effector molecules in immunity and the analyses of their function also provides an excellent example of the way in which sub-families of the Ig superfamily have evolved to generate different ligand binding activities. Comparison of these receptors is interesting as each has two domains and they share a considerable number of amino acids. However, their specificities are distinct (e.g. IgG versus IgE) and their affinities for ligand are spectacularly different – four orders of magnitude, 10^6 M^{-1} versus 10^{10} M^{-1}. However, the interdomain interface is clearly important in FcεRI and FcγRII. In FcγRIII there is a major difference in the region of interaction compared with FcγRII. This makes sense in that FcγRII and FcγRIII are often expressed on the same cell i.e. neutrophils, macrophages. The binding of multivalent complexes to such a cell might most effectively induce a response by engaging both classes of receptor simultaneously, rather than induce receptor competition. The Ig fold which is the fundamental structural unit of most members of the Ig superfamily has been fine tuned to develop a series of receptors of diverse function.

While a huge effort has gone into understanding the interaction which takes place between these receptors and Ig, the definitive co-crystallization has been performed only for FcRn and we eagerly await the solution of the genuine structure of leukocyte Fc receptors.

References

1. Hogarth PM, Hulett MD, Ierino FL, Tate B, Powell MS, Brinkworth RI. Identification of the immunoglobulin binding regions (IBR) of FcγRII and FcεRI. Immunol Rev. 1992;125:21–35.
2. Hulett MD, Hogarth PM. Molecular basis of Fc Receptor function. Adv Immunol. 1994;57: 1–127.
3. Hogarth PM, Hulett MD. Immunoglobulin Fc Receptors. Biomembranes. 1996;3:269–314.
4. Powell MD, Hulett MD, Brinkworth RI, Hogarth PM. Human FcγR-ligand interaction. In: van de Winkel J. Capel P, eds, Human Fc Receptors. Heidelberg: RG Landes; 1996:7–25.
5. Rigby AJ. Hulett MD, Brinkworth RI, Hogarth PM. The structural basis of the interaction of IgE and FcεRI. In: Hamawy MM, ed., IgE Receptor (FcεRI) Function in Mast Cells and Basophils. Heidelberg: RG Landes; 1996:7–32.
6. Hulett MD, McKenzie IFC, Hogarth PM. Chimeric receptors identify immunoglobulin binding regions in human FcγRII and FcεRI. Eur J Immunol. 1993;23:640–5.
7. Hulett MD, Witwort E, Brinkworth RI, McKenzie IFC, Hogarth PM. Identification of the IgG Binding Site for the human low affinity receptor for IgG, FcγRII: enhancement and ablation of binding by site directed mutagenesis. J Biol Chem. 1994;269:15287–93.
8. Hulett MD, Witort E, Brinkworth RI, McKenzie IFC, Hogarth PM. Multiple regions of human FcγRII (CD32) contribute to the binding of IgG. J Biol Chem. 1995;270:21188–94.
9. Hibbs ML, Tolvanen M, Carpen O. Membrane proximal Ig-like domain of FcγRIII (CD16) contains residues critical for ligand binding. J Immunol. 1994;152:4466–74.
10. Tamm A, Kister A, Nolte KU, Gessner JE, Schmidt RE. The IgG binding site of human FcγRIIIB receptor involves CC′ and FG loops of the membrane proximal domain. J Biol Chem. 1996;271:3659–66.
11. Anderson CL, Abraham GN. Characterisation of the Fc receptor for IgG on the human macrophage cell line, U937. J Immunol. 1980;125:2735–41.
12. Anderson CL, Guyre PM, Whitin JC, Ryan DH, Looney RJ. Fanger MW. Monoclonal antibodies to Fc receptors for IgG on human mononuclear phagocytes. Antibody characterisation and induction of superoxide production in a monocyte cell line. J Biol Chem. 1986; 261:12856–64.
13. Dougherty GJ. Selvendran Y, Murdoch S, Palmer DG, Hogg N. The human mononuclear phagocyte high-affinity Fc receptor, FcRI, defined by a monoclonal antibody 10.1. Eur J Immunol. 1987;17:1453–60.
14. Frey J. Engelhardt W. Characterisation and structural analysis of Fcγ receptors of human monocytes, a monoblast cell line (U937) and a myeloblast cell line (HL-60) by a monoclonal antibody. Eur J Immunol. 1987;17:583–91.
15. Peltz GA, Frederick K, Anderson CL, Peterlin BM. Characterization of the human monocyte high affinity Fc receptor (huFcRI). Mol Immunol. 1988;25:243–50.
16. Quilliam AL, Osman N, McKenzie IFC, Hogarth PM. Biochemical characterisation of murine FcγRI. Immunology. 1993;78:358–63.

17. Allen JM, Seed B. Isolation and expression of functional high affinity Fc receptor complementary DNAs. Science. 1989;243:378–81.
18. Sears DW, Osman N, Tate B, McKenzie IFC, Hogarth PM. Molecular cloning and expression of the mouse high affinity Fc receptor for IgG. J Immunol. 1990;144:371–8.
19. Ernst LK, Duchemin AM, Anderson CL. Association of the high affinity receptor for IgG (FcγRI) with the γ subunit of the IgE receptor. Proc Natl Acad Sci USA. 1993;90:6023–7.
20. Scholl PR, Geha RS. Physical association between the high-affinity IgG receptor (FcγRI) and the γ subunit of the high-affinity IgE receptor. Proc Natl Acad Sci USA. 1993;90:6023–9.
21. Masuda M, Roos D. Association of all three types of FcγR (CD64, CD32, CD16) with a γ chain homodimer in cultured human monocytes. J Immunol. 1993;151:7188–95.
22. Gavin AL, Hamilton JA, Hogarth PM. Extracellular mutations of non-obese diabetic mouse FcγRI modify surface expression and ligand binding. J Biol Chem. 1996;271:17091–9.
23. Miller KL, Duchemin AM, Anderson CL. A novel role for the Fc receptor γ subunit: enhancement of FcγR ligand affinity. J Exp Med. 1996;183:2227–32.
24. Letourneur O, Kennedy ICS, Brini AT, Ortaldo JR, O'Shea JJ. Kinet J-P. Characterisation of the family of dimers associated with Fc receptors (FcεRI and FcγRIII). J Immunol. 1991; 147:2652–6.
25. van Vugt MJ. Heijnen AF, Capel PJ et al. FcR-γ chain is essential for both surface expression and function of human FcγRI (CD64) *in vivo*. Blood. 1996;87:3593–9.
26. Fries LF, Hall RP, Lawley TJ. Crabtree GR, Frank MM. Monocyte receptor for the Fc portion of IgG studies with monomeric human IgG_1: normal *in vitro* expression of Fc receptors in HLA-B8/Drw3 subjects with defective Fc-mediated *in vivo* clearance. J Immunol. 1982;129:1041–9.
27. Kurlander RJ. Batker J. The binding of human immunoglobulin G1 monomer and small, covalently cross linked polymers of immunoglobulin G1 to human peripheral blood monocytes and polymorphonuclear leukocytes. J Clin Invest. 1982;69:1–8.
28. Woof JM, Partridge LF, Jeffries R, Burton DR. Localisation of the monocyte-binding region on human immunoglobulin G. Mol Immunol. 1986;23:319–30.
29. Wiener E, Atwal A, Thompson KM, Melamed MD, Gorick B, Hughes-Jones NC. Differences between the activities of human monoclonal IgG_1 and IgG_3 subclasses of anti-D (Rh) antibody in their ability to mediate red cell binding to macrophages. Immunology. 1987;62: 401–8.
30. Jones DH, Looney RJ. Anderson CL. Two distinct classes of IgG Fc receptors on a human monocyte line (U937) defined by differences in binding of murine IgG subclasses at low ionic strength. J Immunol. 1985;135:3348–53.
31. van de Winkel JGJ. Tax WJM, van Bruggen MCJ et al. Characterisation of two Fc receptors for mouse immunoglobulins on human monocytes and cell lines. Scand J Immunol. 1987;26: 663–72.
32. Ceuppens JL, van Vaecjk F. Human T cell activation induced by a monoclonal mouse IgG_3 anti-CD3 antibody (Riv9) requires binding of the Fc part of the antibody to the monocytic 72 kDa high affinity Fc receptor (FcRI). Cell Immunol. 1989;118:136–46.
33. Haagen IA, Geerars AJ. Clark MR, van de Winkel JGJ. Interaction of human monocyte Fcγ receptors with rat IgG_{2b}. A new indicator for the FcγRIIa ($R-H^{131}$) polymorphism. J Immunol. 1995;154:1852–60.
34. Cosio FG, Ackerman SK, Douglas SD, Friend PS, Michael AF. Soluble immune complexes binding to human monocytes and polymorphonuclear leukocytes. Immunology. 1981;44: 773–80.
35. Crowell RE, Du Clos TW, Montoya G, Heaphy E, Mold C. C-reactive protein receptors on the human monocytic cell line U937. Evidence for additional binding to FcγRI. J Immunol. 1991;147:3445–50.
36. Marnell LL, Mold C, Volzer MA, Burlingame RW, Du Clos TW. C-reactive protein binds to FcγRI in transfected Cos cells. J Immunol. 1995;155:2185–90.
37. Robey FA, Jones KD, Tanaka T, Liu TY. Binding of C-reactive protein to chromatin and nucleosome core particles: a possible physiological role of C-reactive protein. J Biol Chem. 1984;259:7311–16.
38. Salonen EM, Vartio T, Hedman K, Vaheri A. Binding of fibronectin by the acute phase reactant C-reactive protein. J Biol Chem. 1984;259:1496–502.
39. Du Clos TW, Zlock LT, Rubin RL. Analysis of the binding of C reactive protein to histones and chromatin. J Immunol. 1988;141:4266–71.

40. Swanson SJ. McPeek MM, Mortensen RF. Characteristics of the binding of human C reactive protein (CRP) to laminin. J Cell Biochem. 1989;40:121–6.
41. Zeller JM, Kubak BM, Gewurz H. Binding sites for C-reactive protein on human monocytes are distinct from IgG Fc receptors. Immunology. 1989;67:51–8.
42. Tebo JM, RF Mortensen. Characterisation and isolation of a C-reactive protein receptor from the monocytic cell line U-937. J Immunol. 1990;144:231–7.
43. van Vugt MJ. van den Herik-Oudijk IE, van de Winkel JGJ. Binding of PE-CY5 conjugates to the human high affinity receptor for IgG (CD64). Blood. 1996;88:2358–61.
44. Unkeless JC, Eisen HN. Binding of monomeric immunoglobulins to Fc receptors of mouse macrophages. J Exp Med. 1975;142:1520–33.
45. Hulett MD, Osman N, McKenzie IFC, Hogarth PM. Chimeric Fc receptors identify functional domains of the murine FcγRI high affinity receptor for IgG. J Immunol. 1991; 147:1863–8.
46. Haeffner-Cavaillon N, Dorrington KJ. Klein M. Studies of the Fcγ receptor of the murine macrophage-like cell line P388D1. II. Binding of human IgG subclass proteins and their proteolytic fragments. J Immunol. 1979;123:1914–19.
47. Haeffner-Cavaillon N, Klein M, Dorrington KJ. Studies of the Fcγ receptor of the murine macrophage-like cell line P388D1. I. The binding of homologous and heterologous immunoglobulin G. J Immunol. 1979;123:1905–13.
48. Porges AJ. Redecha PB, Doebele R, Pan LC, Salmon JE, Kimberly RP. Novel Fcγ receptor I family gene products in human mononuclear cells. J Clin Invest. 1992;90:2101–9.
49. O'Grady JH, Looney RJ. Anderson CL. The valence for ligand of the human monocyte phagocyte 72 kD high affinity IgG Fc receptor is one. J Immunol. 1986;137:2307–10.
50. Koolwijk P, Spierenburg GT, Frasa T, Boot JHA, van de Winkel JGJ, Bast BJ. Interaction between hybrid mouse monoclonal antibodies and the human high-affinity IgG FcR, human FcγRI, on U937: involvement of only one of the mIgG heavy chains in receptor binding. J Immunol. 1989;143:1656–62.
51. Woof JM, Nik Jaafar MI, Jefferies R, Burton DR. The monocyte binding domains on human immunoglobulin G. Mol Immunol. 1984;21:523–7.
52. Partridge LF, Woof JM, Jefferies R, Burton DR. The use of anti-IgG monoclonal antibodies in mapping the monocyte subpopulation in human peripheral blood. Mol Immunol. 1988;23: 1365–72.
53. Shopes B, Weetall M, Holowka D, Baird B. Recombinant human IgG_1-murine IgE chimeric Ig: construction, expression and binding to human Fcγ receptors. J Immunol. 1990;145: 3842–8.
54. Canfield SM, Morrison SL. The binding affinity of human IgG_2 for its high affinity Fc receptor is determined by multiple amino acids in the CH2 domain and is modulated by the hinge region. J Exp Med. 1991;173:1483–91.
55. Chappel MS, Isenman DE, Everett M, Xu YY, Dorrington KJ. Klein MH. Identification of the Fc γ receptor class I binding site in human IgG through the use of recombinant IgG_1/ IgG_2 hybrid and point mutated antibodies. Proc Natl Acad Sci USA. 1991;88:9036–40.
56. Lund J. Winter G, Jones PT et al. Human FcγRI and FcγRII interact with distinct but overlapping sites on human IgG. J Immunol. 1991;147:2657–62.
57. Duncan AR, Woof JM, Partridge LJ. Burton DR, Winter G. Localization of the binding site for the human high affinity Fc receptor for IgG. Nature. 1988;332:563–4.
58. Burton DR, Jefferis R, Partridge LJ. Woof JM. Molecular recognition of antibody by cellular Fc receptor (FcRI). Mol Immunol. 1988;25:1175–81.
59. Chappel MS, Isenman DE, Oomen R, Xu YY, Klein MH. Identification of a secondary FcγRI binding site within a genetically engineered human IgG antibody. J Biol Chem. 1993; 268:25124–31.
60. Lund J. Takahashi N, Pound JD, Goodall M, Nakagawa H, Jefferis R. Oligosaccharide-protein interactions in IgG can modulate recognition by Fcγ receptors. FASEB J. 1995;9: 115–19.
61. Lund J. Takahashi N, Pound JD, Goodall M, Jefferis R. Multiple interactions of IgG with its core oligosaccharide can modulate recognition by complement and human FcγRI and influence the synthesis of its oligosaccharide chains. J Immunol. 1996;157:4963–9.
62. Steel DM, Whitehead AS. The major acute phase reactants: C-reactive protein, serum amyloid P component and serum amyloid A protein. Immunol Today. 1994;15:81–6.
63. Whitehead AS, Rits M, Michaelson J. Molecular genetics of mouse serum amyloid P component (SAP): cloning and gene mapping. Immunogenetics. 1988;28:388–91.

64. Woo P, Korenberg JR, Whitehead AS. Characterisation of genomic and complementary DNA sequences of human C-reactive protein, comparison with the complementary DNA sequence of serum amyloid P component. J Biol Chem. 1985;260:13384–89.
65. Zahedi K, Tebo JM, Siripont J. Klimo GF, Mortensen RF. Binding of human C-reactive protein to mouse macrophages is mediated by distinct receptors. J Immunol. 1989;142:2384–9.
66. Mellman IS, Unkeless JC. Purification of a functional mouse Fc receptor through the use of a monoclonal antibody. J Exp Med. 1980;152:1048–69.
67. Mellman IS, Plutner H, Steinman RM, Unkeless JC, Cohn ZA. Internalization and degradation of macrophage Fc receptors during receptor mediated phagocytosis. J Cell Biol. 1983;96:887–95.
68. Cohen L, Sharp S, Kulczycki A Jr. Human monocytes, B lymphocytes and non-B lymphocytes each have structurally unique Fc receptors. J Immunol. 1983;131:378–83.
69. Pilkington GR, Kraft N, Murdolo V et al. Serological typing of acute leukaemia using the monoclonal antibodies PHM 1,2,3,6, CIKM5 and the rabbit antisera RARC2a (Ad) and RAALLP50. In: Bernard A et al., eds, Leukocyte Typing. Berlin, Heidelberg: Springer-Verlag; 1984:588.
70. Green SA, Plutner H, Mellman I. Biosynthesis and intracellular transport of the mouse macrophage Fc receptor. J Biol Chem. 1985;260:9867–71.
71. Hibbs ML, Hogarth PM, McKenzie IFC. The mouse Ly17 locus identifies a polymorphism of the Fc receptor. Immunogenetics. 1985;22:335–48.
72. Holmes KL, Palfree RGE, Hammerling U, Morse HC. Alleles of Ly17 alloantigen define polymorphism of the murine IgG Fc receptor. Proc Natl Acad Sci USA. 1985;82:7706–10.
73. Rosenfeld SI, Looney RJ. Leddy JP, Phipps DC, Abraham GN, Anderson CL. Human platelet Fc receptor for immunoglobulin G: Identification as a 40,000 molecular weight membrane protein shared by monocytes. J Clin Invest. 1985;76:2317–22.
74. Vaughn M, Taylor M, Mohanakumar T. Characterization of human IgG Fc receptors. J Immunol. 1985;135:4059–65.
75. Looney RJ. Abraham GN, Anderson CL. Human monocytes and U937 cells bear two distinct Fc receptors for IgG. J Immunol. 1986;136:1641–7.
76. van de Winkel JGJ. van Ommen R, Huizinga TWJ et al. Proteolysis induces increased binding affinity of the monocyte type II FcR for human IgG. J Immunol. 1989;143:571–8.
77. Ierino FL, Hulett MD, McKenzie IFC, Hogarth PM. Human FcγRII monoclonal antibodies define structural domains. J Immunol. 1993;150:1794–9.
78. van den Herik-Oudijk IE, Westerdaal NAC, Henriquez NV, Capel PJA, van de Winkel JGJ. Functional analysis of human FcγRII (CD32) isoforms expressed in B lymphocytes. J Immunol. 1994;152:574–85.
79. Karas SP, Rosse WF, Kurlander RJ. Characterisation of the IgG-Fc receptor on human platelets. Blood. 1982;60:1277–82.
80. Kurlander RJ. Haney AF, Gartrell J. Human peritoneal macrophages possess two populations of IgG Fc receptors. Cell Immunol. 1984;86:479–90.
81. Tax WJM, van de Winkel JGJ. Human Fcγ receptor II: A standby receptor activated by proteolysis? Immunol Today. 1990;11:308–10.
82. Koenderman L, Hermans SW, Capel PJA, van de Winkel JGJ. Granulocyte-macrophage colony-stimulating factor induces sequential activation and deactivation of binding via a low-affinity IgG Fc receptor, hFcγRII, on human eosinophils. Blood. 1993;81:2413–19.
83. Walker MR, Lund J. Thompson KM, Jefferies R. Aglycosylation of human IgG_1 and IgG_3 monoclonal antibodies can eliminate recognition by human cells expressing FcγRI and/or FcγRII receptors. Biochem J. 1989;259:1347–53.
84. van de Winkel J. Anderson CL. Biology of human IgG Fc receptors. J Leuk Biol. 1991;49: 511–24.
85. Warmerdam PAM, van de Winkel JGJ. Vlug A, Westerdaal NAC, Capel PJA. A single amino acid in the second Ig-like domain of the human Fcγ receptor II is critical in human IgG_2 binding. J Immunol. 1991;147:1338–43.
86. Warmerdam PAM, van de Winkel JGJ. Grosselin EJ. Capel PJA. Molecular basis for a polymorphism of human FcγRII (CD32). J Exp Med. 1990;172:19–25.
87. Tate BJ. Witort E, McKenzie IFC, Hogarth PM. Expression of the high responder/non-responder human FcγRII: Analysis by PCR and transfection into FcR^- COS cells. Immunol Cell Biol. 1992;70:79–87.

88. Warmerdam PAM, van den Herik-Oudijk IE, Parren PWHI, Westerdaal NAC, van de Winkel JGJ. Capel PJA. Interaction of human FcγRIIb1 (CD32) isoform with murine and human IgG subclasses. Int Immunol. 1993;3:239–47.
89. Unkeless JC, Scigliano E, Freedman VH. Structure and function of human and murine receptors for IgG. Ann Rev Immunol. 1988;6:251–81.
90. Mellman I, Koch T, Healey G et al. Structure and function of Fc receptors on macrophages and lymphocytes. J Cell Sci. 1988;9(suppl.):45–65.
91. Unkeless JC. The presence of two Fc receptors on mouse macrophages:Evidence from a variant cell line and differential trypsin sensitivity. J Exp Med. 1977;145:931–47.
92. Unkeless JC. Characterisation of a monoclonal antibody directed against mouse macrophage and lymphocyte Fc receptors. J Exp Med. 1979;150:580–96.
93. Heuser CH, Anderson CL, Grey HM. Receptors for IgG: subclass specificity of receptors on different mouse cell types and the definition of two distinct receptors on a macrophage cell line. J Exp Med. 1977;145:1316–27.
94. Lopez AF, Strath M, Sanderson CJ. Mouse immunoglobulin isotypes mediating cytotoxicity of target cells by eosinophils and neutrophils. Immunology. 1983;48:503–9.
95. Teillaud JL, Diamond B, Pollock RR, Fajtova V, Scharff MD. Fc receptors on cultured myeloma and hybridoma cells. J Immunol. 1985;134:1774–9.
96. Diamond B, Yelton DE. A new Fc receptor on mouse macrophages binding IgG_3. J Exp Med. 1981;153:514–19.
97. Takizawa F, Adamczewski M, Kinet JP. Identification of the low affinity receptor for immunoglobulin E on mouse mast cells and macrophages as FcγRII and FcγRIII. J Exp Med. 1992;176:469–75.
98. Clark MR, Clarkson SB, Ory PA, Stollman N, Goldstein IM. Molecular basis for a polymorphism involving Fc receptor II on human monocytes. J Immunol. 1989;143:1731–4.
99. Parren PW, Warmerdam PAM, Boeije LC, Capel PJ. van de Winkel JGJ. Aarden LA. Characterization of IgG FcR-mediated proliferation of human T cells induced by mouse and human anti-CD3 monoclonal antibodies: Identification of a functional polymorphism to human IgG_2 anti-CD3. J Immunol. 1992;148:695–701.
100. Shen FW, Boyse EA. An alloantigen selective for B cells Ly17.1. Immunogenetics. 1980;11: 315–17.
101. Davidson WF, Morse III HC, Mathieson BJ. Kozak CA, Shen FW. The B-cell alloantigen Ly17.1 is controlled by a gene closely linked to Ly20 and Ly9 on chromosome 1. Immunogenetics. 1983;17:325–9.
102. Lah M, Quelch K, Deacon NJ. McKenzie IFC, Hogarth PM. Identification of the mouse βFcγRII polymorphism by direct sequencing of amplified genomic DNA. Immunogenetics. 1990;31:202-6.
103. Fleit HB, Wright SD, Unkless JC. Human neutrophil Fcγ receptor distribution and structure. Proc Natl Acad Sci USA. 1982;79: 3275–9.
104. Rumpold H, Kraft D, Obexer G, Bock G, Gebhart W. A monoclonal antibody against a surface antigen shared by human large granular lymphocytes and granulocytes. J Immunol. 1982;129:1458–64.
105. Lanier LL, Le AM, Phillips JH, Warner NL, Babcock GF. Subpopulations of human natural killer cells defined by expression of the leu-7 (HNK-1) and Leu-11 (NK-15) antigens. J Immunol. 1983;131:1789–96.
106. Perussia B, Starr S, Abraham S, Fanning V, Trinchieri G. Human natural killer cells analyzed by B73.1, a monoclonal antibody blocking Fc receptor function. J Immunol. 1983;130:2133–41.
107. Kulczycki A. Human neutrophils and eosinophils have structurally distinct Fcγ receptors. J Immunol. 1984;133:849–54.
108. Simmons D, Seed B. The Fcγ receptor of natural killer cells is a phospholipid-linked membrane protein. Nature. 1988;333:568–70.
109. Edberg JC, Redecha PB, Salmon JE, Kimberly RP. Human FcγRIII (CD16): Isoforms with distinct allelic expression, extracellular domains and membrane linkages on polymorphonuclear and natural killer cells. J Immunol. 1989;143:1642–9.
110. Peltz GA, Grundy HO, Labo RV, Yssel H, Barsh GS, Moore KW. Human FcγRIII cloning, expression, identification of the chromosomal locus of two Fc receptors for IgG. Proc Natl Acad Sci USA. 1989;86:1013–17.

111. Scallon BJ. Scigliano E, Freedman VH et al. A human immunoglobulin G receptor exists in both polypeptide-anchored and phosphoinositol-glycan anchored forms. Proc Natl Acad Sci USA. 1989;86:5079–83.
112. Ravetch JV, Perussia B. Alternative membrane forms of FcγRIII (CD16) on human NK cells and neutrophils: Cell-type specific expression of two genes which differ in single nucleotide substitutions. J Exp Med. 1989;170:481–91.
113. Qiu WQ, De Bruin D, Brownstein BH, Pearse R, Ravetch JV. Organization of the human and mouse low-affinity FcγR genes: Duplication and recombination. Science. 1990;248:732–5.
114. Kurosaki T, Ravetch JV. A single amino acid in the GPI attachment domain determines the membrane topology of FcγRIII. Nature. 1989;342:805–7.
115. Selvaraj P, Carpen O, Hibbs ML, Springer TA. Natural killer cell and granulocyte Fcγ receptor III (CD16) differ in membrane anchor and signal transduction. J Immunol. 1989;143:3283–8.
116. Lanier LL, Ruitenberg JJ. Phillips JH. Functional and biochemical analysis of CD16 antigen on natural killer cells and granulocytes. J Immunol. 1988;141:3478–85.
117. Hibbs ML, Selvaraj P, Carpen O et al. Mechanisms for regulating expression of membrane isoforms of FcγRIII (CD16). Science. 1989;246:1608–11.
118. Kurosaki T, Gander I, Wirthmueller U, Ravetch JV. The β subunit of FcεRI is associated with the FcγRIII on mast cells. J Exp Med. 1992;175:447–451.
119. Lanier LL, Yu G, Phillips JH. Co-association of CD3-z with receptor (CD16) for IgG on human natural killer cells. Nature. 1989;342:803–5.
120. Ra C, Jouvin M-H, Blank U, Kinet J-P. A macrophage Fcγ receptor and the mast cell receptor for immunoglobulin E share an identical subunit. Nature. 1989;341:752–4.
121. Vance BA, Huizinga TWJ. Guyre PM. Functional polymorphism of FcγRIII on human LGL/NK cells. FASEB J. 1992;6:1620A.
122. Anderson CL, Looney RJ. Human leukocyte IgG Fc receptors. Immunol Today. 1986;7:264 6.
123. Huizinga TWJ. van Kemenade F, Koenderman L et al. The 40-kDa Fcγ Receptor (FcRII) on human neutrophils is essential for the IgG induced respiratory burst and IgG-induced phagocytosis. J Immunol. 1989;142:2365 9.
124. Kipps TJ. Parham P, Punt J. Herzenberg LA. Importance of immunoglobulin isotype in human antibody dependent cell mediated cytotoxicity directed by murine monoclonal antibodies. J Exp Med. 1985;161:1 17.
125. Anasetti C, Martin PJ. Morishita Y, Badger CC, Bernstein ID, Hansen JA. Human large granular lymphocytes express high affinity receptors for murine monoclonal antibodies of the IgG_3 subclass. J Immunol. 1987;138:2979 81.
126. Braakman E, van de Winkel JGJ. van Krimpen BA, Janze M, Bolhuis RLH. CD16 on human gd T lymphocytes: Expression, function and specificity for mouse IgG isotypes. Cell Immunol. 1993;143:97 107.
127. Ghirlando R, Keown MB, Mackay GA, Lewis MS, Unkeless JC, Gould HJ. Stoichiometry and thermodynamics of the interaction between the Fc fragment of human IgG_1 and its low-affinity receptor FcγRIII. Biochemistry. 1995;34:13320 7.
128. Tamm A, Schmidt RE. The binding epitopes of human CD16 (FcγRIII) monoclonal antibodies. Implications for ligand binding. J Immunol. 1996;157:1576 81.
129. Barnett-Foster DE, Sjoquist J. Painter RH. The effect of fragment B of staphyloccocal protein A on the binding of rabbit IgG to human granulocytes and monocytes. Mol Immunol. 1982;19:407 12.
130. Sarmay G, Jefferies R, Klein E, Benczur M, Gergely J. Mapping of the functional topography of IgG Fcγ with monoclonal antibodies: Localisation of epitopes interacting with the binding sites of Fc receptor on human K cells. Eur J Immunol. 1985;15:1037 42.
131. Lund J. Tanaka T, Takahashi N, Sarmay G, Arata Y, Jefferis R. A protein structural change in aglycosylated IgG_3 correlates with loss of huFcγRI and huFcγRIII binding and/or activation. Mol Immunol. 1990;27:1145 53.
132. Ravetch JV and Kinet J-P. Fc receptors. Ann Rev Immunol. 1991;9:457 62.
133. Miller L, Blank U, Metzger H, Kinet J-P. Expression of high affinity binding of human immunoglobulin E by transfected cells. Science. 1989;244:334 7.
134. Ra C, Jouvin M-H,, Kinet J-P. Complete structure of the mouse mast cell receptor for IgE (FcεRI) and surface expression of chimeric receptors (rat-mouse-human) on transfected cells. J Biol Chem. 1989;264:15323 7.

135. Kuster H, Zhang L, Brini AT, MacGlashan DWJ. Kinet J-P. The gene and cDNA for the human high affinity immunoglobulin E receptor β chain and expression of the complete human receptor. J Biol Chem. 1992;267:12782–7.
136. Kulczycki A, Metzger H. The interaction of IgE with rat basophilic leukemia cells. II Quantitative aspects of the binding reaction. J Exp Med. 1974;140:1676–95.
137. Ishizaka T, Dvorak AM, Conrad DH, Niebyl JR, Marquett JP, Ishizaka K. Morphologic and immunological characterization of human basophils developed in cultures of cord blood mononuclear cells. J Immunol. 1985;134:532–40.
138. Conrad DH, Wingard JR, Ishizaka T. The interaction of human and rodent IgE with the human basophil IgE receptor. J Immunol. 1983;130:327–33.
139. Keown MB, Ghirlando R, Young RJ et al. Hydrodynamic studies of a complex between the Fc fragment of human IgE and a soluble fragment of the FcεRI alpha chain. Proc Natl Acad Sci USA. 1995;92:1841–5.
140. Hakimi J. Seals C, Kondas JA, Pettine L, Danho W, Kochan J. The α subunit of the human IgE receptor (FcεRI) is sufficient for high affinity IgE binding. J Biol Chem. 1990;265:22079–81.
141. Blank U, Ra C, Kinet J-P. Characterization of truncated α chain products from human, rat and mouse high affinity receptor for Immunoglobulin E. J Biol Chem. 1990;266:2639–46.
142. Ra C, Kuromitsu S, Hirose T, Yasuda S, Furuichi K and Okumura K. Soluble human high-affinity receptor for IgE abrogates the IgE-mediated allergic reaction. Int Immunol. 1993;5:47–54.
143. Haak-Frendscho M, Ridgway J. Schields R, Robbins K, Gorman C, Jardieu P. Human IgE receptor a chain IgG- chimera blocks passive cutaneous anaphylaxis reaction *in vivo*. J Immunol. 1993;151:351–8.
144. Gavin AL, Hulett MD, McKenzie IFC, Hogarth PM. Expression of recombinant soluble FcεRI: function and tissue distribution studies. Immunology. 1995;86:392–8.
145. Riske F, Hakimi J. Mallamaci M et al. High affinity human IgE receptor (FcεRI): Analysis of functional domains of the α subunit with monoclonal antibodies. J Biol Chem. 1991;266:11245–51.
146. Mallamaci MA, Chizzonite R, Griffen M et al. Identification of sites on the human FcεRI α subunit which are involved in binding human and rat IgE. J Biol Chem. 1993;268:22076–83.
147. Robertson MW. Phage and *Escherichia coli* expression of the human high affinity immunoglobulin E receptor alpha sub unit ecto-domain. J Biol Chem. 1993;268:12736–43.
148. McDonnell JM, Beavil AJ. Mackay GA et al. Structure based design and characterisation of peptides that inhibit IgE binding to its high affinity receptor. Nat Struct Biol. 1996;3:419–26.
149. Perez-Montfort R, Metzger H. Proteolysis of soluble IgE receptor complexes: Localization of sites on IgE which interact with the Fc receptor. Mol Immunol. 1982;19:1113–25.
150. Helm BA, Marsh P, Vercelli D, Padlan E, Gould H, Geha R. The mast cell binding site on human immunoglobulin E. Nature. 1988;331:180–3.
151. Helm BA, Kebo D, Vercelli D et al. Blocking of passive sensitization of human mast cells and basophil granulocytes with IgE antibodies by a recombinant human ε-chain fragment of 76 amino acids. Proc Natl Acad Sci USA. 1989;86:9465–9.
152. Basu M, Hakimi J. Dharm E et al. Purification and characterization of human recombinant IgE-Fc fragments that bind to the human high affinity IgE receptor. J Biol Chem. 1993;268:13118–25.
153. Presta L, Shields R, O'Connell L et al. The binding site of human immunoglobulin E for its high affinity receptor. J Biol Chem. 1994;269:26368–73.
154. Helm BA, Sayers I, Higginbottom A et al. Identification of the high affinity receptor binding region in human immunoglobulin E. J Biol Chem. 1996;271:7494–500.
155. Weetall M, Shopes B, Holowka D, Baird B. Mapping of the site of interaction between murine IgE and its high affinity receptor with chimeric Ig. J Immunol. 1990;145:3849–54.
156. Nissim AS, Jouvin MHE, Eshhar Z. Mapping of the high affinity Fcε receptor binding site to the third constant region domain of IgE. EMBO J. 1991;10:101–7.
157. Nissim AS, Schwarzbaum R, Siraganian R, Eshhar Z. Fine specificity of the IgE interaction with the low and high affinity Fcε receptors. J Immunol. 1993;150:1365–74.
158. Morton HC, van Egmond M, van de Winkel JG. Structure and function of human IgA Fc receptors (FcαR). Crit Rev Immunol. 1996;16:423–40.
159. Maliszewski CR, March C J. Schrenborn MA, Gimpel S, Shen L. Expression cloning of a human Fc receptor for IgA. J Exp Med. 1990;172:1665–72.

160. Monteiro RC, Cooper MD, Kubagawa H. Molecular heterogeneity of Fcα receptors detected by receptor specific monoclonal antibodies. J Immunol. 1992;148:1764–70.
161. Morton HC, van den Herik-Oudijk IE, Vossebeld P et al. Functional association between the human myeloid immunoglobulin A Fc receptor (CD89) and FcR gamma chain. Molecular basis for CD89/FcR γ chain association. J Biol Chem. 1995;270:29781–7.
162. Albrechtsen M, Yeaman GR, Kerr MA. Characterisation of the IgA Receptor from human polymorphic nuclear leukocytes. Immunology. 1988;64:205–10.
163. Monteiro RC, Kubagawa H, Cooper MD. Cellular distribution, regulation and biochemical nature of an Fcα Receptor in humans. J Exp Med. 1990;171:597–613.
164. Stewart WW and Kerr MA. The specificity of the human neutrophil IgA receptor (FcαR) determined by measurement of chemiluminescence induced by serum or secretory IgA_1 or IgA_2. Immunology. 1990;71:328–34.
165. Pleass RJ. Andrews PD, Kerr MA, Woof JM. Alternative splicing of the human IgA Fc receptor CD89 in neutrophils and eosinophils. Biochem J. 1996;318:771–7.
166. van Dijk TB, Bracke M, Caldenhoven E, Lammers JWJ. Koendermanl I, Der Groot RP. Cloning and characterisation of FcαRB – a novel Fcα receptor (CD89) isoform expressed in eosinophils and neutrophils. Blood. 1996;88:4229–38.
167. Carayannopoulos L, Hexham JM, Capra JD. Localisation of the binding site for monocyte immunoglobulin (IgA) Fc Receptor (CD89) to the domain boundary between Ca2 and Ca3 in Human IgA_1. J Exp Med. 1996;183:1579–86.
168. Simister NE, Rees AR. Isolation and characterisation of an Fc receptor from neo-natal rat small intestine. Eur J Immunol. 1985;15:733–8.
169. Jakoi ER, Cambier J, Saslow S. Transepithelial transport of maternal antibody: purification of IgG receptor from newborn rat intestine. J Immunol. 1985;135:3360–4.
170. Simister NE, Mostov KE. An Fc receptor structurally related to MHC Class I antigens. Nature. 1989;337:184–7.
171. Ahouse J, Hagerman CL, Mittal P et al. Mouse MHC. Class I-like Fc receptor encoded outside MHC. J Immunol. 1993;151:6076–88.
172. Raghavan M, Wang Y, Bjorkman PJ. Effects of receptor dimerisation on the interaction between the Class I major histocompatibility complex-related Fc receptor and IgG. Proc Natl Acad Sci USA. 1995;92:11200–4.
173. Brambell FWR, Hemmings WA, Morris IG. A theoretical model of γ-globulin catabolism. Nature. 1964;203:1352–5.
174. Junghans RP, Anderson CL. The protection receptor for IgG catabolism is the β-2 microglobulin-containing neo-natal intestinal transport receptor. Proc Natl Acad Sci USA. 1996;93:5512–16.
175. Hobbs SM, Jackson LE, Peppard JV. Binding of subclass of rat immunoglobulin G to detergent-isolated Fc receptor from neo-natal rat intestine. J Biol Chem. 1987;262:8041–6.
176. Huber AH, Kelley RF, Gastinel, LN, Bjorkman PJ. Crystallisation and stoichiometry of binding of a complex between a rat intestinal receptor and Fc. J Mol Biol. 1993;230:1077–83.
177. Burmeister WP, Gastinel LN, Simister NE, Blum ML, Bjorkman PJ. Crystal structure at 2.2Å resolution of the MHC related neonatal Fc receptor. Nature. 1994;372:336–43.
178. Burmeister WP, Huber AH, Bjorkman PJ. Crystal structure of the rat neonatal Fc receptor with Fc. Nature. 1994;372:379–83.
179. Raghavan M, Bonagura VR, Morrison SL, Bjorkman PJ. Analysis and pH dependence of the neo-natal Fc receptor/immunoglobulin G interaction using antibody and receptor variants. Biochem. 1995;34:1649–57.
180. Kim JK, Tsen MF, Ghetie V, Ward ES. Localisation of the site of the murine IgG_1 molecule that is involved in binding to the murine intestinal Fc receptor. Eur J Immunol. 1994;24:2429–34.
181. Mostov KE, Friedlander M, Blobel G. The receptor for transepithelial transport of IgA and IgM contains multiple immunoglobulin-like domains. Nature. 1984;308:37–43.
182. Mostov KE. Transepithelial transport of immunoglobulins. Ann Rev Immunol. 1994;12:63–84.
183. Frutiger S, Hughes GJ, Hanly WC, Kingette M, Jaton JC. The amino-terminal domain of rabbit secretory component is responsible for noncovalent binding to immunoglobulin A dimers. J Biol Chem. 1986;261:16673–81.
184. Bakos MA, Kurosky A, Goldblum RM. Characterisation of a critical binding site for human polymeric Ig on secretory component. J Immunol. 1991;147:3419–26.

185. Bakos MA, Kurosky A, Czerwinski EW, Goldblum RM. A conserved binding site on the receptor for polymeric Ig is homologous to CD R1 of Ig V kappa domains. J Immunol. 1993;951:1346–52.
186. Coyne RS, Siebrecht M, Peitsch MC, Casanova JE. Mutational analysis of polymeric immunoglobulin receptor/ligand interactions. Evidence for the involvement of multiple complementary determining regions (CDR)-like loops in receptor domain I. J Biol Chem. 1994;269:31620–5.
187. Piskunch JF, Blanchard MH, Youngman KR, France JA, Kaetzel CS. Molecular cloning of mouse polymeric Ig receptor. Functional regions are conserved among five mammalian species. J Immunol. 1995;154:1735–47.
188. Bakos MA, Kurosky A, Woodard CS, Denney RM, Goldblum RM. Probing the topography of free and polymeric Ig-bound human secretory component with monoclonal antibodies. J Immunol. 1991;146:162–8.
189. Edelman GM, Cunningham BA, Gall WE, Gottlieb PD, Rutishauser U, Waxdal MJ. The covalent structure of an entire gamma G immunoglobulin molecule. Proc Natl Acad Sci USA. 1969;63:78–83.

3
Fc receptor genetics and the manipulation of genes in the study of FcR biology

T. TAKAI and J. V. RAVETCH

Over the last several years, a number of mouse lines deficient in specific Fc receptor (FcR) genes, and those with exogenous FcR genes have been developed. They include transgenic mice strains with a transgene for human FcγRIα [1], FcεRIα [2], FcRγ [3], or FcγRIIIα [4], and knock-out mice strains lacking the gene for FcεRIα [5], FcεRII [6–8], FcRγ [9], FcγRIIB [10], or FcγRIII [11]. These mice have been highly useful for analysing the functions of individual FcRs or their subunit molecule in the immune system. This review will focus mainly on the findings derived from such mice, emphasizing aspects of positive and negative regulatory functions for FcRs in the immune system.

FcR GENETICS: AN OVERVIEW

Considerable progress has been made in the last decade in defining the genetic organization, molecular structures, and detailed functions of FcRs through molecular biological techniques [reviewed in References 12–18]. Most of the ligand-binding a subunits of FcRs belong to the immunoglobulin (Ig) superfamily and constitute a group of complex but highly related molecules (Table 3.1). Eight genes have been identified for the human FcγRs: three genes for the high affinity IgG receptor FcγRI (FcγRIA, FcγRIB, and FcγRIC) [19] and five genes for the low affinity IgG receptors FcγRII (FcγRIIA, FcγRIIB, and FcγRIIC) [20] and FcγRIII (FcγRIIIA and FcγRIIIB) [21]. The low affinity FcγR genes and the genes for the FcεRIα and γ subunits, are clustered on

J.G.J. van de Winkel and P.M. Hogarth (eds.), The Immunoglobulin Receptors and their Physiological and Pathological Roles in Immunity. 37–48.

Table 3.1 Genes and transcripts of Fc receptors and some related molecules

Species	*CD*	*Gene*	*Chromosome*[a] *Human*	*Mouse (cM)*	*Transcripts*
Human	CD64	FcγRIA	1q21		FcγRIa
		FcγRIB	1q21		
		FcγRIC	1q21		
Mouse		FcγRI		3 (45.2)	FcγRI
Human	CD32	FcγRIIA	1q23		FcγRIIa1, sFcγRIIa2
		FcγRIIB	1q23		FcγRIIb1, FcγRIIb2, FcγRIIb3
		FcγRIIC	1q23		FcγRIIc
Mouse		FcγRII		1 (92)	FcγRIIb1, FcγRIIb1′, FcγRIIb2, sFcγRIIb3
Human	CD16	FcγRIIIA	1q23		FcγRIIIa
		FcγRIIIB	1q23		FcγRIIIb, sFcγRIIIb
Mouse		FcγRIII		1 (92.3)	FcγRIII
Human		FcεRIα	1q23		FcεRIα
Mouse		FcεRIα		1 (94.2)	FcεRIα
Human		FcεRIβ	11q13		FcεRIβ
Mouse		FcεRIβ		19 (8.0)	FcεRIβ
Human		FcRγ	1q23		FcRγ
Mouse		FcRγ		1 (93.3)	FcRγ
Human		CD3ζ	1q22-23		CD3ζ, CD3η
Mouse		CD3ζ		1 (87.2)	CD3ζ, CD3η
Human	CD23	FcεRIIa	19pRIIa		FcεRIIa, sFcεRII
		FcεRIIb			FcεRIIb
Mouse		FcεRII		8 (0.4)	FcεRII, sFcεRII
Human	CD89	FcαR	19q13.4		FcαRa1, FcαRa2, FcαRa3
Human		poly-IgR	1q31-q41		poly-IgR
Mouse		poly-IgR		?	poly-IgR
Human		FcRn	19q13.3		FcRn
Mouse		FcRn		7 (20.0)	FcRn
Human	CD158	p58 KIR	19q13.4		p58 KIR

[a]These data are available at http://www.genome.ad.jp/.

[b]s: soluble receptor

chromosome 1q23. This region of 1q23 is syntenic to mouse chromosome 1, where single genes for these receptors are found [22,23]. The three FcγRI genes map to chromosome 1q21. The mouse FcγRI gene has been mapped on chromosome 3. A locus on chromosome 11q, encoding the β subunit of FcεRI has been suggested to be a candidate gene for atopy [24]. A single amino acid substitution, Ile181Leu, in the fourth transmembrane domain is found segregating with the atopic phenotype.

A receptor for IgA, FcαR, was found on human monocytoid cell line U937 [25] and other cells including glomerular mesangial cells [26] and monocytes [27]: the gene has been mapped on chromosome 19q13.4. A novel class of FcγR,

Fcγ2R, was cloned from a cattle alveolar macrophage library [28], which has ability to bind erythrocytes sensitized with IgG_2, but not IgG_1. Homology search indicated a greater level of similarity with human FcαR than with any other FcRs: the percentage of identical amino acids was 41% [28]. Similarities with human FcγRI, FcγRII, FcγRIII, FcεRI, or bovine FcγRII were less than 28%. Analysis of genetic divergence indicated that the Fcγ2R gene and the human FcαR gene probably evolved from a common ancestor, which is not shared by other FcγRs [28]. More recently, it was reported that human killer-cell inhibitory receptors (KIRs) found on NK cells as well as T cells show meaningfully high homology to human FcαR both at the nucleotide and amino-acid sequence levels [29]. Chromosomal location of the human p58 KIR has been mapped to 19q13.4. Molecular characteristics of murine KIR-like molecules, gp 49 [30–33] and p91 [34], have also been reported. The genes for human KIRs and murine KIR-like molecules, human FcαR, and bovine Fcγ2R may have evolved from a common ancestor.

FcR TRANSGENIC MICE

Mononuclear phagocytes are critical antigen presenting cells during the initiation of specific humoral immune responses. Antigen presenting cells capture and internalize antigen and provide co-stimulatory signals for activation of helper T cells, which are then capable of stimulating production of antibody-forming cells. To evaluate a role of FcγRI in antigen presentation *in vivo*, Heijnen *et al.* [1] generated human FcγRI transgenic mice. The animals immunized with an anti-human FcγRI antibody produced antigen-specific antibody responses, demonstrating that human FcγRI on myeloid cells is highly active in mediating enhanced antigen presentation *in vivo*.

FcεRI–IgE interaction is highly species specific, and rodent FcεRI does not bind human IgE. Fung-Leung *et al.* [2] generated transgenic mice expressing the human FcεRIα. Interestingly, the transgenic human FcεRIα was complexed with the mouse β and γ chains. Their data demonstrate that the human FcεRIα alone not only confers the specificity in human IgE binding, but also can reconstitute a functional receptor by coupling with the mouse β and γ chains to trigger mast cell activation and degranulation in a whole animal system [2].

The human FcγRIIIA is expressed in NK cells and macrophages and is absent in neutrophils, whereas FcγRIIIB is expressed only in neutrophils. To determine the molecular basis for this differential expression, Li *et al.* [4] generated transgenic mice of human FcγRIIIA and FcγRIIIB. These transgenic mice showed faithful reconstitution of this human pattern of cell type specificity. To determine the *cis* acting sequence elements that confer this specificity, they created promoter swap transgenic mice in which the putative regulatory sequences have been exchanged. The promoter swap transgenic mice that carry IIIA 5′ flanking sequences express FcγRIII in macrophages and NK cells. In contrast, promoter swap transgenic mice that contain IIIB 5′ sequences express FcγRIII in neutrophils only. They concluded that the elements conferring the cell type-specific expression of the human FcγRIII genes locate within the 5′ flanking sequences and first intron [4].

The FcγRIIIα chain forms a complex with a FcRγ homodimer [35,36]. Initially identified as a subunit of FcεRI, FcRγ is also a subunit of the FcγRI, and in a subset of T cells, functions as a subunit of the TCR complex [37–41]. It is also critical for transducing signals into the cell interior which results in cellular activation [42,43] through tyrosine kinase Lyn- and Syk-activation pathway mediated by its immunoreceptor tyrosine-based activation motif (ITAM) [44]. Structural and functional similarity between FcRγ and CD3ζ suggests that these proteins are members of a family of signal-transducing proteins [35,45]. Although FcRγ is reported to be expressed in early thymocytes, its potential role in thymocyte development remains controversial. The observation that T cell development is unaffected in mice lacking FcRγ (see below) argues that its function is not critical for T cell maturation [9]. However, recent experiments indicate that FcγRs on fetal thymocytes may transduce developmentally important signals [46]. In addition, it has been reported that FcRγ-deficient mice show an impaired peripheral T-cell development [47]. TCRγδ$^+$ intraepithelial T cells in reproductive organs (r-IEL) were dramatically decreased, suggesting that the development of r-IEL is FcRγ-dependent, possibly due to the predominant usage of FcRγ homodimers in the TCR complex [47].

Flamand *et al.* [3] generated transgenic mice in which FcRγ was overexpressed at all stages of ontogeny. Overexpression of FcRγ inhibited the maturation of T cells as well as NK cells. The developmental effects were transgene dose related and correlated with markedly delayed maturation of fetal CD4 CD8 FcγRII/III$^+$ thymocytes, cells thought to include the progenitors of both T and NK cells. These results suggest that the CD3ζ and FcRγ chains serve distinctive functions in thymocyte development and that FcRs may play an important role in regulating the differentiation of early progenitor cells within the thymus [3].

ABOLITION OF EFFECTOR FUNCTIONS IN FcR-DEFICIENT MICE

FcεRIα deletion

Mast cells and basophils play a prominent role in anaphylaxis upon activation by IgE and allergen. These cells express at least three types of receptors for IgE: they are FcεRI, FcγRII, and FcγRIII. FcεRI binds monomeric IgE, whereas the two low affinity IgG receptors bind both IgG and IgE immune complexes [48]. To evaluate the relative contribution of these IgE receptors to the genesis of *in vivo* anaphylaxis, Dombrowicz *et al.* [5] have generated FcεRIα-deficient mice by gene targeting. The development and differentiation of mast cells does not seem to be affected by the absence of a FcεRIα gene. IgE-mediated cutaneous and systemic anaphylaxis were tested in these mice [5]. Murine anti-DNP IgE was injected into the ear or i.v., and then challenged i.v. with antigen. The deposition of fibrin in the injected ear, or the rectal temperature of mice was measured for the severity of passive cutaneous anaphylaxis (PCA) or of systemic anaphylaxis, respectively. The FcεRIα-deficient mice were resistant to the cutaneous and systemic anaphylaxis triggered by IgE.

FcRγ deletion

Takai *et al.* [9] have generated a mouse strain genetically deficient in FcRγ. The ablation of this chain resulted in the almost complete loss of the effector responses on NK cells, macrophages, and mast cells due to the loss of at least FcγRI, FcγRIII and FcεRI on these cells [9]. These FcRγ-deficient mice did not respond to various experimental induction protocols of inflammatory reactions [9,49,50]. These results demonstrate the FcR as a key molecule which triggers a variety of inflammatory cascades.

Loss of phagocytic functions in FcRγ-deficient mice

FcR-mediated phagocytosis is accompanied by a variety of transmembrane signaling events. Greenberg *et al.* [51] found that tyrosine phosphorylation is an early event after Fc receptor ligation in mouse inflammatory macrophages. Murine macrophages express FcγRI, II, III on their surface. FcR-mediated phagocytosis was assessed by the internalization of IgG-opsonized sheep red blood cells (SRBC) using thioglycollate-elicited peritoneal macrophages from the FcRγ-knockout mice [9]. In macrophage from the mutant mice, although rosetting activity to IgG-opsonized SRBC was still observed, phagocytic activity was totally absent. This result confirm that FcγRIII on macrophages is essential for manifesting phagocytic activity against IgG-opsonized target cells. The positive contribution of FcγRII to antibody-mediated phagocytosis [52,53] was not observed [9].

Ablation of IgE-induced type I anaphylaxis in FcRγ-deficient mice

Bone marrow-derived cultured mast cells (BMMCs) isolated from FcRγ-deficient mice lack IgE binding, as a result of the loss of FcεRI [9]. Consistent with the loss of these *in vitro* mast cell functions, PCA reaction mediated via FcεRI and IgE on mast cells *in vivo* was ablated in the FcRγ-deficient mice [9], agreeing with the observations from FcεRIα-deficient mice described above [5]. Then, is the murine IgE/FcεRI system the sole triggering system of anaphylactic response *in vivo*? Mast cells and basophils can be activated *in vitro* and *in vivo* by non-IgE stimuli [54]. Oettgen *et al.* [55] have reported that IgE knockout mice were able to mount systemic anaphylaxis at the similar magnitude to that of wild-type mice on antigen challenge, suggesting that type I hypersensitivity reaction is inducible upon crosslinking between FcγRII/III on mast cells by IgG immune complexes as well as crosslinking of FcεRI via IgE and antigen, in agreement with the phenotypes of FcγRIIB-deficient mice [10] described below.

Type II hypersensitivity and FcRγ knockout

Clynes and Ravetch [49] showed that FcRγ-deficient mice were resistant to the development of experimental autoimmune haemolytic anaemia induced by rabbit anti-mouse red blood cell IgG antibodies. This resistance is primarily a consequence of ineffective erythrophagocytosis, resulting from the lack of FcγRs on phagocytes. They also found that FcRs were critical to the induction of experimental autoimmune thrombocytopenia induced by mouse anti-platelet antibodies. These results suggest that FcRs play an important role in the pathogenesis of antibody-induced type II hypersensitivity [49].

Type III hypersensitivity and FcγR

Type III hypersensitivity response is caused by the deposition of immune complex, and the most characteristic response of this type is the Arthus reaction. Experimental induction of the Arthus reaction is achieved by injection of antibody into the skin and administration of the antigen intravenously followed by assessment of inflammation such as oedema, haemorrhage, and neutrophil infiltration after 2- to 24-h (reverse passive Arthus reaction). This induction scheme is interpreted that the local formation of immune complex, complement activation via the classical pathway, and triggering of the inflammatory cascade. Sylvestre and Ravetch [50] re-evaluated this cascade using FcRγ-deficient mice and found that the Arthus reaction is not able to be induced without FcγR. Mice were injected i.v. with OVA and then given i.d. injections of anti-OVA IgG. The FcRγ knockout mouse showed a distinct reduction in oedema, haemorrhage, and neutrophil infiltration. Therefore, the inflammatory deficit observed in the mutant mice can be attributed to the lack of FcRs in these animals [50].

Sylvestre *et al.* [56] performed the Arthus reaction in complement C3- and C4-deficient mice, and found that the magnitude of the reaction was comparable to that of wild-type animals, confirming that the complement system is not essential and that FcγRs are the primary initiator of the Arthus reaction. Using differential reconstitution of the mast cell-deficient mouse strain W/W^v with mast cells derived from wild-type or FcRγ-deficient mice, Sylvestre and Ravetch [57] further demonstrated that the Arthus reaction in FcRγ-deficient mice was shown to be reconstituted by such mast-cell transfer. Therefore, it was confirmed that the tissue mast cells from FcRγ-deficient mice are primary candidate of this unresponsiveness to type III hypersensitivity, and the effector molecule which is responsible for the initiation of this type of reaction is FcγRIII. This notion has been verified by Hazenbos *et al.* [11]: they have generated mice deficient for the FcγRIIIα in which NK cell-mediated antibody-dependent cytotoxicity and phagocytosis of IgG_1-coated particles by macrophages are lost. Strikingly, these mice lack IgG-mediated mast cell degranulation, and are resistant to IgG-dependent PCA, and exhibit an impaired Arthus reaction. These data support the notion that FcγRIII play a dominant role in the type III hypersensitive response.

ELIMINATION OF NEGATIVE IMMUNE REGULATION IN FcγRIIB-DEFICIENT MICE

Immune regulation via FcγRIIB

FcγRIIB is unique among other FcRs both structurally and functionally. This receptor molecule does not associate with a FcRγ homodimer and possesses a characteristic amino acid sequence, immunoreceptor tyrosine-based inhibition motif (ITIM) [58]. When B cells are stimulated with anti-Ig $F(ab')_2$, the cells proliferate efficiently due to crosslinking of surface Ig, but intact anti-Ig antibody fails to do so. This phenomenon has been interpreted that FcγRIIB molecules on B cells bind the Fc portion of the antibody and form co-crosslinks between antigen receptor and FcγRIIB via anti-Ig antibody. This observation suggested that FcγRIIB on B cells might inhibit antibody production in immune response *in vivo* [59]. Amigorena *et al.* [60] and Muta *et al.* [61] transfected FcγRII-deficient A20 B lymphoma cell line, IIA1.6, with FcγRIIB and observed that anti-Ig antibody which crosslinks surface FcγRIIB with membrane Ig, inhibits Ca^{2+} mobilization in that cell line. It was shown that FcγRIIB modulates membrane Ig-induced Ca^{2+} mobilization by inhibiting Ca^{2+} influx, without changing the pattern of tyrosine phosphorylation [61,62]. The 13-amino-acid sequence including ITIM in the cytoplasmic domain of FcγRIIB was both necessary [60] and sufficient [61] for this effect. The inhibitory motif in FcγRIIB controls lymphocyte activation by inhibiting a Ca^{2+} signaling pathway triggered through ITAMs of Ig-α and Ig-β as a result of recruitment of novel SH2-containing proteins that interact with the tyrosine-phosphorylated ITIM [61]. Cambier and colleagues [63] identified one of the SH2-containing proteins as a cytoplasmic protein tyrosine phosphatase, SHP-1. Daëron *et al.* [58] have reported that the inhibition signalling through FcγRIIB is a more general mechanism of immunosuppression in a variety of cells including B cells, T cells and mast cells, showing that FcγRIIB inhibit FcεRI- and T-cell-receptor-dependent cell activation as well as B-cell-receptor pathway. Recently, Ono *et al.* [64] have shown that the inhibitory signaling by FcγRIIB does not require SHP-1, in mast cells, and results in the recruitment of the SH2-domain-containing inositol polyphosphate 5-phosphatase, SHIP [65], to the tyrosine-phosphorylated 13-amino-acid inhibitory motif of FcγRIIB in both B cells and mast cells.

Augmented antibody response in FcγRIIB-deficient mice

Crosslinking stimulation of surface IgM on splenic B cells using intact anti-μ IgG resulted in the marked proliferative response in B cells from FcγRIIB-deficient mice but not in those cells from wild-type mice, confirming the role of FcγRIIB on B cells as a negative regulator of signalling through antigen receptor *in vitro* [10]. The FcγRIIB-deficient mice showed higher antibody titres than those of wild-type mice when they were immunized with thymus-dependent (TD) such as SRBC and thymus-independent antigens. The number of anti-SRBC IgM and IgG plaque-forming cells in splenocytes from FcγRII-deficient mice were higher than those of wild-type mice in the secondary immune response. Therefore, it was shown that FcγRIIB negatively regulate antibody production *in vivo* [10].

Down regulation of anaphylactic response through FcγRIIB

BMMCs from FcγRIIB-deficient mice were stimulated by crosslinking FcεRI with IgE and anti-IgE or FcγRII/III with 2.4G2 [66] and anti-rat IgG, and estimated for their degranulation *in vitro* [10]. Interestingly, although the mast cells from wild-type mice did not degranulate upon crosslinking of surface FcγRIIB and FcγRIII with 2.4G2 and anti-rat IgG, mast cells from the mutant animals responded significantly on the same crosslinking stimulus [10]. *In vitro* observations by Daëron and colleagues [54,58] are consistent with these data. IgG-induced PCA reaction in FcγRIIB-deficient mice was augmented several-fold over that in wild-type mice [10]. These results provide interesting suggestions on the onset of allergic responses. It is clear that IgE antibody works very efficiently as a triggering molecule through its binding to the high-affinity receptor for IgE, FcεRI on immune cells such as mast cells and basophils. Likewise, the IgG immune complex-induced positive switching through FcγRIII is also present but is attenuated in normal situations because FcγRIIB is co-expressed simultaneously with FcγRIII on most effector cells, and this receptor delivers abortive or negative signaling upon co-engagement with FcγRIII. In FcγRIIB-deficient mice, this negative switching does not operate so that FcγRIII-mediated signaling cascade will work effectively.

FcεRII deletion

Since structure and functional aspects of FcεRII will be discussed in other chapters, this section will only introduce the results of FcεRII-deficient animals briefly. FcεRII-deficient mouse lines have been generated independently by three groups [6–8]. The results obtained are consistent in that B- and T-cell development is normal in such animals, but exhibit inconsistency in antibody responses to some TD antigens. Although the origin of the inconsistency is not known, possible explanations would be that these mice strains were different in their genetic background and that they received different immunization protocols such as in an adjuvant employed.

The genetic background of FcεRII-deficient mouse generated by Yu *et al.* [6] is C57BL/6. B- and T-cell development of the targeted mouse was normal. Immune responses to the helminth *Nippostrongylus brasiliensis* were unaffected. In contrast, immunization with TD antigens such as DNP-OVA, NP-OVA or KLH in alum with *Bordetella pertussis* leads to increased and sustained specific IgE and IgG_1 antibody titres compared with controls. Formation of germinal centers after administration of DNP-OVA in alum with *B. pertussis* is normal. From these observations, they proposed that murine FcεRII may act as a negative feedback component of IgE regulation [6]. The second and the third FcεRII-deficient mouse strains generated by Stief *et al.* [7] and Fujiwara *et al.* [8], respectively, have the genetic background of 129/B6 hybrid. The lack of FcεRII did not cause profound changes in lymphocyte compartments including thymocytes, peripheral T cells, and B-1 and B-2 cells [7]. In a polyclonal Ig production *in vivo* after infection with *N. brasiliensis*, serum IgE and IgG_1 levels of FcεRII-deficient mice were similar to those of wild-type mice in both primary and secondary immune responses [7]. The knockout mice displayed normal lymphocyte differentiation and could mount normal antibody responses, including IgE responses upon immunization with TD antigens such as DNP-OVA in complete Freund's adjuvant and infection with *N. brasiliensis* [8]. Antigen-specific IgE-mediated enhancement of antibody response was impaired in the knockout mice [8].

CONCLUDING REMARKS

Using FcRγ-, FcεRIα- or FcγRIIIα-deficient mice, the mechanisms of initiating type I, II, and type III hypersensitivity cascades have been evaluated. It was shown that the binding of IgE to FcεRI and the binding of IgG immune complexes to FcγRIII on effector cells are pivotal events which lead to anaphylaxy and Arthus reaction, respectively. On the other hand, regulation of these activation events through inhibition molecules [10,58] is a potent means of maintaining homeostasis and controlling inappropriate effector cell activation.

References

1. Heijnen, IAFM, van Vugt MJ, Fanger NA et al. Antigen targeting to myeloid-specific human FcγRI/CD64 triggers enhanced antibody responses in transgenic mice. J Clin Invest. 1996; 97:331–8.
2. Fung-Leung W-P, Sousa-Hitzler JD, Ishaque A et al. Transgenic mice expressing the human high-affinity immunoglobulin (Ig) E receptor α chain respond to human IgE in mast cell degranulation and in allergic reactions. J Exp Med. 1996;183:49–56.
3. Flamand V, Shores EW, Tran T et al. Delayed maturation of CD4 CD8 FcγRII/III$^+$ T and natural killer cell precursors in FcεRIγ transgenic mice. J Exp Med. 1996;184:1725–35.
4. Li M, Wirthmueller U, Ravetch JV. Reconstitution of human FcγRIII cell type specificity in transgenic mice. J Exp Med. 1996;183:1259–63.
5. Dombrowicz D, Flamand V, Brigman KK, Koller BH, Kinet J-P. Abolition of anaphylaxis by targeted disruption of the high affinity immunoglobulin E receptor a chain gene. Cell. 1993;75:969–76.
6. Yu P, Kosco-Vilbois M, Richards M, Kohler G, Lamers MC. Negative feedback regulation of IgE synthesis by murine CD23. Nature. 1994;369:753–6.
7. Stief A, Texido G, Sansig G, Eibel H, Gros GL, van der Putten H. Mice deficient in CD23 reveal its modulatory role in IgE production but no role in T and B cell development. J Immunol. 1994;52:3378–90.
8. Fujiwara H, Kikutani H, Suematsu S et al. The absense of IgE-antibody-mediated augmentation of immune responses in CD23-deficient mice. Proc Natl Acad Sci USA. 1994;91:6835–9.
9. Takai T, Li M, Sylvestre D, Clynes R, Ravetch JV. FcR γ chain deletion results in pleiotrophic effector cell defects. Cell. 1994;76:519–29.
10. Takai T, Ono M, Hikida M, Ohmori H, Ravetch JV. Augmented humoral and anaphylactic responses in FcγRII-deficient mice. Nature. 1996;379:346–9.

11. Hazenbos WLW, Gessner JE, Hofhuis FMA et al. Impaired IgG-dependent anaphylaxis and Arthus reaction in FcγRIII (CD16) deficient mice. Immunity. 1996;5:181–8.
12. Ravetch JV, Anderson CL. FcγR family: proteins, transcripts, and genes. In: Metzger H, ed., Fc Receptors and the Action of Antibodies. Washington D.C.: American Society for Microbiology; 1990:211–35.
13. Ravetch JV, Kinet J-P. Fc receptors. Annu Rev Immunol. 1991;9:457–92.
14. Metzger H. The receptor with high affinity for IgE. Immunol Rev. 1992;125:37–48.
15. van de Winkel JGJ, Capel PJA. Human IgG Fc receptor heterogeneity: molecular aspects and clinical implications. Immunol Today. 1993;14:215–21.
16. Beaven MA, Metzger H. Signal transduction by Fc receptors: the FcεRI case. Immunol Today. 1993;14:222–6.
17. Ravetch JV. Fc receptors: Rubor redux. Cell. 1994;78:553–60.
18. Takai T. Multiple loss of effector cell functions in FcRγ-deficient mice. Intern Rev Immunol. 1996;13:369–81.
19. Ernst LK, van de Winkel JG, Chiu IM, Anderson CL. Three genes for the human high affinity Fc receptor for IgG (FcγRI) encode four distinct transcription products. J Biol Chem. 1992;267:15692–700.
20. Qiu WQ, de Bruin D, Brownstein BH, Pearse R, Ravetch JV. Organization of the human and mouse low-affinity FcγR genes: duplication and recombination. Science. 1990;248:732–5.
21. Ravetch JV, Perussia B. Alternative membrane forms of FcγRIII on human NK cells and neutrophils: cell type specific expression of two genes which differ in single nucleotide substitutions. J Exp Med. 1989;170:481–97.
22. Oakey RJ, Howard TA, Hogarth PM, Tani K, Seldin MF. Chromosomal mapping of the high affinity Fcγ receptor gene. Immunogenetics. 1992;35:279–82.
23. Seldin MF, Prins JB, Rodrigues N, Todd JA, Meisler MH. Encyclopedia of the mouse genome III: mouse chromosome 3. Mammal Genome. 1993;4:1–10.
24. Shirakawa T, Li A, Dubowitz M et al. Association between atopy and variants of the β subunit of the high-affinity immunoglobulin E receptor. Nature Genet. 1994;7:125–30.
25. Maliszewski CR, March CJ, Schoenborn MA, Gimpel S, Shen L. Expression cloning of a human Fc receptor for IgA. J Exp Med. 1990;172:1665–72.
26. Gomez-Guerrero C, Gonzalez E, Egido J. Evidence for a specific IgA receptor in rat and human mesangial cells. J Immunol. 1993;151:7172–81.
27. Shen L, Collins JE, Schoenborn MA, Maliszewski CR. Lipopolysaccharide and cytokine augmentation of human monocyte IgA receptor expression and function. J Immunol. 1994; 152:4080–6.
28. Zhang G, Young JR, Tregaskes CA, Sopp P, Howard CJ. Identification of a novel class of mammalian Fcγ receptor. J Immunol. 1995;155:1534–41.
29. Wagtmann N, Biassoni R, Cantoni C et al. Molecular clones of the p58 NK cell receptor reveal immunoglobulin-related molecules with diversity in both the extra- and intracellular domains. Immunity. 1995;2:439–49.
30. Arm, JP, Gurish MF, Reynolds DS et al. Molecular cloning of gp49, a cell-surface antigen that is preferentially expressed by mouse mast cell progenitors and is a new member of the immunoglobulin superfamily. J Biol Chem. 1991;266:15966–73.
31. Katz HR, Vivier E, Castells MC, McCormick MJ, Chambers JM, Austen KF. Mouse mast cell gp49B1 contains two immunoreceptor tyrosine-based inhibition motifs and suppresses mast cell activation when coligated with the high-affinity Fc receptor for IgE. Proc Natl Acad Sci USA. 1996;93:10809–14.
32. Wang LL, Mehta IK, LeBlanc PA, Yokoyama WM. Mouse natural killer cell express gp49B1, a structural homologue of human killer inhibitory receptors. J Immunol. 1997;158:13–7.
33. Rojo S, Burshtyn DN, Long EO, Wagtman N. Type I transmembrane receptor with inhibitory function in mouse mast cells and NK cells. J Immunol. 1997;158:9–12.
34. Hayami K, Fukuta D, Nishikawa Y et al. Molecular cloning of a novel murine cell-surface glycoprotein homologous to killer cell inhibitory receptors. J Biol Chem. [in press].
35. Ra C, Jouvin M-HE, Kinet J-P. Complete structure of the mouse mast cell receptor for IgE (FcεRI) and surface expression of chimeric receptors (rat-mouse-human) on transfected cells. J Biol Chem. 1989;264:15323–7.
36. Kurosaki T, Gander I, Ravetch JV. A subunit common to an IgG Fc receptor and the T-cell receptor mediates assembly through different interactions. Proc Natl Acad Sci USA. 1991;88: 3837–41.

37. Orloff DG, Ra C, Frank SJ, Klausner RD, Kinet J-P. Family of disulphide-linked dimers containing the ζ and η chains of the T-cell receptor and the γ chain of Fc receptors. Nature. 1990;347:189–91.
38. Vivier E, Rochet N, Kochan JP, Preski DH, Schlossman SF. Structural similarity between Fc receptors and T cell receptors. Expression of the γ subunit of FcεRI in human T cells, natural killer cells and thymocytes. J Immunol. 1991;147:4263–70.
39. Koyasu S, D'Adamio L, Arulanandam ARN, Abraham S, Clayton LK, Reinherz E. T cell receptor complexes containing FcεRIg homodimers in lieu of CD3ζ and CD3η components: a novel isoform expressed on large granular lymphocytes. J Exp Med. 1992;175:203–9.
40. Ernst LK, Duchemin A-M, Anderson CL. Association of the high-affinity receptor for IgG (FcγRI) with the γ subunit of the IgE receptor. Proc Natl Acad Sci USA. 1993;90:6023–7.
41. Scholl PR, Geha RS. Physical association between the high-affinity IgG receptor (FcγRI) and the γ subunit of the high-affinity IgE receptor (FcεRIg). Proc Natl Acad Sci USA. 1993;90: 8847–50.
42. Wirthmueller U, Kurosaki T, Murakami MS, Ravetch JV. Signal transduction by FcγRIII (CD16) is mediated through the γ chain. J Exp Med. 1992;175:1381–90.
43. Paolini R, Renard V, Vivier E et al. Different roles for the FcεRI γ chain as a function of the receptor context. J Exp Med. 1995;181:247–55.
44. Reth M. Antigen receptor tail clue. Nature. 1989;338:383–4.
45. Kuster H, Thompson H, Kinet J-P. Characterization and expression of the gene for the human Fc receptor γ subunit. Definition of a new gene family. J Biol Chem 1990;265:6448–52.
46. Sandor M, Galon J, Takacs L et al. An alternative Fcγ-receptor ligand: potential role in T-cell development. Proc Natl Acad Sci USA. 1994;91:12857–61.
47. Park SY, Arase H, Wakizaka K et al. Differential contribution of the FcRγ chain to the surface expression of the T cell receptor among T cells localized in epithelia: analysis of FcRγ-deficient mice. Eur J Immunol. 1995;25:2107–10.
48. Takizawa F, Adamczewski M, Kinet J-P. Identification of the low affinity receptor for immunoglobulin E on mouse mast cells and macrophages as FcγRII and FcγRIII. J Exp Med. 1992;176:469–76.
49. Clynes R, Ravetch JV. Cytotoxic antibodies trigger inflammation through Fc receptors. Immunity. 1995;3:21–6.
50. Sylvestre DL, Ravetch JV. Fc receptors initiate the Arthus reaction: redefining the inflammatory cascade. Science. 1994;265:1095–8.
51. Greenberg S, Chang P, Silverstein SC. Tyrosine phosphorylation is required for Fc receptor-mediated phagocytosis in mouse macrophages. J Exp Med. 1993;177:529–34.
52. Joiner KA, Fuhrman SA, Miettinen HM, Kasper LH, Mellman I. *Toxoplasma gondii*: fusion competence of parasitophorous vacuoles in Fc receptor-transfected fibroblasts. Science. 1990;249:641–6.
53. Daëron M, Malbec O, Latour S, Bonnerot C, Segal DM, Fridman WH. Distinct intracytoplasmic sequences are required for endocytosis and phagocytosis via murine FcγRII in mast cells. Intern Immunol. 1993;5:1393–401.
54. Daëron M, Bonnerot C, Latour S, Fridman WH. Murine recombinant FcγRIII but not FcγRII, trigger serotonin release in rat basophilic leukemia cells. J Immunol. 1992;149:1365–73.
55. Oettgen HC, Martin TR, Wynshaw-Boris A, Deng C, Drazen JM, Leder P. Active anaphylaxis in IgE-deficient mice. Nature. 1994;370:367–70.
56. Sylvestre D, Clynes R, Ma M, Warren H, Carroll MC, Ravetch JV. Immunoglobulin G-mediated inflammatory responses develop normally in complement-deficient mice. J Exp Med. 1996;184:2385–92.
57. Sylvestre DL, Ravetch JV. A dominant role for mast cell Fc receptors in the Arthus reaction. Immunity. 1996;5:387-90.
58. Daëron M, Latour S, Malbec O et al. The same tyrosine-based inhibition motif, in the intracytoplasmic domain of FcγRIIB, regulates negatively BCR-, TCR-, and FcR-dependent cell activation. Immunity. 1995;3:635–46.
59. Chan PL, Sinclair NR St C. Regulation of the immune response. V. An analysis of the function of the Fc portion of antibody in suppression of an immune response with respect to interaction with components of the lymphoid system. Immunology. 1971;21:967–81.
60. Amigorena S, Bonnerot C, Drake JR et al. Cytoplasmic domain heterogeneity and functions of IgG Fc receptors in B lymphocytes. Science. 1992;256:1808–12.

61. Muta T, Kurosaki T, Misulovin Z, Sanchez M, Nussenzweig MC, Ravetch JV. A 13-amino-acid motif in the cytoplasmic domain of FcγRIIB modulates B-cell receptor signalling. Nature. 1994;368:70–3.
62. Diegel ML, Rankin BM, Bolen JB, Dubois PM, Kiener PA. Cross-linking of Fcγ receptor to surface immunoglobulin on B cells provides an inhibitory signal that closes the plasma membrane calcium channel. J Biol Chem. 1994;15:11409–16.
63. D'Ambrosio D, Hippen KL, Minskoff SA et al. Recruitment and activation of PTP-1C in negative regulation of antigen receptor signaling by FcγRIIB1. Science. 1995;268:293–7.
64. Ono M, Bolland S, Tempst P, Ravetch JV. Role of the inositol phosphatase SHIP in negative regulation of the immune system by the receptor FcγRIIB. Nature. 1996;383:263–6.
65. Damen JE, Liu L, Rosten P et al. The 145-kDa protein induced to associate with Shc by multiple cytokines is an inositol tetraphosphate and phosphatidylinositol 3,4,5-trisphosphate 5-phosphatase. Proc Natl Acad Sci USA. 1996;93:1689–93.
66. Unkeless JC. Characterization of a monoclonal antibody directed against mouse macrophage and lymphocyte Fc receptors. J Exp Med. 1979;150:580–96.

4
Commentary

R. G. LYNCH

During the past 25 years there has been an unprecedented growth in fundamental knowledge about the structures and functions of Fc receptors. It has been a time of great excitement and discovery. The present volume records some of the history of these developments and provides a comprehensive summary of the current state of knowledge.

In the 1970s when the concept of an Fc receptor was beginning to take root, a prominent immunologist commented to me that 'Fc receptors couldn't be very important, because you found them everywhere you looked'. In retrospect, the broad range of cells that express Fc receptors, the pleiotropic functions associated with these receptors, and certain aspects of the variation in experimental data between laboratories, might have been considered clues pointing to a substantial structural heterogeneity of Fc receptors. Nonetheless, it was still somewhat surprising to see the extensive degree of structural diversity that has been identified by investigations of Fc receptor genes and gene products.

The remarkable ability of recombinant DNA technology to dissect and resolve molecular structures redirected the focus of Fc receptor investigations towards determining the molecular basis of Fc receptor functions, a theme that is present in many of the chapters of this volume. In their scholarly review Gavin *et al.* present a comprehensive and detailed review of the structural basis of Ig-ligand interactions with FcγRI, FcγRII, FcγRIII, FcεRI, FcεRII, FcαRI, FcRn and pIgR; they also provide an extensive, and valuable bibliography for this subject. Mapping the structure–function connections for FcR and their ligands is a promising area of investigation because of its potential to lead to therapeutic interventions in disease. As reviewed by Gavin *et al.*, there is a substantial and growing body of knowledge about the molecular determinants of some Ig-FcR

J.G.J. van de Winkel and P.M. Hogarth (eds.), The Immunoglobulin Receptors and their Physiological and Pathological Roles in Immunity. 49–50.

interactions and this is proving valuable. However, we are just beginning to identify some of the structural determinants of the constitutive and inducible interactions that occur between Fc receptors and their associated molecules, and there is essentially no information available about the structural basis of interactions between Fc receptors and their alternative, non-Ig ligands.

Recombinant DNA technology made if feasible to determine gene structure, and from it to infer protein structure. Knowledge of gene structure, coupled with remarkable advances in technology, made it feasible to experimentally control the expression of genes in eukaryotic cells. Initially cell lines, and later strains of mice were engineered for over-expression or non-expression of Fc receptor genes and their associated molecules. These approaches generated a wealth of knowledge that provided insights into the determinants of production, assembly, surface membrane insertion, and physiological and pathological functions of several Fc receptors. Interesting observations, some of which challenge previous paradigms, have come from investigations in FcR-transgenic mice and FcR-deficient mice. This area of research is succinctly reviewed in chapter 3 by Takai and Ravetch who have also contributed many of the key observations. Of particular interest are investigations in the FcRγ-deficient mouse that have identified a key role for the FcRγ chain in a variety of inflammatory reactions, including Types I, II, and III hypersensitivity reactions, and which also suggest that the γ-chain plays a role in T cell development. The pleiotropic functions linked to the γ-chain reflects that the γ-chain is a subunit of FcεRI, FcγRI, FcγRIII, FcαRI and of the TCR complex in a subset of T cells. Also of interest are the mice deficient in FcγRIIB, which provides a useful model to further investigate the inhibitory functions ascribed to this unique, ITIM-containing IgG Fc receptor. As additional strains of genetically engineered mice are developed they will continue to provide valuable tools to address fundamental questions about the role of Fc receptors in development and disease.

5
The polymeric immunoglobulin receptor

C. KAETZEL and K. MOSTOV

INTRODUCTION

IgA is the primary immunoglobulin found in a wide range of epithelial and mucosal secretions, including milk, saliva, intestinal secretions, bile, respiratory secretions, and tears [1]. Over 30 years ago it was observed that IgA isolated from such secretions contains an extra polypeptide of about 70 kDa, known as secretory component (SC) [2]. SC is synthesized by the epithelial cell and associates with dimeric IgA (dIgA) as it is transported across the epithelial cell. It was subsequently discovered that SC is a proteolytic fragment of an integral membrane protein, known as the polymeric immunoglobulin receptor (pIgR) [3]. The basic model for the transepithelial transport of dIgA by the pIgR is shown in Figure 5.1. The pIgR is synthesized as an integral membrane protein in the rough endoplasmic reticulum and then travels to the Golgi apparatus. In the last station of the Golgi, known as the trans-Golgi network (TGN), the pIgR is sorted into vesicles that deliver it to the basolateral surface of the epithelial cell. At that surface the pIgR can bind to dIgA that is produced by plasma cells in the lamina propria underlying the epithelium. The pIgR and bound dIgA are then endocytosed and delivered to endosomes. The receptor and ligand move through a series of endocytotic and transcytotic vesicles and are ultimately delivered to the apical plasma membrane. There the extracellular, ligand binding portion of the pIgR is cleaved and released together with the dIgA into external secretions. This cleaved fragment is the SC.

J.G.J. van de Winkel and P.M. Hogarth (eds.), The Immunoglobulin Receptors and their Physiological and Pathological Roles in Immunity. 51–62.

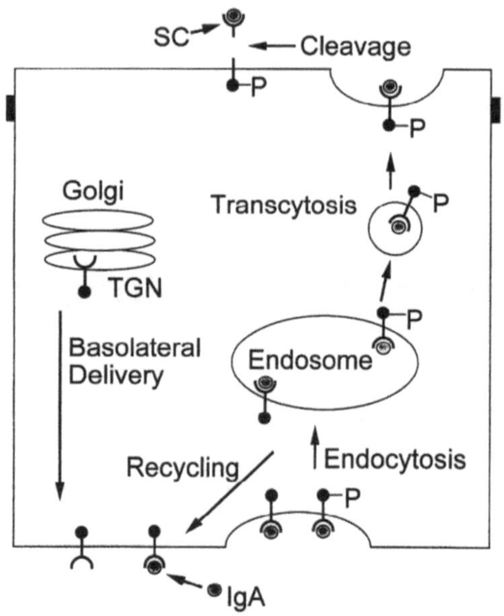

Figure 5.1 Pathway of pIgR through an epithelial cell. A simplified epithelial cell is illustrated, with the apical surface at the top, and basolateral surface at the bottom

STRUCTURAL FEATURES OF THE POLYMERIC IMMUNOGLOBULIN RECEPTOR

Cell surface pIgR has been observed in most human secretory epithelia, including intestine, bronchus, salivary glands, gallbladder, renal tubules, pancreas, lacrimal glands, sweat glands, lactating mammary glands and uterus [reviewed in Reference 4]. While expression of pIgR in human liver is restricted primarily to the bile duct epithelium, rodents and lagomorphs express high levels of pIgR in hepatocytes and actively transport dIgA from blood to bile. The primary structure of pIgR has been determined from five mammalian species by molecular cloning of complementary DNA [reviewed in Reference 5]. Alignment of the amino acid sequences of human, mouse, rat, bovine and rabbit pIgR revealed a number of structural features that are highly conserved. The basic structure of pIgR comprises an N-terminal extracellular region of approximately 560 amino acids, a 23-amino acid membrane-spanning region, and a 103-amino acid C-terminal cytoplasmic region (Figure 5.2). The extracellular region of pIgR is the ligand-binding portion of the molecule conferring specificity for polymeric Ig. This region contains five domains with homology to Ig variable domains and a sixth non-Ig-like domain connecting to the transmembrane region. Interspecies homology is greatest in domain 1, especially within a 23-amino acid segment that has been shown to be critical for Ig

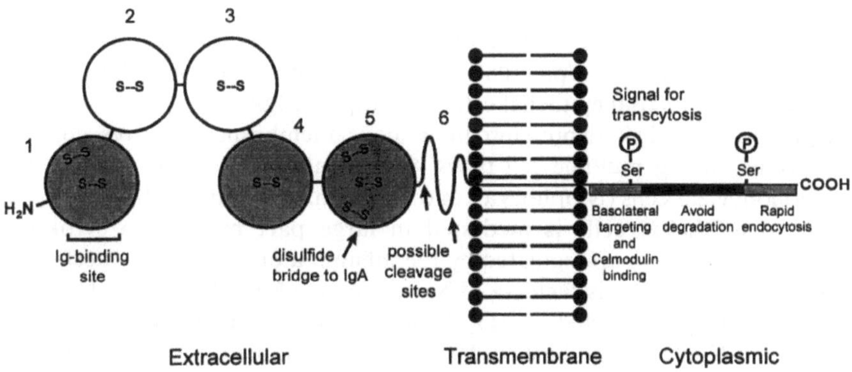

Figure 5.2 Conserved structural features of the polymeric Ig receptor

binding [6]. Domains 2 and 3 are encoded by a single exon, which is frequently spliced out in rabbit pIgR mRNA [7] but not in other species. The functional roles of domains 2 and 3 are unclear, because the alternatively spliced 'short' form of rabbit pIgR is competent for binding and transcytosis of polymeric Ig [8].

Each of the five Ig-like domains contains an internal disulphide bond, characteristic of Ig homology units, and conserved in all known species of pIgR (Figure 5.2). A second internal disulphide bond is found in all species in domains 1 and 5, and in some species in domains 2, 3 and 4. A third highly conserved disulphide bond is found in domain 5. In human secretory IgA, the 'extra' disulphide bond in domain 5 has been shown to rearrange to form a disulphide bond with cysteine residues in one of the α heavy chains of dIgA [9]. Disulphide bonding between pIgR and dIgA has been shown to occur late in the transcytotic pathway [10], and normal transcytosis of the pIgR–dIgA complex can occur under conditions where disulphide bonding is prevented [10,11]. Therefore, the significance of disulphide bonding between pIgR and dIgA may not be to facilitate transcytosis, but rather to prevent dissociation of the secretory IgA complex in the often harsh milieu of external secretions.

Other conserved structural features of pIgR include two potential proteolytic cleavage sites [5,12]; an α-helical transmembrane region and intracellular targeting signals in the cytoplasmic region that interact with cytoplasmic proteins to direct pIgR through the transcytotic pathway (see below).

REGULATION OF pIgR SYNTHESIS

Regulation by cytokines and hormones

A unique feature of pIgR is that it makes only one trip across epithelial cells before being cleaved and released at the apical surface. Because there is a 1:1 stoichiometry between SC and dIgA in secretory IgA, it follows that one molecule of pIgR must be produced by the epithelial cell for every molecule of dIgA that is transported. Upregulation of pIgR expression would thus increase the capacity of mucosal epithelial cells to transport dIgA.

Regulation of pIgR expression in mucous membranes involves complex interactions among multiple cell types. Communication between lamina propria mononuclear cells (lymphocytes and macrophages), intraepithelial lymphocytes and epithelial cells is mediated in large part by soluble cytokines. Expression of pIgR is up-regulated by the inflammatory cytokines interferon-γ (IFN-γ), tumour necrosis factor-α (TNF-α), interleukin (IL)1-α and IL-β in a variety of cell types, including HT-29 human colon carcinoma cells [13–16] (Blanch and Kaetzel, unpublished data), human endometrial cells [17], human keratinocytes [18], rat uterus [19,20] and rat lacrimal gland [21]. Although the effects of the Th1-type cytokine IFN-γ and the Th2-type cytokine IL-4 are often antagonistic, with respect to pIgR these cytokines are synergistic. Transforming growth factor-β has also been found to elevate pIgR levels in human and rat intestinal epithelial cell lines [16,22].

A widely used model for studying pIgR regulation in human intestinal epithelium is the HT-29 human colon carcinoma cell line, where pIgR is up-regulated by recombinant IFN-γ, TNF-α, and IL-4 [reviewed in Reference 16]. To determine the physiological significance of these observations, further experiments were conducted using natural cytokines produced by freshly isolated human intestinal lamina propria mononuclear cells activated *in vitro* [16]. Supernatants from these activated cells dramatically induced expression of pIgR by HT-29 cells, to levels far exceeding those achieved with recombinant cytokines. Antibody-mediated neutralization of these cytokines suggested roles for both IFN-γ and TNF-α, with IFN-γ acting as the central regulator. IFN-γ and TNF-α producing cells are abundant in the human intestine, and infection or immunization with pathogenic organisms such as *Salmonella typhimurium*, *Listeria monocytogenes*, *Vibrio cholerae* and cytomegalovirus has been shown to enhance intestinal production of these cytokines [reviewed in Reference 23].

Expression of pIgR is also regulated by hormones, in a cell-type specific manner. In human uterus, pIgR levels vary significantly during the menstrual cycle, with the highest expression seen during the luteal phase [24,25]. In rat uterus, pIgR expression is elevated by estrogen and antagonized by progesterone [26,27]. In contrast, pIgR expression in rabbit mammary gland is up-regulated by prolactin and antagonized by estrogen and progesterone [28]. In the male reproductive tract, including the prostate and seminal vesicles, pIgR expression is up-regulated by androgens [29]. Interestingly, androgens also up-regulate pIgR expression in the lacrimal gland of male rats; SC output in tears is five-fold higher than in female rats [30].

Mechanisms of pIgR gene regulation

Investigations into the molecular mechanisms of pIgR regulation in HT-29 cells demonstrated that IFN-γ and TNF-α up-regulate pIgR mRNA levels by a mechanism involving *de novo* protein synthesis [12,31–33]. These observations suggest that IFN-γ and TNF-α may induce synthesis of one or more factors that enhance transcription of the *PIGR* gene. Recently, genomic DNA has been isolated encompassing the 5′-flanking and exon 1 regions of the human, rat and mouse *PIGR* genes ([32,34–36] (H. de Boer, personal communication). The promoter-proximal region of the human *PIGR* gene contains three motifs homologous to the interferon stimulated response element [ISRE; reviewed in Reference 37]. Two of the ISRE motifs are in the 5′-flanking region and one is in exon 1. The exon 1 ISRE is perfectly conserved in the human, rat and mouse *PIGR* genes, supporting the concept that this ISRE may be functionally significant. In contrast, the upstream ISREs are unique to the human *PIGR* gene. Another common feature of the human, rat and mouse *PIGR* genes is the presence of two 10-bp inverted repeats in the 5′-flanking region, which may play a structural role in the transcriptional activity of the *PIGR* promoter.

Basal transcriptional activity of the human and rat *PIGR* genes has been tested using transient transfection of chimeric reporter plasmids in which segments of the pIgR promoter were fused to the firefly luciferase gene [32,35,36]. A segment spanning approximately 210 bp of 5′-flanking region and a portion of exon 1 was found to be optimal for basal transcription of the human and rat *PIGR* genes in liver and intestinal epithelial cells. To test for regulation of the human PIGR promoter by IFN-γ and TNF-α, these reporter plasmids were transiently transfected into HT-29 cells, which were then cultured in the presence or absence of IFN-γ or TNF-α [12,32]. Analysis of luciferase activity in the transfected cells demonstrated that the promoter region from nt −280 to +29 was necessary and sufficient for induction of transcription by both IFN-γ and TNF-α. Deletion of the exon 1 ISRE did not affect induction of transcription by IFN-γ, but abolished induction by TNF-α. Further truncation of the promoter fragment from the 5′ end to nt −95 (which deleted both upstream ISREs) led to a loss of inducibility by both IFN-γ and TNF-α. These results suggested that the exon 1 ISRE is necessary for regulation of the proximal promoter by TNF-α, but that all three ISREs may mediate IFN-γ regulation.

Binding of nuclear proteins to the ISRE motifs in the PIGR promoter

The interferon regulatory factor (IRF) family of transcription factors includes IRF-1, IRF-2, interferon-stimulated gene factor 3 (ISGF3)-γ, and several lymphoid-specific factors [reviewed in Reference 38]. IRF-1 has been associated with transcriptional activation and cell cycle arrest while IRF-2 has been associated with transcriptional repression and promotion of cell growth/oncogenesis. Stimulation of cells with IFN-γ, TNF-α or IL-1 dramatically upregulates IRF-1 synthesis, leading to activation of many inducible genes containing ISRE sites [38,39]. Electrophoretic mobility shift assays demonstrated that the *PIGR* upstream ISREs bound nuclear proteins from both unstimulated and cytokine-stimulated HT-29 cells [32]. In contrast, the exon 1 ISRE bound nuclear proteins only after stimulation with IFN-γ or TNF-α. Experiments with antibodies against IRF-1, IRF-2 and ISGF-3-γ demonstrated that IRF-1 binds to the exon 1 ISRE, but indicated that the upstream ISREs may bind novel proteins. Inhibition of protein synthesis with cycloheximide blocked the IFN-γ induced binding of IRF-1 to the exon 1 ISRE, suggesting that IRF-1 is newly synthesized by HT-29 cells in response to IFN-γ. A role for IRF-1 was confirmed independently by demonstrating that transient expression of IRF-1 in HT-29 cells was sufficient for upregulation of the *PIGR* promoter (Blanch and Kaetzel, manuscript in preparation). These results suggest that both IFN-γ and TNF-α induce synthesis of the IRF-1 transcription factor in human intestinal epithelial cells, which in turn upregulates transcription of the *PIGR* gene by binding to the exon 1 ISRE.

pIgR TRAFFICKING AND SIGNALS

The pIgR has been an extremely useful model system to elucidate the pathways and mechanisms of polarized membrane traffic in epithelial cells [reviewed in References 40–42]. Most of our understanding of pIgR trafficking has come from two experimental systems. The first is rat liver, where expression of pIgR by hepatocytes allows transcytosis of dIgA from blood to bile. In rat hepatocytes the pIgR is directed first to the sinusoidal surface, which is equivalent to the basolateral surface. Transcytosis is to the bile canalicular surface, equivalent to the apical surface, where the SC and dIgA are released into the bile [reviewed in Reference 43]. A great advantage of rat liver is that highly purified subcellular fractions can be isolated. For instance, it has been possible to isolate a fraction highly enriched in vesicles that appear to be involved in the transcytosis of the pIgR [44]. As a second experimental system we have expressed the cloned cDNA for pIgR in Madin-Darby canine kidney (MDCK) cells, which form a tight, well-polarized monolayer when cultured on permeable supports. The transfected pIgR behaves in the MDCK cells as *in vivo*, and serves as a useful model to study protein trafficking.

Basolateral sorting signal of the pIgR

In a systematic effort to understand the signals that direct the pIgR during its complex pathway through the cell, we have made a series of mutations in the cytoplasmic domain of the pIgR. We have shown that basolateral sorting requires a sorting signal in the cytoplasmic domain of the membrane protein [40,45]. The pIgR spans the membrane once and contains a carboxyl-terminal, cytoplasmic domain of 103 amino acids. The 17 amino acids of this cytoplasmic domain that lie closest to the membrane comprise a signal that is necessary and sufficient for targeting from the TGN to the basolateral surface. This 17-residue segment can be transplanted to a heterologous reporter molecule and direct the delivery of this molecule to the basolateral surface [reviewed in Reference 40].

Analysis of the 17-residue basolateral sorting signal of pIgR by alanine scanning mutagenesis indicates that three residues appear to be crucial for basolateral sorting [46]. Individual point mutations of His^{656}, Arg^{657}, or Val^{660} to Ala largely blocked basolateral targeting. In addition, the three-dimensional solution structure of a synthetic peptide corresponding to this 17 residue signal has been determined by nuclear magnetic resonance spectroscopy. One of the three crucial residues (Val^{660}) is in position 3 of a type I β turn, while the other two crucial residues (His^{656} and Arg^{657}) precede the turn. This β turn is particularly interesting because signals for clathrin-mediated endocytosis (and probably for several other sorting events) generally contain a type I β turn [40].

The pathway of transcytosis

pIgR transcytosis can be divided into three steps [47]. In step 1, pIgR undergoes clathrin-mediated endocytosis from the plasma membrane to basolateral early endosomes. In step 2, a microtubule-dependent event transports pIgR to a tubular compartment located beneath the center of the apical plasma membrane (the apical recycling compartment). In step 3, pIgR is delivered from the apical recycling compartment to the apical plasma membrane. The organization of the apical recycling compartment around the centriole is dependent on microtubules [47]. Protein traffic to both the apical and basolateral surfaces is slowed to varying degrees by nocodazole, suggesting that microtubules play a role in delivery to both surfaces [48].

Interaction of pIgR with viruses

Many viruses infect epithelial cells, and the pIgR/dIgA system has been shown to interact with these viruses in two highly significant and surprising ways. First, antiviral dIgA that is being transcytosed by the pIgR can neutralize virus inside the epithelial cell [49,50]. Neutralization of intracellular virus is a function previously associated only with the cellular immune system, not with soluble antibodies. This phenomenon was originally observed in MDCK cells infected with either Sendai or influenza virus, using antibodies directed against membrane glycoproteins (haemagglutinin-neuraminidase) found on the surface of virions. One possible explanation for these results is that newly synthesized viral glycoproteins that move from the TGN to the cell surface pass through an endosomal compartment that contains transcytosing dIgA. This concept is supported by the observation that dIgA antibodies against internal viral proteins did not inhibit viral replication. Intracellular neutralization of virus has recently been found to be a major mechanism for protection against rotavirus infection *in vivo* [51]. Mice bearing tumours secreting dIgA specific for the inner capsid protein VP6 were protected, while mice bearing tumours secreting dIgA specific for the outer membrane hemagglutinin VP4 were not protected. In this case the mechanism is less clear, because the protective dIgA is directed against an internal antigen of the virion, not a surface glycoprotein. One possible clue comes the very recent report that transcytosing dIgA passes through a compartment containing the Rab1a protein [52]. This protein is classically associated with the endoplasmic reticulum (ER) and transport between ER and the Golgi. This surprising result suggests that a portion of dIgA internalized by the pIgR is not transcytosed, but instead transported to the ER, and from there back to the cytosol. This could provide a mechanism for delivery of antiviral dIgA to the cytosol and thus viral neutralization. From these two contrasting examples it appears that the mechanism of viral neutralization by intracellular dIgA may vary among viruses depending on their mode of replication, and perhaps also on the characteristics of the infected cell. A second type of interaction of dIgA and viruses comes from work showing that antibodies to Epstein-Barr virus can bind to the virus in the nasopharynx [53]. The dIgA–virus complex can then be internalized via the pIgR on the nasopharyngeal epithelium, leading to infection of these otherwise refractory epithelial cells. Therefore, the pIgR can cause tissue-specific spread of a virus.

REGULATION OF pIgR TRAFFIC

Transcytosis of the pIgR is regulated at multiple levels and provides an excellent model to study the regulation of membrane traffic [41,54]. The internalization of pIgR from the basolateral plasma membrane (step 1 of transcytosis) is regulated by phosphorylation [55]. The pIgR contains two tyrosine-based internalization signals, which resemble internalization signals found in other rapidly endocytosed receptors [56]. However, phosphorylation of Ser^{726}, which is distantly located from both of these tyrosine-based signals, is required for rapid endocytosis of the pIgR.

From both the biosynthetic and endocytotic pathways, the pIgR basolateral sorting signal helps direct the pIgR to the basolateral surface. This sorting signal can be thought of as a 'basolateral retrieval signal'. If this basolateral retrieval signal were permanently active, the pIgR might continuously recycle to the basolateral surface, and never be transcytosed to the apical surface. However, we have found at three mechanisms that promote transcytosis. First, Ser^{664}, located in the 17 residue basolateral sorting signal, is phosphorylated [57]. We suggest that the non-phosphorylated pIgR is initially targeted to the basolateral surface. Once reaching that surface (or perhaps after endocytosis) the pIgR is phosphorylated on Ser^{664}, thereby weakening the basolateral sorting signal and allowing the pIgR to be transcytosed.

The second control mechanism is binding of the ligand, dIgA [58]. Although a significant rate of transcytosis occurs when the pIgR is not bound to its ligand, binding of dIgA augments this rate of transcytosis. This stimulation does not depend on phosphorylation of Ser^{664}. In fact, because of the low baseline rate of transcytosis of the Ala^{664} mutant without dIgA bound, binding of dIgA to this mutant gives a proportionately greater stimulation of transcytosis. However, maximal transcytosis of pIgR is achieved only when dIgA is bound and phosphorylation of Ser^{664} occurs. That dIgA binding to the pIgR stimulates transcytosis suggests that the pIgR is capable of transducing a signal across the plasma membrane to the cytoplasmic sorting machinery. Binding of dIgA to the pIgR very rapidly causes the tyrosine-phosphorylation of several proteins, including a phosphatidylinositol-specific phospholipase Cγ1 (PLCγ1). This enzyme causes hydrolysis of phosphatidylinositol-4,5-bis-phosphate (PIP2) to diacylglyceride and inositol 1,4,5-trisphosphate (IP3). The diacylglyceride in turn leads to activation of protein kinase C (PKC). Activation of PKC by phorbol esters stimulates transcytosis, so it is likely that activation of PKC by dIgA binding to pIgR also stimulates transcytosis [59,60]. The production of IP3 probably causes the release of Ca^{2+} from intracellular stores and an increase in intracellular free Ca^{2+} ($[Ca^{2+}]_i$). Artificially increasing $[Ca^{2+}]_i$ with the drug thapsigargin rapidly stimulates transcytosis, so it is likely that the increase in $[Ca^{2+}]_i$ caused by dIgA binding to pIgR also stimulates transcytosis [60]. Binding of Ca^{2+} to calmodulin (CaM) could stimulate transcytosis in several possible ways. CaM binds directly to the 17 residue basolateral sorting signal of the pIgR [61]. We suggest that CaM binding may regulate transcytosis. The stimulation of transcytosis by dIgA binding acts on step 3 of transcytosis, i.e. the delivery from the apical recycling compartment to the apical plasma membrane [62].

A third mechanism for regulating transcytosis is comprised of extracellular signals. Binding of bradykinin or cholinergic drugs to their receptors stimulates transcytosis $[Ca^{2+}]_i$. Activation of the heterotrimeric Gs protein also stimulates transcytosis [63].

Acknowledgements

This work was supported by National Institutes of Health grants CA-51998, AI-26449, DK43999, AI25144, AI-36953, and AI-39161 and the University of Kentucky Research Fund.

References

1. Mostov KE. Transepithelial transport of immunoglobulins. Annu Rev Immunol. 1994;12: 63–84.
2. Tomasi TB Jr., Tan EM, Solomon A, Prendergast RA. Characteristics of an immune system common to certain external secretions. J Exp Med. 1965;121:101–24.
3. Mostov KE, Kraehenbuhl J-P, Blobel G. Receptor-mediated transcellular transport of immunoglobulin: synthesis of secretory component as multiple and larger transmembrane forms. Proc Natl Acad Sci USA. 1980;77:7257–61.
4. Mostov KE, Kaetzel CS. Immunoglobulin transport and the polymeric immunoglobulin receptor. In: Ogra PL, Mestecky J, Lamm ME, Strober W, McGhee JR, Bienenstock J, eds, Handbook of Mucosal Immunology, 2nd edn. San Diego, CA: Academic Press; 1998: in press.

5. Piskurich JF, Blanchard MH, Youngman KR, France JA, Kaetzel CS. Molecular cloning of the mouse polymeric Ig receptor: Functional regions of the molecule are conserved among five mammalian species. J Immunol. 1995;154:1735–47.
6. Bakos M-A, Kurosky A, Goldblum RM. Characterization of a critical binding site for human polymeric Ig on secretory component. J Immunol. 1991;147:3419–26.
7. Deitcher DL, Mostov KE. Alternate splicing of rabbit polymeric immunoglobulin receptor. Mol Cell Biol. 1986;6:2712–5.
8. Schaerer E, Verrey F, Racine L, Tallichet C, Reinhardt M, Kraehenbuhl J-P. Polarized transport of the polymeric immunoglobulin receptor in transfected rabbit mammary epithelial cells. J Cell Biol. 1990;110:987–98.
9. Fallgreen-Gebauer E, Gebauer W, Bastian A et al. The covalent linkage of secretory component to IgA. Structure of sIgA. Biol Chem Hoppe-Seyler. 1993;374:1023–8.
10. Chintalacharuvu KR, Tavill AS, Louis LN, Vaerman J-P, Lamm ME, Kaetzel CS. Disulfide bond formation between dimeric immunoglobulin A and the polymeric immunoglobulin receptor during hepatic transcytosis. Hepatology. 1994;19:162–73.
11. Tamer CM, Lamm ME, Robinson JK, Piskurich JF, Kaetzel CS. Comparative studies of transcytosis and assembly of secretory IgA in Madin-Darby canine kidney cells expressing human polymeric Ig receptor. J Immunol. 1995;155:707–14.
12. Kaetzel CS, Blanch VB, Hempen PM, Phillips KM, Piskurich JF, Youngman KR. The polymeric Ig receptor: structure and synthesis. Biochem Soc Trans. 1997;25:475–80.
13. Sollid LM, Kvale D, Brandtzaeg P, Markussen G, Thorsby E. Interferon-gamma enhances expression of secretory component, the epithelial receptor for polymeric immunoglobulins. J Immunol. 1987;138:4303–6.
14. Kvale D, Brandtzaeg P, Lovhaug D. Up-regulation of the expression of secretory component and HLA molecules in a human colonic cell line by tumour necrosis factor-alpha and gamma interferon. Scand J Immunol. 1988;28:351–7.
15. Phillips JO, Everson MP, Moldoveanu Z, Lue C, Mestecky J. Synergistic effect of IL-4 and IFN-gamma on the expression of polymeric Ig receptor (secretory component) and IgA binding by human epithelial cells. J Immunol. 1990;145:1740–4.
16. Youngman KR, Fiocchi C, Kaetzel CS. Inhibition of IFN-gamma activity in supernatants from stimulated human intestinal mononuclear cells prevents up-regulation of the polymeric Ig receptor in an intestinal epithelial cell-line. J Immunol. 1994;153:675–81.
17. Menge AC, Mestecky J. Surface expression of secretory component and HLA class II DR antigen on glandular epithelial cells from human endometrium and two endometrial adenocarcinoma cell lines. J Clin Immunol. 1993;13:259–64.
18. Nihei Y, Maruyama K, Endo Y, Sato T, Kobayashi K, Kaneko F. Secretory component (polymeric immunoglobulin receptor) expression on human keratinocytes by stimulation with interferon-gamma and differences in response. J Dermatol Sci. 1996;11:214–22.
19. Wira CR, Bodwell JE, Prabhala RH. In vivo response of secretory component in the rat uterus to antigen, IFN-gamma, and estradiol. J Immunol. 1991;146:1893–9.
20. Prabhala RH, Wira CR. Cytokine regulation of the mucosal immune system: In vivo stimulation by interferon-gamma of secretory component and immunoglobulin A in uterine secretions and proliferation of lymphocytes from spleen. Endocrinology. 1991;129:2915–23.
21. Kelleher RS, Hann LE, Edwards JA, Sullivan DA. Endocrine, neural, and immune control of secretory component output by lacrimal gland acinar cells. J Immunol. 1991;146:3405–12.
22. McGee DW, Aicher WK, Eldridge JH, Peppard JV, Mestecky J, McGhee JR. Transforming growth factor-beta enhances secretory component and major histocompatibility complex class I antigen expression on rat IEC-6 intestinal epithelial cells. Cytokine. 1991;3:543–50.
23. VanCott JL, Staats HF, Pascual DW et al. Regulation of mucosal and systemic antibody responses by T helper cell subsets, macrophages, and derived cytokines following oral immunization with live recombinant *Salmonella*. J Immunol. 1996;156:1504–14.
24. Sullivan DA, Richardson GS, MacLaughlin DT, Wira CR. Variations in the levels of secretory component in human uterine fluid during the menstrual cycle. J Steroid Biochem. 1984;20:509–13.
25. Bjercke S, Brandtzaeg P. Glandular distribution of immunoglobulins, J chain, secretory component, and HLA-DR in the human endometrium throughout the menstrual cycle. Hum Reprod. 1993;8:1420–5.
26. Richardson J, Kaushic C, Wira CR. Estradiol regulation of secretory component: expression by rat uterine epithelial cells. J Steroid Biochem Mol Biol. 1993;47:143–9.

27. Kaushic C, Richardson JM, Wira CR. Regulation of polymeric immunoglobulin A receptor messenger ribonucleic acid expression in rodent uteri: effect of sex hormones. Endocrinology. 1995;136:2836–44.
28. Rosato R, Jammes H, Belair L, Puissant C, Kraehenbuhl J-P, Djiane J. Polymeric-Ig receptor gene expression in rabbit mammary gland during pregnancy and lactation: evolution and hormonal regulation. Mol Cell Endocrinol. 1995;110:81–7.
29. Stern JE, Gardner S, Quirk D, Wira CR. Secretory immune system of the male reproductive tract: effects of dihydrotestosterone and estradiol on IgA and secretory component levels. J Reprod Immunol. 1992;22:73–85.
30. Sullivan DA, Bloch KJ, Allansmith MR. Hormonal influence on the secretory immune system of the eye: Androgen regulation of secretory component levels in rat tears. J Immunol. 1984;132:1130–5.
31. Piskurich JF, France JA, Tamer CM, Willmer CA, Kaetzel CS, Kaetzel DM. Interferon-gamma induces polymeric immunoglobulin receptor mRNA in human intestinal epithelial cells by a protein synthesis dependent mechanism. Mol Immunol. 1993;30:413–21.
32. Piskurich JF, Youngman KR, Phillips KM et al. Transcriptional regulation of the human polymeric immunoglobulin receptor gene by IFN-gamma. Mol Immunol. 1997; 34:75–91.
33. Krajci P, Tasken K, Kvale D, Brandtzaeg P. Interferon-gamma stimulation of messenger RNA for human secretory component (poly-Ig receptor) depends on continuous intermediate protein synthesis. Scand J Immunol. 1993;37:251–6.
34. Krajci P, Kvale D, Tasken K, Brandtzaeg P. Molecular cloning and exon-intron mapping of the gene encoding human transmembrane secretory component (the poly-Ig receptor). Eur J Immunol. 1992;22:2309–15.
35. Verrijdt G, Swinnen J, Peeters B, Verhoeven G, Rombauts W, Claessens F. Characterization of the human secretory component gene promoter. Biochim Biophys Acta. 1997;1350:147 54.
36. Fodor E, Feren A, Jones A. Isolation and genomic analysis of the rat polymeric immunoglobulin receptor gene terminal domain and transcriptional control region. DNA Cell Biol. 1997;16:215 25.
37. Tanaka N, Kawakami T, Taniguchi T. Recognition DNA sequences of interferon regulatory factor 1 (IRF-1) and IRF-2, regulators of cell growth and the interferon system. Mol Cell Biol. 1993;13:4531 8.
38. Taniguchi T, Harada H, Lamphier M. Regulation of the interferon system and cell growth by the IRF transcription factors. J Cancer Res Clin Oncol. 1995;121:516–20.
39. Fujita T, Reis LF, Watanabe N, Kimura Y, Taniguchi T, Vilcek J. Induction of the transcription factor IRF-1 and interferon-beta mRNAs by cytokines and activators of second-messenger pathways. Proc Natl Acad Sci USA. 1989;86:9936–40.
40. Mostov K, Apodaca G, Aroeti B, Okamoto C. Plasma membrane protein sorting in polarized epithelial cells. J Cell Biol. 1992;116:577 83.
41. Mostov KE, Cardone MH. Regulation of protein traffic in polarized epithelial cells. BioEssays. 1995;17:129–38.
42. Rodriguez-Boulan E, Powell SK. Polarity of epithelial and neuronal cells. Annu Rev Cell Biol. 1992;8:395–427.
43. Brown WR, Kloppel TM. The liver and IgA: immunological, cell biological and clinical implications. Hepatology. 1989;9:763–84.
44. Sztul E, Kaplin A, Saucan L, Palade G. Protein traffic between distinct plasma membrane domains: Isolation and characterization of vesicular carriers involved in transcytosis. Cell. 1991;64:81–9.
45. Casanova JE, Apodaca G, Mostov KE. An autonomous signal for basolateral sorting in the cytoplasmic domain of the polymeric immunoglobulin receptor. Cell. 1991;66:65 75.
46. Aroeti B, Kosen PA, Kuntz ID, Cohen FE, Mostov KE. Mutational and secondary structural analysis of the basolateral sorting signal of the polymeric immunoglobulin receptor. J Cell Biol. 1993;123:1149–60.
47. Apodaca G, Katz LA, Mostov KE. Receptor-mediated transcytosis of IgA in MDCK cells is via apical recycling endosomes. J Cell Biol. 1994;125:67–86.
48. Breitfeld PP, McKinnon WC, Mostov KE. Effect of nocodazole on vesicular traffic to the apical and basolateral surfaces of polarized MDCK cells. J Cell Biol. 1990;111:2365–73.
49. Mazanec MB, Kaetzel CS, Lamm ME, Fletcher D, Nedrud JG. Intracellular neutralization of virus by immunoglobulin A antibodies. Proc Natl Acad Sci USA. 1992;89:6901–5.

50. Mazanec MB, Coudret CL, Fletcher DR. Intracellular neutralization of influenza virus by immunoglobulin A anti-hemagglutinin monoclonal antibodies. J Virol. 1995;69:1339–43.
51. Burns JW, Siadat-Pajouh M, Krishnaney AA, Greenberg HB. Protective effect of rotavirus VP6-specific IgA monoclonal antibodies that lack neutralizing activity. Science. 1996;272:104–7.
52. Jin M, Saucan L, Farquhar MG, Palade GE. Rab1a and multiple other Rab proteins are associated with the transcytotic pathway in rat liver. J Biol Chem. 1996;271:30105–13.
53. Sixbey JW, Yao QY. Immunoglobulin A-induced shift of Epstein-Barr virus tissue tropism. Science. 1992;255:1578–80.
54. Bomsel M, Mostov K. Role of heterotrimeric G proteins in membrane traffic. Mol Biol. Cell. 1992;3:1317–28.
55. Okamoto CT, Song W, Bomsel M, Mostov KE. Rapid internalization of the polymeric immunoglobulin receptor requires phosphorylated serine 726. J Biol Chem. 1994;269:15676–82.
56. Okamoto CT, Shia S-P, Bird C, Mostov KE, Roth MG. The cytoplasmic domain of the polymeric immunoglobulin receptor contains two internalization signals that are distinct from its basolateral sorting signal. J Biol Chem. 1992;267:9925–32.
57. Casanova JE, Breitfeld PP, Ross SA, Mostov KE. Phosphorylation of the polymeric immunolglobulin receptor required for its efficient transcytosis. Science. 1990;248:742–5.
58. Song W, Bomsel M, Casanova J, Vaerman J-P, Mostov K. Stimulation of transcytosis of the polymeric immunoglobulin receptor by dimeric IgA. Proc Natl Acad Sci USA. 1994;91:163–6.
59. Cardone MH, Smith BL, Song W, Mochly-Rosen D, Mostov KE. Phorbol myristate acetate-mediated stimulation of transcytosis and apical recycling in MDCK cells. J Cell Biol. 1994;124:717–27.
60. Cardone MH, Smith BL, Mennitt PA, Mochly-Rosen D, Silver RB, Mostov KE. Signal transduction by the polymeric immunoglobulin receptor suggests a role in regulation of receptor transcytosis. J Cell Biol. 1996;133:997–1005.
61. Chapin SJ, Enrich C, Aroeti B, Havel RJ, Mostov KE. Calmodulin binds to the basolateral targeting signal of the polymeric immunoglobulin receptor. J Biol Chem. 1996;271:1336–42.
62. Song W, Apodaca G, Mostov K. Transcytosis of the polymeric immunoglobulin receptor is regulated in multiple intracellular compartments. J Biol Chem. 1994;269:29474–80.
63. Bomsel M, Mostov KE. Possible role of both the alpha and beta gamma subunits of the heterotrimeric G protein, Gs, in transcytosis of the polymeric immunoglobulin receptor. J Biol Chem. 1993;268:25824–35.

6
Multiple roles of FcRn

N. E. SIMISTER

TRANSMISSION OF IgG FROM MOTHER TO YOUNG

FcRn in neonatal rat and mouse gut

FcRn is named for the source from which it was first purified, the intestinal epithelium of neonatal rats. Long before the receptor was characterized it was known that rats and mice acquire maternal antibodies by suckling [reviewed in Reference 1]. These antibodies provide passive immunity during the period when the capacity to mount an effective antibody response develops [1]. Neonates lose the ability to take up antibodies around the age of weaning [1]. During the nursing period, IgG antibodies are transmitted. In contrast, very little IgM, IgA, IgD or IgE is taken up from the gut [1,2]. All subclasses of rat and mouse IgG are transported [reviewed in Reference 3], as are IgG immune complexes [4]. As well as selecting between immunoglobulin classes, the transport process discriminates between antibodies from different species: some rat, mouse, guinea pig, hamster, rabbit, cattle, sheep and human antibodies are transported; chicken antibodies are not [1]. The selectivity of immunoglobulin transport and the competition between antibodies for transmission suggest that an IgG receptor is a part of the transport mechanism [1]. The Fc fragment of IgG is also transported, indicating that the receptor recognizes a site in this region [1].

IgG is transported across the proximal third of the small intestine of suckling rats [5]. IgG binds the apical plasma membrane of epithelial cells on the upper parts of the villi in the proximal small intestine [5]. IgG also binds to isolated epithelial cells [6] and brush borders [7] from this region of the gut. IgG binding sites behave as specific and saturable receptors, with an affinity for human IgG of 10–50 nM [3]. These receptors bind Fc, but not Fab or $F(ab')_2$ [8]. Fc

J.G.J. van de Winkel and P.M. Hogarth (eds.), The Immunoglobulin Receptors and their Physiological and Pathological Roles in Immunity. 63–71.

receptors are detected by electron microscopy [9] or by radioligand binding [10] only in the proximal small intestine and only before weaning. The coincidence in time and place of specific IgG binding sites and IgG transport implies that these Fc receptors mediate the intestinal transmission of maternal antibodies [5,9].

IgG binds epithelial cells from suckling rats [11] and mice [12] at pH 6, but not at pH 7.4. IgG also binds the basolateral membranes of isolated epithelial cells with this pH preference [6]. The pH in the duodenum and jejunum of neonatal rats is 6–6.5 [13], so IgG obtained by suckling can bind receptors on the luminal surface of epithelial cells. However, an acid pH is not needed for IgG transport across segments of neonatal rat proximal small intestine maintained *in vitro*, suggesting that IgG may bind its receptor after uptake, in acidified endosomes [14]. Two proteins from detergent extracts of neonatal rat proximal small intestine, with relative molecular weights of about 50 000 and 12 000, are eluted from immobilized IgG at pH 7.4 or 8.0 after binding at pH 6.0 or 6.5 [15,16]. The 50 and 12 kDa molecules are cross-linked in brush borders by dithiobis (succinimidyl propionate), and are therefore subunits of one protein, rather than two proteins that fortuitously copurify [17]. These proteins are isolated from immobilized Fc but not $F(ab')_2$ [16], and are obtained in amounts detectable by silver staining after SDS–PAGE only from the proximal part of the small intestine and only before weaning [16]. The co-expression of the 50 and 12 kDa proteins where IgG is transported, the sensitivity to pH of their ability to bind IgG, and their specificity for Fc, suggest that they are subunits of the transporting Fc receptor [15,16].

The 12 kDa subunit of the neonatal Fc receptor (FcRn) is β_2-microglobulin (β_2m) [17], which is also the small subunit of class I MHC proteins. The 50 kDa subunit is an integral membrane glycoprotein sharing about 28% amino acid identity with the extracellular part of class I MHC α chains [17]. The similarity of FcRn to class I MHC proteins suggests that the large and small subunits exist as a dimer, and this is confirmed by the structure obtained by X-ray crystallography [18]. The major structural difference between the FcRn and classical class I MHC α chains is the groove between the helices of the α1 and α2 domains is closed in FcRn and contains no peptides [18]. The α chain of mouse FcRn has 91% amino acid identity with that of the rat [19]. Mouse FcRn α chain is encoded outside the MHC on chromosome 7 [19]. Intestinal epithelial cells from neonatal mice homozygous for a targeted disruption of the β_2m gene lack the normal ability to bind IgG [20]. These mice acquire no maternal IgG by suckling [21]. This implies that FcRn is the only transporter of IgG from mother to young in the mouse. While it is possible that another FcR that contains β_2m transports IgG, this is unlikely because Southern blots show that there are no close relatives of FcRn α chain [22].

FcRn in fetal yolk sac of rats and mice

Although rats and mice acquire most maternal IgG by suckling, they also receive antibodies *in utero*. IgG begins to be transported from the maternal circulation to fetal mice between the eleventh and twelfth days of gestation [23]. Similarly, maternal IgG can be detected in sera of rats at 17 days gestational age [1]. Routes of transmission exist across the yolk sac splanchnopleur, placenta and fetal intestine, but their relative importance is not known [1]. FcRn is present in tissues that transport IgG prenatally. By 20 days of gestation pH-dependent Fc receptors can be detected on rat intestinal epithelial cells [24]. At 19–23 days gestational age FcRn can be detected in rat yolk sac endoderm by a monoclonal antibody against the α chain [25], and FcRn α chain mRNA is transcribed in mouse yolk sac by 18 days [19]. FcRn cDNA cloned from a mouse yolk sac library from day 18 is identical to that encoding intestinal FcRn (C.M. Story and N.E. Simister, unpublished). FcRn does not bind IgG at the neutral pH [25] of the uterine fluid, but could bind in the vesicles where it is detected in yolk sac endoderm, if these are acidified endosomes [25]. Mice that lack β_2m are born without maternal IgG [21]. This suggests that FcRn is not only available when prenatal IgG transmission occurs, but, with the caveat noted above, that it is essential for transport of IgG from mother to fetus [21].

FcRn in human placenta

Humans obtain maternal IgG only before birth, across the placenta. Fetal IgG synthesis is low [26], and most IgG in the blood of newborns is maternal [27]. IgG is acquired in preference to IgA, IgM or other maternal serum proteins [28]. The IgG concentration in fetal serum rises sharply between 22 and 26 weeks of gestation [26]. By the time of birth fetal IgG_3 and IgG_4 typically equal maternal levels, fetal IgG_1 concentration may exceed the mother's by a little, and fetal IgG_2 is often lower [29,30].

IgG from maternal blood crosses two cell layers to enter the fetal circulation: the syncytiotrophoblast, which is the outermost layer of the chorionic villi and is in contact with maternal blood, and the endothelium of the fetal blood vessels. These layers are separated by a loose connective tissue, the stroma. The relatively low level of IgG_2 in fetal sera appears to result from its inefficient transport through the endothelium, not the syncytiotrophoblast [31]. In the syncytiotrophoblast IgG is detected by immuno-electron microscopy in coated pits, coated vesicles, non-coated vesicles, tubulovesicular bodies (which may be early endosomes), and multivesicular bodies [32,33]. IgG is located in vesicles and multivesicular bodies within endothelial cells, but is not found in the paracellular clefts between cells [32,33].

Several IgG binding proteins are present in the syncytiotrophoblast [30]. Placental alkaline phosphatase (PLAP) binds IgG with low affinity [34]. PLAP is also expressed in the trophoblastic layer of amniochorion [35], which is exposed to maternal IgG in the uterine decidua, but does not contain maternal IgG [36], arguing against a role for PLAP in IgG uptake. Annexin II also binds IgG [37] and appears to be exposed on the surface of syncytiotrophoblast [38]. One monoclonal anti-FcγRIII, Leu-11b, consistently stains syncytiotrophoblast from term placenta [39–41], but several others do not [41,42]. The appearance of the Leu-11b-reactive material on western blots is not typical of FcγRIII [41]. It therefore remains uncertain whether FcγRIII or a cross-reactive protein is expressed on syncytiotrophoblast. Another syncytiotrophoblast FcR detected by IgG binding is of higher affinity than annexin II or PLAP, and distinct in its subclass specificity from FcγRIII and FcRn [43]. FcRn is also expressed in the placenta. FcRn α chain mRNA [44] and protein [44–46] are present in syncytiotrophoblast. β_2m co-localizes with FcRn α chain in syncytiotrophoblast [used in Reference 45] and co-purifies with it from a placental extract [46]. The latter observations conflict with immunohistochemical studies in which no β_2m was detected in syncytiotrophoblast [47]. This discrepancy might be due to loss of β_2m from frozen sections [used in Reference 47] but not paraformaldehyde-fixed tissue [45]. Of these IgG-binding proteins, FcRn is the only one with a known function in antibody transport.

Human FcRn α chain is very similar to those of the rat and mouse, with 69% nucleotide identity with rat FcRn over the open reading frame, and 65% predicted amino acid identity [48]. Like rat and mouse FcRns, human FcRn binds IgG at pH 6, but not at pH 7.5 [48]. Human FcRn does not bind IgG at the neutral pH of maternal blood at the apical plasma membrane of syncytiotrophoblast but could bind at a pH near 6 in an early endosome. Consistent with this possibility, FcRn is detected not at the syncytiotrophoblast plasma membrane but in cytoplasmic granules [45]. The question of how IgG is delivered to endosomes remains open. IgG follows the same pathway to endosomes in syncytiotrophoblast as a fluid phase marker [32], so it is possible that IgG is initially taken up unbound. Alternatively, or additionally, IgG might be taken up by one of the IgG binding proteins discussed above, or by a novel receptor. IgG carried by FcRn to the basal membrane of the syncytiotrophoblast would be released at the neutral pH of the intercellular fluid of the stroma.

It is not known how IgG is transported from the stroma to the lumen of fetal blood vessels. One study found no FcRn in the endothelium [44], but occasional [45] and low level expression [46] have been reported. FcγRII is present [33,39,41,42] but has little, if any, affinity for monomeric IgG [49]. Another receptor may therefore be partly, if not wholly, responsible for IgG transport into the fetal capillaries.

PROTECTION OF IgG FROM DEGRADATION

Inefficiency of transport from mother to offspring

IgG transport from mother to offspring is generally inefficient [1]. Much of the IgG taken up by human placental syncytiotrophoblast is degraded and not transmitted intact [50]. Similarly, a fraction of the IgG taken up by epithelial cells from the proximal small intestine of neonatal rats is degraded, not transported [1,14]. This fraction increases with the amount of IgG taken up, suggesting that a limited number of Fc receptors protect IgG from intracellular degradation [1].

Protection of IgG from catabolism

Selective protection of IgG from degradation is not limited to its occurrence during the transport of antibodies from mother to young. IgG also enjoys protection from catabolism compared with other classes of immunoglobulins, and therefore has the longest survival time in the circulation [51]. Immunoglobulins are broken down by several mechanisms, some general and others preferring particular classes [reviewed in Reference 52]. Likewise, immunoglobulins are catabolized at a variety of sites, including the liver, intestine, and kidney, but most IgG catabolism occurs at low levels at sites broadly distributed through the body [52]. The capillary endothelium is a plausible place for this widespread catabolism because of its access to circulating IgG and its extensive surface area [52].

IgG catabolism is related to its serum concentration, rising at high concentrations to a level similar to those of other immunoglobulin classes. This suggests that a saturable receptor protects IgG from catabolism [53]. Fc fragments have a similar half-life to intact IgG, and much longer survival than Fab fragments [54], suggesting that the receptor that rescues IgG from catabolism binds a site in the Fc region. Two lines of evidence show that this site overlaps that recognized by FcRn. First, IgG protection [55] and binding to rat FcRn [56] are both blocked by *Staphylococcus aureus* protein A (SpA), or its low molecular weight fragment B. The latter observation is consistent with crystal structures of Fc complexes with SpA [57] and rat FcRn [58], which predict that both receptors bind to amino acid residues 252–254, 308–311 and 433–436 (EU numbering) in the C_H2–C_H3 region of IgG. Second, mutations in these regions cause both rapid catabolism [59] and impaired interaction with FcRn [60,61].

Comparisons of the saturability and specificity of the mechanisms that protect IgG from catabolism and transport it from mother to offspring had earlier suggested an underlying similarity [62]. Evidence that FcRn itself protects IgG from catabolism has come from studies of transgenic mice deficient in β_2-microglobulin, and thus without functional FcRn. In addition to lacking maternal IgG, $\beta_2m^{/}$ mice have lower endogenous IgG levels than littermates possessing functional FcRn [21]. Adult $\beta_2m^{/}$ mice also have less serum IgG than heterozygotes, despite having similar numbers of B cells [63]. $\beta_2m^{/}$ mice clear mouse IgG subclasses from the circulation more rapidly than do wild-type mice [64–66]. In contrast, Fc fragments of mouse IgG_1 mutated in the catabolic site [64] and chicken IgY [66], which does not bind FcRn [13], are degraded equally rapidly by β_2-microglobulin knockout and normal mice. Increased catabolism accounts for the decreased serum IgG concentrations in $\beta_2m^{/}$ mice, without any change in the rate of IgG synthesis [65]. Accelerated catabolism of autoantibodies may explain the decreased severity of the systemic lupus erythematosus-like autoimmune syndrome in MRL-lpr/lpr mice that lack β_2m, compared with normal MRL-lpr/lpr mice [67]. Because it is unlikely that there are other β_2m-dependent FcR receptors [22], these studies suggest that FcRn protects IgG from degradation.

FcRn is expressed at potential sites of IgG clearance. FcRn α chain mRNA [64] and protein (E.J. Israel, D.F. Wilsker and N.E. Simister, unpublished) are present in endothelial cell lines. FcRn α chain mRNA is also detectable in adult rat liver, kidney and intestine [17], mouse liver [64], and human liver, kidney and intestine [48]. FcRn is present on adult rat hepatocytes [68], and on human intestinal epithelial cells [69]. FcRn is scarce in these tissues compared with the neonatal rat or mouse intestine [17,64], but its potential capacity to protect IgG is considerable because of the very large number of expressing cells.

The mechanism of this protection may indeed by very similar to that of IgG transport [62]. Because IgG does not bind FcRn at the extracellular pH, immunoglobulins must be delivered by fluid phase endocytosis or by other receptors to an acidified compartment containing FcRn, probably sorting early endosomes. IgG that binds FcRn there will be sorted through recycling early endosomes to the plasma membrane, and released at neutral pH. In contrast, immunoglobulins in excess of FcRn molecules, and isotypes that cannot bind FcRn, will remain in the fluid phase in sorting early endosomes and be trafficked by way of late endosomes to lysosomes, where they will be degraded.

CONCLUSIONS

FcRn is a protein whose one function allows it to play several roles. This function appears to be to transport IgG from endosomes to the plasma membrane, avoiding lysosomal degradation. Maternal IgG is delivered to endosomes in the neonatal intestine or fetal yolk sac of mice and rats, and in human placental syncytiotrophoblast. IgG may be brought to endosomes by FcRn itself, by fluid phase endocytosis, or by other receptors. Some of this IgG will be transported across the cells by FcRn, ultimately to the neonatal or fetal circulation. FcRn is expressed in these transporting cells at high levels for a limited time. FcRn constitutively expressed at low levels in endothelial cells, hepatocytes and elsewhere may likewise return IgG from endosomes to the cell surface and to the circulation, extending the window in which antibodies of particular specificities remain available to the immune system.

Acknowledgements

This work was supported by NIH grants HD27691 and HD01146.

References

1. Brambell FWR. The Transmission of Passive Immunity from Mother to Young. Amsterdam: North-Holland; 1970.
2. Waldmann TA, Jones EA. The role of cell surface receptors in the transport and catabolism of immunoglobulins. Ciba Found Symp. 1973;9:5–17.
3. Simister NE. Transport of monomeric antibodies across epithelia. In: Metzger H, ed., Fc Receptors and the Action of Antibodies. Washington, DC: American Society for Microbiology; 1990:57–73.

4. Abrahamson DR, Powers A, Rodewald R. Intestinal absorption of immune complexes by neonatal rats: a route of antigen transfer from mother to young. Science. 1979;206:567–9.
5. Rodewald R. Selective antibody transport in the proximal small intestine of the neonatal rat. J Cell Biol. 1970;45:635–40.
6. Wild AE, Richardson LJ. Direct evidence for pH-dependent Fc receptors on proximal enterocytes of suckling rat gut. Experientia. 1979;35:838–40.
7. Wallace KH, Rees AR. Studies on the immunoglobulin-G Fc-fragment receptor from neonatal rat small intestine. Biochem J 1980;188:9–16.
8. Rodewald R. Intestinal transport of peroxidase-conjugated IgG fragments in the neonatal rat. In: Hemmings WA, ed., Maternofoetal Transmission of Immunoglobulins. London: Cambridge University Press; 1976:137–49.
9. Rodewald R. Intestinal transport of antibodies in the newborn rat. J Cell Biol. 1973;58:189–211.
10. Simister NE, Rees AR. Properties of immunoglobulin G-Fc receptors from neonatal rat intestinal brush borders. Ciba Found Symp. 1983;95:273–86.
11. Jones EA, Waldman TA. The mechanism of intestinal uptake and transcellular transport of IgG in the neonatal rat. J Clin Invest. 1972;51:2916–27.
12. Mackenzie NM, Keeler KD. A flow microfluorimetric analysis of the binding of immunoglobulins to Fcγ receptors of the neonatal mouse jejunal epithelium. Immunology. 1984;51:529–33.
13. Rodewald R. pH-dependent binding of immunoglobulins to intestinal cells of the neonatal rat. J Cell Biol. 1976;71:666–70.
14. Benlounes N, Chedid R, Thuillier F, Desjeux JF, Rousselet F, Heyman M. Intestinal transport and processing of immunoglobulin G in the neonatal and adult rat. Biol Neonate. 1995;67:254–63.
15. Rodewald R, Kraehenbuhl J-P. Receptor-mediated transport of IgG. J Cell Biol. 1984;99: 159s–64s.
16. Simister NE, Rees AR. Isolation and characterization of an Fc receptor from neonatal rat small intestine. Eur J Immunol. 1985;15:733–8.
17. Simister NE, Mostov KE. An Fc receptor structurally related to MHC class I antigens. Nature. 1989;337:184–7.
18. Burmeister WP, Gastinel LN, Simister NE, Blum ML, Bjorkman PJ. Crystal structure at 2.2 Å resolution of the MHC-related neonatal Fc receptor. Nature. 1994;372:336–43.
19. Ahouse JJ, Hagerman CL, Mittal P et al. A mouse MHC class-I-like Fc receptor encoded outside the MHC. J Immunol. 1993;151:6076–88.
20. Zijlstra M, Bix M, Simister NE, Loring JM, Raulet DH, Jaenisch R. β_2-Microglobulin deficient mice lack CD4 8^+ cytolytic T cells. Nature. 1990;344:742–6.
21. Israel EJ, Patel VK, Taylor SF, Marshak-Rothstein A, Simister NE. Requirement for a β_2-microglobulin-associated Fc receptor for acquisition of maternal IgG by fetal and neonatal mice. J Immunol. 1995;154:6246–51.
22. Kandil E, Noguchi M, Ishibashi T, Kasahara M. Structural and phylogenetic analysis of the MHC class I-like Fc receptor gene. J Immunol. 1995;154:5907–18.
23. Morphis LG, Gitlin D. Maturation of the maternofoetal transport system for human gamma-globulin in the mouse. Nature. 1970;228:573.
24. Griffin P, Wild AE. Ontogeny of the Fc gamma receptor in rat small intestine. J Reprod Immunol. 1987;11:287–306.
25. Roberts DM, Guenthert M, Rodewald R. Isolation and characterization of the Fc receptor from the fetal yolk sac of the rat. J Cell Biol. 1990;111:1867–76.
26. Gitlin D, Biasucci A. Development of γG, γA, γM, βIC/βIA, C'1 esterase inhibitor, ceruloplasmin, transferrin, hemopexin, haptoglobin, fibrinogen, plasminogen, α1-antitrypsin, orosomucoid, β-lipoprotein, α2-macroglobulin, and prealbumin in the human conceptus. J Clin Invest. 1969;48:1433–46.
27. Mellbye OJ, Natvig JB. Presence and origin of human IgG subclass proteins in newborns. Vox Sang. 1973;24:206–15.
28. Gitlin D, Kumate J, Urrusti J, Morales C. The selectivity of the human placenta in the transfer of plasma proteins from mother to fetus. J Clin Invest. 1964;43:1938–51.
29. Malek A, Sager R, Schneider H. Maternal-fetal transport of immunoglobulin G and its subclasses during the third trimester of human pregnancy. Am J Reprod Immunol. 1994;32: 8–14.

30. Simister NE, Story CM. Human placental Fc receptors and the transmission of antibodies from mother to fetus. J Reprod Immunol. 1997;37:1–23.
31. Mongan LC, Ockleford CD. Behavior of two of IgG subclasses in transport across the human placenta. J Anat. 1996;188:43–51.
32. Leach L, Eaton BM, Firth JA, Contractor SF. Immunocytochemical and labelled tracer approaches to uptake and intracellular routing of immunoglobulin-G (IgG) in the human placenta. Histochem J. 1991;23:444–9.
33. Bright NA, Ockleford CD, Anwar M. Ontogeny and distribution of Fc gamma receptors in the human placenta. Transport or immune surveillance? J Anat. 1994;184:297–308.
34. Makiya R, Stigbrand T. Placental alkaline phosphatase has a binding site for the human immunoglobulin-G Fc portion. Eur J Biochem.1992;205:341–5.
35. Johnson PM, Molloy CM. Localization in human term placental bed and amniochorion of cells bearing trophoblast antigens identified by monoclonal antibodies. Am J Reprod Immunol. 1983;4:33–7.
36. Bright NA, Ockleford CD. Heterogeneity of Fc gamma receptor-bearing cells in human term amniochorion. Placenta. 1994;15:247–55.
37. Kristoffersen EK, Ulvestad E, Bjørge L, Aarli Å, Matre R. Fc gamma receptor activity of human placental annexin II. Scand J Immunol. 1994;40:237–42.
38. Kristoffersen EK, Matre R. Surface annexin II on placental membranes of the fetomaternal interface. Am J Reprod Immunol. 1996;36:141–9.
39. Kristoffersen EK, Ulvestad E, Vedeler CA, Matre R. Fcγ receptor heterogeneity in the human placenta. Scand J Immunol. 1990;32:561–4.
40. Kameda T, Koyama M, Matsuzaki N, Taniguchi T, Saji F, Tanizawa O. Localization of three subtypes of Fcγ receptors in human placenta by immunohistochemical analysis. Placenta. 1991;12:15–26.
41. Wainwright SD, Holmes CH. Distribution of Fcγ receptors on trophoblast during human placental development: an immunohistochemical and immunoblotting study. Immunology. 1993;80:343–51.
42. Sedmak DD, Davis DH, Singh U, van de Winkel JG, Anderson CL. Expression of IgG Fc receptor antigens in placenta and on endothelial cells in humans. An immunohistochemical study. Am J Pathol. 1991;138:175–81.
43. Esterman AL, Dancis J, Lee JD, Rindler MJ. Two mechanisms for IgG uptake in cultured human trophoblast: evidence for a novel high affinity Fc receptor. Pediatr Res. 1995;38:1–6.
44. Simister NE, Story CM, Chen H-L, Hunt JS. An IgG transporting Fc receptor expressed in the syncytiotrophoblast of human placenta. Eur J Immunol. 1996;26:1527–31.
45. Kristoffersen EK, Matre R. Co-localization of the neonatal Fc gamma receptor and IgG in human placental term syncytiotrophoblasts. Eur J Immunol. 1996;26:1668–71.
46. Leach JL, Sedmak DD, Osborne JM, Rahill B, Lairmore MD, Anderson CL. Isolation from human placenta of the IgG transporter, FcRn, and localization to the syncytiotrophoblast: implications for maternal-fetal antibody transport. J Immunol. 1996;157:3317–22.
47. Faulk WP, Temple A. Distribution of β_2 microglobulin and HLA in chorionic villi of human placentae. Nature. 1976;262:799–802.
48. Story CM, Mikulska JE, Simister NE. A major histocompatibility complex class I-like Fc receptor cloned from human placenta: possible role in transfer of immunoglobulin G from mother to fetus. J Exp Med. 1994;180:2377–81.
49. Karas SP, Rosse WF, Kurlander RJ. Characterization of the IgG-Fc receptor on human platelets. Blood. 1982;60:1277–82.
50. Contractor SF, Eaton BM, Stannard PJ. Uptake and fate of exogenous immunoglobulin G in the perfused human placenta. J Reprod Immunol. 1983;5:265–73.
51. Waldmann TA, Strober W. Metabolism of immunoglobulins. Prog Allergy. 1969;13:1–110.
52. Mariani G, Strober W. Immunoglobulin metabolism. In: Metzger H, ed., Fc Receptors and the Action of Antibodies. Washington, D.C.: American Society of Microbiology; 1990:94–177.
53. Fahey JL, Robinson AG. Factors controlling serum immunoglobulin concentration. J Exp Med. 1963;118:845–68.
54. Spiegelberg HL, Weigle WO. The catabolism of homologous and heterologous 7S gamma globulin fragments. J Exp Med. 1965;121:323–38.
55. Dima S, Medesan C, Mota G, Moraru I, Sjoquist J, Ghetie V. Effect of protein A and its fragment B on the catabolic and Fc receptor sites of IgG. Eur J Immunol. 1983;13:605–14.

56. Raghavan M, Chen MY, Gastinel LN, Bjorkman PJ. Investigation of the interaction between the class I MHC-related Fc receptor and its immunoglobulin G ligand. Immunity. 1994;1: 303–15.
57. Diesenhofer J. Crystallographic refinement and atomic models of a human Fc fragment and its complex with fragment B of protein A from *Staphylococcus aureus* at 2.9- and 2.8-Å resolution. Biochemistry. 1981;20:2361–70.
58. Burmeister WP, Huber AH, Bjorkman PJ. Crystal structure of the complex between the rat neonatal Fc receptor and Fc. Nature. 1994;372:379–83.
59. Kim JK, Tsen MF, Ghetie V, Ward ES. Identifying amino acid residues that influence plasma clearance of murine IgG_1 fragments by site-directed mutagenesis. Eur J Immunol. 1994;24: 542–8.
60. Kim JK, Tsen MF, Ghetie V, Ward ES. Localization of the site of the murine IgG_1 molecule that is involved in binding to the murine intestinal Fc receptor. Eur J Immunol. 1994;24: 2429–34.
61. Raghavan M, Bonagura VR, Morrison SL, Bjorkman PJ. Analysis of the pH dependence of the neonatal Fc receptor/immunoglobulin G interaction using antibody and receptor variants. Biochemistry. 1995;34:1649–57.
62. Brambell FWR, Hemmings WA, Morris IG. A theoretical model of γ-globulin catabolism. Nature. 1964;203:1352–5.
63. Spriggs MK, Koller BH, Sato T et al. Beta 2-microglobulin , $CD8^+$ T-cell-deficient mice survive inoculation with high doses of vaccinia virus and exhibit altered IgG responses. Proc Natl Acad Sci USA. 1992;89:6070–4.
64. Ghetie V, Hubbard JG, Kim JK, Tsen MF, Lee Y, Ward ES. Abnormally short serum half-lives of IgG in beta 2-microglobulin-deficient mice. Eur J Immunol. 1996;26:690–6.
65. Junghans RP, Anderson CL. The protection receptor for IgG catabolism is the beta2-microglobulin-containing neonatal intestinal transport receptor. Proc Natl Acad Sci USA. 1996;93:5512–6.
66. Israel EJ, Wilsker DF, Hayes KC, Schoenfeld D, Simister NE. Increased clearance of IgG in mice that lack β_2-microglobulin: possible protective role of FcRn. Immunology. 1996;89:573–8.
67. Christianson GJ, Blankenburg RL, Duffy TM et al. Beta 2-microglobulin dependence of the lupus-like autoimmune syndrome of MRL-lpr mice. J Immunol. 1996;156:4932–9.
68. Blumberg RS, Koss T, Story CM et al. An MHC class I-related receptor for IgG on rat hepatocytes. J Clin Invest. 1995;95:6246–51.
69. Israel EJ, Taylor S, Wu Z et al. Expression of the neonatal Fc receptor, FcRn, on human intestinal epithelial cells. Immunology. 1997;92:69–74.

7
Commentary
The polymeric immunoglobulin receptor and the Brambell receptor: non-haematopoietic FcRs, with much to compare

R. P. JUNGHANS

The foregoing chapters cogently summarize the biology and molecular biology of the IgA and IgG transport systems, which play crucial roles in defence against infection. Whereas IgG transmission is confined to the perinatal period and is essential to the survival of newborn animals, the continuous release of IgA into secretions provides mucosae with the first barrier to infection throughout life. The juxtaposition of these chapters is most felicitous, due to the many opportune contrasts and comparisons to be made. Further, the recent confirmation of a long-postulated role of the receptor for IgG transmission in also protecting IgG from catabolism stimulates additional interesting comparisons. These are summarized in Table 7.1.

THE RECEPTORS AND THEIR ANTIBODIES

Between the late 1940s and the early 1960s, Brambell and colleagues defined the IgG receptor that transmits immunity from mother to young [1–3], the first FcγR described, and they subsequently implicated the action of this or a similar receptor in protection of IgG from catabolism [4]. [For a historical overview, see Reference 5]. In the following, we refer to this receptor generically as the Brambell receptor (FcRB). Where discrimination is warranted for the tempo-

J.G.J. van de Winkel and P.M. Hogarth (eds.), The Immunoglobulin Receptors and their Physiological and Pathological Roles in Immunity. 73–82.

Table 7.1 Comparison of IgA and IgG and their transport receptors

	IgA and PIgR	*IgG and FcRB*
Form in blood	Monomer (H_2L_2) (87%) Dimer $(H_2L_2)_2$-J (13%)	Monomer H_2L_2)
Concentration in blood	2 mg/ml	10 mg/ml
Daily synthesis	65 mg/kg/d	33 mg/kg/d
Plasma half-life	6 d	23 d
Concentration in secretions (human)		
Colostrum	17 mg/ml (96% dIgA)	0.4 mg/ml
Milk	1 mg/ml (96% dIgA)	0.04 mg/ml
Jejunal fluid	0.3 mg/ml (96% dIgA)	0.3 mg/ml
Saliva	0.3 mg/ml (96% dIgA)	0.05 mg/ml
Form in transport	Dimer, oligomer	Monomer
Absorption transport?	No	Yes
Secretion transport?	Yes	No
Transport receptor	pIgR	FcRB (FcRn)
Protection receptor	–	FcRB (FcRp)
Related to other FcR?	No	No
Binding	Non-covalent then covalent	Non-covalent
Mechanism of binding/release	Affinity to IgA+J/proteolysis	Low pH/neutral pH
Surface binding?	Yes	No (except neonatal intestine)
Tissues for transport	Mucosae (saliva, tears, bile, etc.)	Yolk sac, intestine, placenta (FcRn)
Tissues for protection		Vasc endothelium (mainly) (FcRp)
Net transport direction	Blood to lumen	Lumen to blood (FcRn) Blood to blood (FcRp)
Duration of expression	Throughout life	Antenatal, neonatal (FcRn) Throughout life (FcRp)
Impact of loss of transport (*inferred from IgA deficients)	Benign or increased broncho-pulmonary infections; not lethal*	Lethal to newborn in nature
Impact of loss of protection from catabolism		Lower IgG levels, increased infections

rally and functionally distinct transport and protection activities of the FcRB, we follow modern convention and employ the term FcRn, for neonatal, to apply broadly to both antenatal and neonatal transmissions, and the term FcRp, for protection, to specify its role to mediate the long *in vivo* survival of IgG. A key advance in the recognition of a receptor for IgA transport was by Mostov *et al.* in 1980 [6] when they identified the secretory component (SC) as a fragment of an integral membrane protein of the secretory cell.

IgG and IgA are structurally very similar, each having three constant domains and one variable domain, and they have similar monomeric masses (150 and 160 kDa), with increased glycosylation and the C terminal polymerization region (tailpiece) of IgA providing the slight increase in size. By contrast, IgE and IgM have an extra constant domain that serves as an extended hinge. IgA and IgG each have FcR on haematopoietic cells, and these receptors share homologies [7,8]. Further, the haematopoietic FcR γ chain that mediates cellular signalling for FcγR and FcεR also associates with the FcαRI and transmits signals for this receptor [9–11]. (Other IgA FcR are less well defined [8].) Both IgG and IgA have been shown to mediate intracellular antiviral activities [12–14]; this has in some instances been speculated to be linked to the pIgR transport of IgA, as discussed in Chapter 5, but the mechanisms in either case are still not entirely clear.

IgG exists in one form only, as an H_2L_2 monomer. IgA exists as mainly monomer, mIgA, in blood (>80%) with a lower fraction of dimer, dIgA [$(H_2L_2)_2$-J] and lesser quantities of higher forms, which are grouped collectively as dIgA. The mIgA and dIgA forms arise from different plasma cells, in which dimerization is promoted by concurrent expression of J chain in the IgA plasma cells. The production of IgA and J chain are regulated independently by cytokines [15,16]. In the basal lamina of the intestine, for example, the IgA plasma cells are predominantly of the J chain type, yielding dIgA. At the basolateral membrane of the secretory epithelial cell, this dIgA is bound by the polymeric Ig receptor (pIgR), for which J chain is required. The bound dIgA-pIgR complex undergoes endocytosis and transcytoplasmic transport to the apex or lumenal side of the cell. The means to release IgA at the apical surface is unique: by proteolytic cleavage of the pIgR itself that yields a covalent complex of the so-called 'secretory component' (SC) of the pIgR with the dIgA. The resultant apical flow yields principally dimeric IgA [$(H_2L_2)_2$-J-SC] in the secretions [17], except for the minor fraction of monomeric IgA (like IgG) which enters secretions by passive filtration [18].

Receptor-dependent IgG transport appears to occur only in the setting of transmission of immunity from mother to young, and this of course is in the reverse direction to that of IgA (Figure 7.1). This transport, mediated by the FcRB (FcRn), occurs at different times perinatally, depending upon the mechanism of transport and on the species (Table 7.2). In humans there is no neonatal transport, but only antenatal, primarily by placenta. This is similarly reflected in the near absence of IgG in human milk (Table 7.1), whereas species that transmit IgG neonatally via intestinal absorption (groups II and III in Table 7.2) have high maternal IgG concentrations in colostrum and milk [5]. Inasmuch as non-transmission is lethal in natural environments, it seems likely that the evolutionary pressure for creating the FcRB was for transmission of immunity and, further, that its primordial expression was in yolk sac, where it is present in birds and probably in earlier phyla [5].

As described in Chapter 6, it is intriguing how this receptor (FcRB) has migrated to different sites (Table 7.1) to ensure efficient transmission of immunity to young concurrent with adapting to the exigencies of species evolution. Its further migration to a different function and location in protection from catabolism for IgG, albeit employing similar mechanisms, is a truly spectacular evolutionary economy. In contrast to the pIgR and the FcRn expression of the Brambell receptor, the IgG protection function appears to mediate no net transport at all. Pharmacokinetic arguments have been invoked

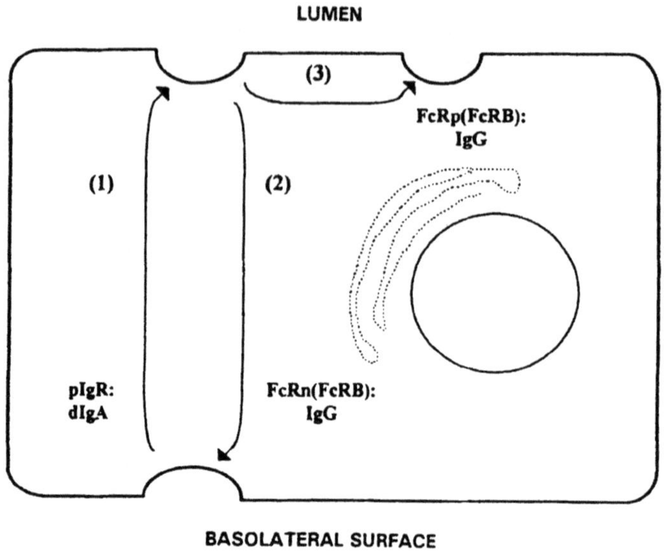

Figure 7.1 The transport paths for IgA and IgG. (1) pIgR transfers IgA from basolateral surface to lumen; (2) FcRB as transport receptor (FcRn) transfers IgG from lumen to basolateral surface; (3) FcRB as protection receptor (FcRp) transfers IgG from lumen back to lumen (plasma). (Figure provided by courtesy of Dr J. Coronella)

[5] to implicate the recycling of IgG back to the lumenal surface of the vascular endothelial cell, with only blood to blood transport (Figure 7.1). It has been known for many years that plasma proteins are catabolized in close contact with the vascular space [19], and an endothelial site for catabolism of plasma proteins (including IgA) – hence the site for protection of IgG by FcRp – accords with this view. This activity accounts for the longer survival of IgG compared with IgA (Table 7.1); in fact, IgG is the longest surviving of all plasma proteins. Other expressions of the FcRB, as in adult hepatocytes [20], appear to be related to its protective (FcRp) rather than transport (FcRn) functions (infra). Interestingly, the pIgR has also been assigned a protective role in one sense, in that the SC is thought to protect secretory IgA from protease activity in the secretory fluids [17]. From the relatively benign course of IgA deficiency in

Table 7.2 Transmission of passive immunity

	Expression of IgG transmission	
Species	*Antenatal*	*Neonatal*
I. Rabbit, guinea pig, primates, humans	++	–
II. Ruminants (cow, sheep), horse, pig	–	+++
III. Mouse, rat, carnivores (dog, cat)	+	++

humans, however, one may speculate that loss of the pIgR would not lead to serious infections, perhaps due to the continued presence of IgG by passive filtration in secretions as well (Table 7.1); one caveat is that there is increased IgM transport by the pIgR in IgA deficiency [21] that would also be lost in a pIgR knockout. This warrants direct investigation.

RECEPTOR STRUCTURES AND BIOCHEMISTRY

In contrast to the haematopoietic FcR, the molecular structures of these transport receptors are very different from each other and from other FcR. The pIgR has five Ig-like domains, of which 1 and 5 make predominant contacts with dIgA and of which domains 2 and 3 may be dispensible [22]. During transcellular migration, a disulphide bond forms between the pIgR and dIgA, but this is not essential to transport [23]. Further details are supplied Chapter 5. Cloning of the FcRB proved it to be an MHC class I-related structure with a β_2-microglobulin subunit [24]. Only the Brambell receptor has been examined by X-ray crystallography. Co-crystals of FcRB with Fc ligand revealed a non-functional peptide binding groove not involved in Fc interaction, and an antiparallel orientation for IgG Fc binding with contacts from both the heavy chain and the β_2m light chain of the FcRB to the CH2–CH3 interface of the Fc [25], distal to the site for binding by the haematopoietic FcγRs [26]. These Fc sites prominently include histidine residues that mediate the pH-dependent binding of the FcRB, as anticipated to a remarkable degree by other studies with judiciously designed Fc mutants that showed the importance of these histidines to both transmission of IgG and protection from catabolism [27,28].

The pIgR necessarily binds dIgA at neutral pH, and is stable to the acidic milieu of the endosomal transport compartment. By contrast, it was shown in the early 1970s by Jones and Waldmann [29] that ligand (IgG Fc) did not bind to FcRB at neutral physiological pH but binds only under acidic conditions (e.g. pH 6.0), which was later reproduced with purified receptor protein [30]. Hence, the FcRB does not bind IgG on the cell surface, which is generally at neutral pH, but does so only after passive endocytosis of the bulk IgG-containing fluid phase and vesicle acidification – as transport receptor (FcRn) in the placenta and yolk sac, and as protection receptor (FcRp) in the vascular endothelium. In fact, the FcRB is clearly absent from the surface of yolk sac [31]; precise information is not available for the other sites. The sole exception to this generalization is in the intestine of the neonatal rodent, where the reduced pH of the intestinal fluids promotes binding to receptor which is present on the surface at detectable levels. Whether such mucosal binding of IgG by the intestinal FcRn is actually important is uncertain [see Reference 5 for discussion]: even in intestine $\approx$99% of the receptor is in intracellular vesicles rather than on the membrane [32]. The dominant expression of the FcRB is as the protection receptor (FcRp) throughout life, in which surface binding of IgG by the FcRp would be truly counterproductive, pulling IgG into the endosome, when the principal role of the receptor is to put IgG out of the endosome.

RECEPTOR TRAFFICKING AND EXPRESSION

The release mechanism of the IgA is by proteolysis of the associated pIgR as discussed above. By contrast, the release of IgG by the Brambell receptor is by pH reneutralization, either on the distal side of the transport cell, or on return to the vascular lumen by the endothelial cell (Figure 7.1). Whereas the mechanism of release of IgA clearly precludes receptor recycling and reutilization, recycling is possible, even probable, for the Brambell receptor [33,34]. In at least one setting, however, the pIgR receptor with IgA still attached will undergo a return flow from the lumen to the basolateral surface. Such 'unreleased' secretory IgA has been shown to facilitate antigen sampling by M cells in the Peyer's patches of the intestine [35,36]. A reverse flow of IgA-pIgR is also compatible with the observed enhancement by IgA of viral infection of nasopharyngeal epithelium [37]. Similarly, IgG has been shown to transport antigen across the neonatal rodent intestine via the Brambell receptor as FcRn [38,39]. As for FcRp, the Brambell receptor has also been suggested to mediate antigen-harvesting from monomeric IgG immune complexes in plasma: termed 'differential catabolism', antigen is separated from antibody within vascular endothelial vesicles, with antibody returning to circulation 'cleansed' of antigen [40,41].

In terms of the ultrastructural trafficking in native tissues, relatively more is known about the FcRB (as FcRn), largely from the structural studies of Rodewald and co-workers [31,32,34,42–43], whereas the gene regulation and molecular determinants of trafficking of the pIgR is far advanced relative to the state of knowledge of the FcRB. The transcellular pathway of FcRn in neonatal rodent intestine was documented from lumen to coated pits to tubular vesicles to endosomes and thence to the lateral intercellular surface of the enterocyte where membrane fusion releases the vesicle contents and liberates the bound IgG – essentially in accord with Brambell's initial model [1,2]. Other studies recently showed FcRB (as FcRp) in endothelial cell lines [44,45] and in vascular endothelium [45], but detailed ultrastructural information is not yet available. Similar ultrastructural studies of pIgR in rat hepatocytes trace a comparable reverse flow from the basolateral to apical surfaces [46,47].

The protein cytoplasmic motifs for sorting of the pIgR have been extensively investigated in reconstructed *in vitro* models that showed critical amino acids in a small membrane proximal segment with characteristic secondary structure features [48,49]. The constitutive transcytosis of the pIgR is accelerated by binding of dIgA and attendent phosphorylation events [50]. There are also motifs in the cytoplasmic domain of the FcRB suggested to interact with coated pits [51], but the further cytoskeletal interactions and their regulation have not been elucidated. pIgR gene expression is sensitive to a number of cytokines and hormones, and the role of nuclear transcription factors have been suggested, as outlined by in Chapter 5. Cytokine response motifs have been noted in the 5′ DNA flanking sequence of the FcRB [52], but no biological testing of FcRB gene regulation has yet been performed. Further, the mechanisms regulating the tissue and temporal expression of the FcRB are totally unknown. Expression appears to be constitutive within the yolk sac splanchnopleur and the placental syncytiotrophoblast as FcRn and in the vascular endothelium as FcRp, but the expression in the rodent intestine, which begins antenatally, abuptly ceases around the time of weaning (after 16 days in mice and 18 days in rats) [3]. This cessation of receptor expression can be hastened by administration of corticosteroids, which induces a maturation of the proximal intestine that is accompanied by diverse morphologic and biochemical alterations [2].

If there is a generalizable feature that can be cited to link all sites of expression of the FcRB, it is the following: all are cells that are active in fluid phase endocytosis. Yolk sac and placenta are extremely active endocytically, as is proximal intestine of neonatal rodent until day 16–18, after which endocytic activity abruptly ceases and FcRB expression is lost. Similarly, the vascular endothelium and renal proximal tubule are the most endocytically active tissues in adults, and have the highest expression of FcRB [45, personal communication]. Smaller amounts of FcRB are expressed in other tissues, such as liver hepatocytes [20], but this likely parallels the basal endocytic activity of this tissue: the FcRB plays no role in transport to bile [P. Telleman and R. Junghans, in preparation]. This connection suggests that the coupling of FcRB expression to the endocytic activities of a cell provides a generalized protection function for IgG wherever endocytosis is significant. Hence, one may speculate that there are shared regulatory elements for FcRB expression and endocytosis, whether perinatal or in the adult animal. There are also clear differences between the transport (FcRn) and protection (FcRp) settings. The cellular interactions with the FcRB perinatally are such as to direct the receptor (with IgG) transcellularly, whereas IgG is apparently recycled to the lumenal surface of the endothelial cell during protection with no net transport. Whether this difference represents distinct signals for different cytoskeletal interactions or merely different ultrastructural architectures of the endocytically active cells can only be speculated at this point.

CONCLUSIONS

The regulation of IgA and IgG distribution and concentration is intrinsic to the effectiveness of these important and abundant immunoglobulin classes. The many biologically interesting similarities and differences between these two receptor systems that participate majorly in this regulation also suggest areas of future research. For example, investigation of the promoter regulatory elements of the FcRB akin to prior studies on the pIgR would be very informative and could reveal connections with features that also regulate endocytosis. Understanding this regulation could additionally shed light on diseases associated with disturbed IgG catabolism [5,53,54]. Similarly, a molecular dissection comparable to the pIgR studies is warranted to explore the cytoplasmic domains of the FcRB that interact with cytoskeleton and direct the receptor in transport and recycling. While noting the dramatic progress of work on pIgR regulation, its substantial complexities have already been alluded to in Chapter 5 and this work will undoubtedly continue. On the other hand, ultrastructural studies of the pIgR analogous to those on the FcRB would be useful to assess the occupancy, kinetics and distribution of pIgR in different cellular compartments under different conditions of transcytosis and activation. Whereas the studies with cell lines have been enormously informative in dissecting features of the pIgR molecular function, these ultrastructural studies would best be performed with native tissues. Ultrastructure analysis of non-hepatocyte tissues for pIgR trafficking (hepatocytes mediate little dIgA transport in humans) and of vascular endothelium for FcRB (FcRp) would be very informative. Co-crystallization of IgA Fc and pIgR would be highly instructive on the nature of the interactions mediated by this unusual receptor; much fundamental information was gained in the corresponding FcRB:Fc studies. Finally, knockout studies of pIgR are needed to assess the importance of this receptor in health; and knockout of the heavy chain of the FcRB is needed to separate the impact of IgG concentration regulated by FcRp from that of MHC class I-related functions in autoimmune diseases, which is not addressed in the light chain (β_2m) knockout that does not separate these effects. The insights provided in the foregoing chapters and the contrasts available from them contribute important perspectives on these receptor systems that will help to guide the future directions of research.

Acknowledgements

I wish to thank several professional colleagues for their prior reviews of this commentary: the authors of the foregoing chapters, Drs Charlotte Kaetzel, Keith Mostov, and Neil Simister, as well as Drs Clark Anderson, Victor Ghetie, Daniel Sedmak, and Jan van de Winkel. I am grateful to Ms. Tara Harris for assistance in preparing the manuscript. I am supported by a Research Career Development Award from the National Cancer Institute.

References

1. Brambell FWR, Halliday RAM, Morris IG. Interference by human and bovine serum protein fractions with the absorption of antibodies by suckling rats and mice. Proc R Soc Lond B. 1958;149:1–11.
2. Brambell FWR. The transmission of immunity from mother to young and the catabolism of immunoglobulins. Lancet. 1966;ii:1087–93.
3. Brambell FWR. The Transmission of Passive Immunity from Mother to Young. Amsterdam: North Holland Publishing Co.; 1970.
4. Brambell FWR, Hemmings WA, Morris IG. A theoretical model of γ-globulin catabolism. Nature. 1964;203:1352–5.
5. Junghans RP. Finally! The Bramball receptor (FcRB): mediator of transmission of immunity and protection from catabolism for IgG. Immunol Res. 1997;16:29–57.
6. Mostov KE, Kraehenbuhl JP, Blobel G. Receptor-mediated transcellular transport of immunoglobulins: synthesis of secretory component as multiple and larger transmembrane forms. Proc Natl Acad Sci USA. 1980;77:7257–61.
7. van de Winkel JGJ, Capel PJA. Human IgG Fc receptor heterogeneity: molecular aspects and clinical implications. Immunol Today. 1993;14:215–21.

8. Morton HC, van Egmond M, van de Winkel JGJ. Structure and function of human IgA Fc receptors (FcαR). Crit Rev Immunol. 1996;16:423–40.
9. Pfefferkorn LC, Yeaman GR. Association of IgA-Fc receptors (FcαR) with FcεRI γ2 subunits in U937 cells. Aggregation induces the tyrosine phosphorylation of γ2. J Immunol. 1994;153: 3228–36.
10. Morton HC, van den Herik-Oudijk IE, Vossebeld P et al. Functional association between the human myeloid IgA Fc receptor (CD89) and FcR γ chain. Molecular basis for CD89/FcR γ chain association. J Biol Chem. 1995;270:29781–7.
11. Saito K, Suzuki K, Matsuda H, Okimura K, Ra C. Physical association of Fc receptor γ homodimer with IgA receptor. J Allergy Clin Immunol. 1995;96:1152–60.
12. Levine B, Harwick JM, Trapp BD, Crawford TO. Antibody-mediated clearance of alphavirus infection from neurons. Science. 1991;254:856–60.
13. Mazanec MB, Kaetzel CS, Lamm ME, Fletcher D. Nedrud JG. Intracellular neutralization of virus by immunoglobulin A antibodies. Proc Natl Acad Sci USA. 1992;89:6901–5.
14. Dietschold B, Kao M, Zheng YM et al. Delineation of putative mechanisms involved in antibody-mediated clearance of rabies virus from the central nervous system. Proc Natl Acad Sci USA. 1992;89:7252–6.
15. Randall TD, Parkhouse RME, Corley JB. J chain synthesis and secretion of hexameric IgM is differently regulated by LPS and IL5. Proc Natl Acad Sci USA. 1992;89:962–6.
16. Matsui K, Nakanishi K, Cohen DJ et al. B cell response pathways regulated by IL5 and IL2. J Immunol. 1989;142:2918–23.
17. Kerr MA. The structure and function of human IgA. Biochem J. 1990;271:285–96.
18. Dive C, Heremans JF. Nature and origin of the proteins of bile. I. A comparative analysis of serum and bile proteins in man. Eur J Clin Invest. 1974;4:235–9.
19. McFarlane AS. The behavior of I^{131}-labeled plasma proteins *in vivo*. Ann NY Acad Sci. 1957; 70:19–25.
20. Blumberg RS, Koss T, Story CM et al. A major histocompatibility complex class I-related Fc receptor for IgG on rat hepatocytes. J Clin Invest. 1995;95:2397–402.
21. Fernandes FR, Nagao AT, Mayer MP, Zelante F, Carneiro-Sampaio MM. Compensatory levels of salivary IgM anti-*Streptococcus mutans* antibodies in IgA-deficient patients. J Invest Allergol Clin Immunol. 1995;5:151–5.
22. Piskurich JF, Blanchard MH, Youngman KR, France JA, Kaetzel CS. Molecular cloning of mouse polymeric Ig receptor: Functional regions of the molecule are conserved among five mammalian species. J Immunol. 1995;154:1735–47.
23. Chintalacharuvu KR, Tavill AS, Louis LN, Vaerman JP, Lamm ME, Kaetzel CS. Disulfide bond formation between dimeric immunoglobulin A and the polymeric immunoglobulin receptor during hepatic transcytosis. Hepatology. 1994;19:1620–73.
24. Simister NE, Mostov KE. An Fc receptor structurally related to MHC class I antigens. Nature. 1989;337:184–7.
25. Burmeister WP, Huber AH, Bjorkman PJ. Crystal structure of the complex of rat neonatal Fc receptor with Fc. Nature. 1994;372:83.
26. Jefferis R, Lund J, Pound J. Molecular definition of interaction sites on human IgG for Fc receptors (huFc gamma R). Mol Immunol. 1990;27:1237–40.
27. Kim JK, Tsen MF, Ghetie V, Ward ES. Identifying amino acid residues that influence plasma clearance of murine IgG_1 fragments by site-directed mutagenesis. Eur J Immunol. 1994;24: 542–8.
28. Kim JK, Tsen MF, Ghetie V, Ward ES. Localization of the site of the murine IgG1 molecule that is involved in binding to the murine intestinal Fc receptor. Eur J Immunol. 1994;24: 2429–34.
29. Jones EA, Waldmann TA. The mechanism of intestinal uptake and transcellular transport of IgG in the neonatal rat. J Clin Invest. 1972;51:2916–27.
30. Raghavan M, Gastinel LN, Bjorkman PJ. The class I major histocompatibility complex related Fc receptor shows pH-dependent stability differences correlating with immunoglobulin binding and release. Biochemistry. 1993;32:8654–60.
31. Roberts DM, Guenthert M, Rodewald R. Isolation and characterization of the Fc receptor from the fetal yolk sac of rat. J Cell Biol. 1990;111:1867–76.
32. Berryman M, Rodewald R. β_2-microglobulin co-distributes with the heavy chain of the intestinal IgG-Fc receptor throughout the transepithelial transport pathway of the neonatal rat. J Cell Sci. 1995;108:2347–60.

33. Hemmings WA. Protein selection in the yolk-sac splanchnopleur of the rabbit: the total uptake estimated as loss from the uterus. Proc R Soc B. 1957;148:76–83.
34. Rodewald R. Distribution of immunoglobulin G receptors in the small intestine of the young rat. J Cell Biol. 1980;85:18–32.
35. Neutra MR, Kraehenbuhl JP. Transepithelial transport and mucosal defense. I. The role of M cells. Trends Cell Biol. 1992;2:134–8.
36. Weltzin R, Kraehenbuhl JP, Neutra MR. Monoclonal IgA adheres to luminal surfaces of M cells in Peyer's patch epithelium. J Cell Biol. 1987;105:234a.
37. Gan YJ, Chodosh J, Morgan A, Sixbey JW. Epithelial cell polarization is a determinant in the infectious outcome of immunoglobulin A-mediated entry by Epstein-Barr virus. J Virol. 1997;71:519–26.
38. Abrahamson DR, Powers A, Rodewald R. Intestinal absorption of immune complexes by neonatal rats: a route of antigen transfer from mother to young. Science. 1979;206:567–9.
39. Peppard JV, Jackson LE, Hall JG, Robertson D. The transfer of immune complexes from the lumen of the small intestine to the bloodstream in sucking rats. Immunology. 1984;53:385–93.
40. Junghans RP, Waldmann TA. Metabolism of Tac (IL2Rα): Physiology of cell surface shedding and renal elimination, and suppression of catabolism by antibody binding. J Exp Med. 1996; 271:10453–60.
41. Junghans RP, Anderson CL. The protection receptor for IgG catabolism is the β_2-microglobulin-containing intestinal transport receptor. Proc Natl Acad Sci USA. 1996;93: 5512–6.
42. Rodewald R. Selective antibody transport in the proximal small intestine of the young rat. J Cell Biol. 1970;45:635–40.
43. Rodewald R. Intestinal transport of antibodies in the newborn rat. J Cell Biol. 1973;58:189–211.
44. Ghetie V, Hubbard JG, Kim JK, Ysen MF, Lee Y, Ward ES. Abnormally short serum half-lives of IgG in β2-microglobulin-deficient mice. Eur J Immunol. 1996;26:690–6.
45. Sedmak DD, Rahill BM, Osborne JM et al. *In vitro* and *in vivo* expression of the IgG protection receptor, FcRp, by vascular endothelial cells. J Immunol. 1998; in press.
46. Takahashi I, Nakane PK, Brown WR. Ultrastructural events in the translocation of polymeric IgA by rat hepatocytes. J Immunol. 1982;128:1181–7.
47. Geuze HJ, Slot JW, Strous GJ et al. Intracellular receptor sorting during endocytosis: comparative immunoelectron microscopy of multiple receptors in rat liver. Cell. 1984;37: 195–204.
48. Aroeti B, Kosen PA, Kuntz ID, Cohen PE, Mostov KE. Mutational and secondary structural analysis of the basolateral sorting signal of the polymeric immunoglobulin receptor. J Cell Biol. 1993;123:1149–60.
49. Mostov KE. Transepithelial transport of immunoglobulins. Ann Rev Immunol. 1994;12:63–84.
50. Song W, Bomsel M, Casanova J, Vaerman JP, Mostov KE. Stimulation of transcytosis of the polymeric immunoglobulin receptor by dimeric IgA. Proc Natl Acad Sci USA. 1994;91:163–6.
51. Ahouse JJ, Hagerman CL, Mittal P et al. Mouse MHC class I-like Fc receptor encoded outside the MHC. J Immunol. 1993;151:6076–88.
52. Kandil E, Noguchi M, Ishibashi T, Kasahara M. Structural and phylogenetic analysis of the MHC class I-like Fc receptor gene. J Immunol. 1995;154:5907–18.
53. Waldmann TA, Strober W. Metabolism of immunoglobulins. Progr Allergy. 1969;13:1–110.
54. Mariani G, Strober W. Immunoglobulin metabolism. In: Metzger H, ed., Fc Receptors and the Action of Antibodies. American Society for Microbiology. 1990:94–177.

8
Signals initiated by the high affinity FcR for IgE

O. H. CHOI and P. G. HOLBROOK

INTRODUCTION

The high affinity receptor for immunoglobulin E (FcεRI) is present on mast cells and basophils and, in conjunction with immunoglobulin E, functions as the cell surface initiator of various physiological responses that mediate certain types of allergic reaction. These responses include degranulation with release of histamine and the biosynthesis and release of arachidonic acid metabolites and cytokines. Since mast cells and basophils are difficult to obtain in large quantities, the rat basophilic leukemia cell line (RBL-2H3) is often employed as a mast cell model. RBL-2H3 cells express FcεRI, release histamine and have proved particularly useful in studies of signal transduction [reviewed in Reference 1]. The presence of FcεRI on Langerhans cells, monocytes and eosinophils has recently been reported [2].

RECEPTOR

Classification

FcεRI is one member of a family of multimeric receptors of the immune system that includes receptors for immunoglobulin G (FcγRI, FcγRII and FcγRIII) [3] and the receptor for T cell (TCR) [4] and B cell antigen (BCR). Some of the individual subunits of these receptors appear to have evolved from common precursors by gene duplication and in some instances are interchangeable among receptor family members. A low affinity receptor for immunoglobulin E (FcεRII or CD23) is not a member of this family [3] and will be discussed elsewhere in this volume.

J.G.J. van de Winkel and P.M. Hogarth (eds.), The Immunoglobulin Receptors and their Physiological and Pathological Roles in Immunity. 83–93.

Structure

Careful attempts to purify FcεRI demonstrated that it was a tetramer composed of three distinct proteins [5]. The cDNA of the α, β and γ subunits of FcεRI from rat, mouse, and human have been cloned and their amino acid sequences deduced [6,7]. The FcεRI receptor most commonly exists as a complex of one α, one β and two γ subunits although complexes of one α and two γ subunits occur on human monocytes [8].

The human α subunit has a single transmembrane region, a short intracellular C-terminal sequence and an extracellular region that contains the immunoglobulin E binding domain and the N-terminus. Two disulphide bonds between cysteine residues of the α subunit form two immunoglobulin-related regions that mediate high-affinity binding of monomeric IgE. The human β subunit has four transmembrane regions. Its C and N termini are both located intracellularly. The human γ subunit has an extracellular N-terminus, a transmembrane region, and an extended C-terminal domain. Two of these γ subunits are linked by disulphide bonds between cysteine residues located at the extracellular aspect of their transmembrane regions.

Signalling motifs

A tyrosine-containing motif common to the cytoplasmic tail regions of components of the FcεRI, TCR and BCR was described by Reth [9] and termed the antigen recognition activation motif or ARAM [10]. As the importance of this motif to activation of nonreceptor tyrosine kinases became apparent [11], it was redesignated the immunoreceptor tyrosine-based activation motif or ITAM [12]. FcεRI contains three ITAMs: one in the intracellular, C-terminal region of the β subunit and one in each C-terminus of the γ subunits. ITAMs vary in structure and phosphorylated peptides containing ITAM motifs of the β and γ subunits of FcεRI have been shown to bind proteins that contain Src homology 2 (SH2) domains with variable affinity [13].

SIGNAL TRANSDUCTION

The signalling complex

Clustering of FcεRI and tyrosine phosphorylation are the earliest events leading to mast cell activation [14]. Clustering occurs when multivalent antigen links two or more monomeric IgE molecules bound to their receptor. An essential role for ITAMS in early FcεRI signalling is supported by two lines of experimental evidence. First, truncation of regions of the cytoplamic domains of β and γ that contain ITAMS results in loss of downstream function in P815 cells [15]. Second, a chimeric protein consisting of the extracellular domain of the α chain of the interleukin 2 receptor and the cytoplasmic domain of FcεRI γ supports serotonin release from RBL cells when cross-linked with antibody [16]. Following receptor clustering, the β and γ subunits of FcεRI are phosphorylated on tyrosine residues in a time- and antigen concentration-dependent manner [17].

An accumulation of data on immunoreceptors supports the general notion that phosphorylation of ITAMs promotes assembly of cytoplasmic proteins into signalling complexes [18] (Figure 8.1). A model specific to FcεRI signalling has been proposed [14]. The Src family tyrosine kinase, Lyn is associated with the β subunit of the FcεRI in its basal state and phosphorylates β and γ subunits of adjacent cross-linked receptor. This promotes recruitment, phosphorylation and activation of Syk of the Syk/ZAP-70 family of tyrosine kinases. Studies comparing the $\alpha\beta\gamma_2$ complex to $\alpha\gamma_2$ in a reconstitution system support this model and suggest further that β functions to amplify activation by γ [19]. Syk activation is mediated by a so-called activation loop phosphorylation chain reaction initiated by Lyn but also dependent upon other molecules of Syk [20]. The activation loop is present in all known tyrosine kinases and has a tyrosine residue that occupies the active site until it is phosphorylated thus freeing the site for access by substrates.

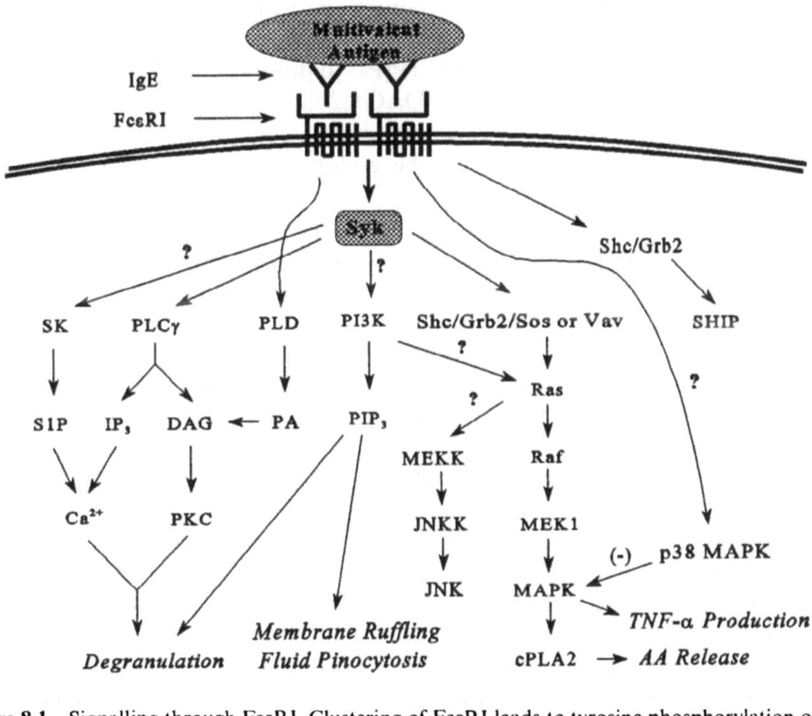

Figure 8.1 Signalling through FcεRI. Clustering of FcεRI leads to tyrosine phosphorylation of β and γ subunits of receptors by a Src kinase Lyn. This promotes tyrosine phosphorylation and activation of Syk. Known downstream signals of Syk are indicated with arrow. Possible downstream signals for which there is no experimental evidence are indicated with a question mark. Abbreviations can be found in the text

Syk, in turn, activates at least two different signalling pathways (Figure 8.1) [1,21]: the phospholipase Cγ (PLCγ) pathway and, indirectly the mitogen activated protein (MAP) kinase/ phospholipase A_2 (PLA_2) pathway, perhaps via Shc [22]. Phosphatidylinositol-3-kinase (PI3K) may represent a third pathway that is ubiquitously activated by immunoreceptors [23]. FcεRI is inactivated and may be regulated by serine/threonine and tyrosine phosphatases [14].

Changes in cytosolic calcium

FcεRI is expressed in electrically non-excitable cells. Thus, only Ca^{2+} signalling mechanisms in non-excitable cells will be discussed. The increase in cytosolic free Ca^{2+} concentration ($[Ca^{2+}]_i$) observed in cells expressing Ca^{2+}-mobilizing receptors is due to Ca^{2+} released from intracellular stores as well as Ca^{2+} influx from outside of the cell [reviewed in Reference 24].

Intracellular calcium release through PLC and sphingosine kinase

Stimulation of G-protein coupled receptors and receptors that utilize tyrosine kinases, such as FcεRI activates PLCβ and PLCγ, respectively. PLC cleaves phosphatidylinositol 4,5-bisphosphate (PIP_2) to produce inositol 1,4,5-trisphosphate (IP_3) and *sn*-1,2-diacylglycerol (DAG) [25]. DAG activates protein kinase C (PKC). IP_3 mediates Ca^{2+} release from intracellular stores by binding to an IP_3-gated Ca^{2+}-channel located on the endoplasmic reticulum [reviewed in Reference 25].

Although IP_3 is produced in antigen-stimulated RBL-2H3 cells, studies in RBL-2H3 cells transfected with the m1 muscarinic receptor (RBL-m1) point to an additional mechanism for release of Ca^{2+}. Whereas there is a good correlation between IP_3 production and the Ca^{2+} response in RBL-m1 cells stimulated with the muscarinic agonist carbachol, this is not the case when FcεRI is cross-linked with antigen [26,27]. The competitive inhibitor of sphingosine kinase (SK), DL-*threo*-dihydrosphingosine (DHS), attenuates the FcεRI -mediated increase in $[Ca^{2+}]_i$ as well as the production of sphingosine 1-phosphate (S1P) suggesting that S1P in addition to IP_3 might be a second messenger for Ca^{2+} mobilization via these receptors [27]. The balance between IP_3 and S1P may vary with different types of cells and stimulants.

Introduction of S1P into a permeabilized DDT_1 MF-2 smooth muscle cell line [28] and Swiss 3T3 fibroblasts [29] has been shown to release Ca^{2+} from intracellular Ca^{2+} stores. S1P is probably generated in the endoplasmic reticulum and activates release of stored calcium [28]. As an S1P analogue, sphingosylphosphorylcholine (SPC) has also been used as a pharmacological tool to study Ca^{2+} release. An SPC-gated Ca^{2+} channel from RBL-2H3 cells has been characterized [30] and recently cloned [31]. It is not yet clear whether this channel is gated by S1P [reviewed in Reference 32].

Although it is not yet clear how SK is activated, the following evidence indicates that Syk might be involved: (1) pretreatment of cells with genestein, a tyrosine kinase inhibitor, partially inhibits SK [27]; (2) expression of dominant negative Syk inhibits the Ca^{2+} response mediated by FcεRI [33]; (3) a Syk negative clone of RBL-2H3 cells is not able to promote a Ca^{2+} response [34].

Calcium influx

Stimulated Ca^{2+} influx mechanisms in electrically non-excitable cells are not well understood at present. The sustained Ca^{2+} signal, that in some cases can last for hours, is largely dependent on influx of extracellular Ca^{2+} in RBL-2H3 cells and is similar to that in other electrically unresponsive cells and triggered via either FcεRI or m1 receptors [26]. Putney has proposed that Ca^{2+} entry occurs in response to depletion of intracellular Ca^{2+} stores and this so-called capacitative calcium entry has been demonstrated for other cells including mast cells [35–37]. Ca^{2+} influx in RBL-2H3 cells is similar to that in other electrically unresponsive cells and triggered via either FcεRI or m1 receptors [26]. Patch clamp techniques have been used to identify a Ca^{2+} release-activated Ca^{2+} current termed I_{crac} in mast cells [38] and a similar current is shown to be inactivated by PKC in RBL-2H3 cells [39]. This Ca^{2+} influx appears to be mediated by the release of a small messenger CIF (calcium influx factor) that is released from intracellular organelles upon depletion of intracellular Ca^{2+} store [40]. In addition to capacitative Ca^{2+} entry in antigen-stimulated RBL-2H3 cells, a small, initial increase in $[Ca^{2+}]_i$ has been observed in the apparent absence of intracellular Ca^{2+} release suggesting that a bona fide receptor-mediated Ca^{2+} influx mechanism is operational in these cells [41]. Thus, it is possible that multiple mechanisms account for Ca^{2+} influx stimulated by FcεRI cross-linking.

Activation of MAPK and $cPLA_2$

Stimulation of RBL-2H3 cells with antigen apparently activates several forms of mitogen-activated protein (MAP) kinase, including p42 MAP kinase (also known as extracellular signal-regulated kinase (ERK)) [42,43], stress-activated p38 MAP kinase [42,43] and c-jun amino-terminal kinase (JNK) [44]. Among these, p42 MAP kinase has been studied most extensively. Analogous to activation of the p42 MAP kinase in other types of cells, activation of these kinases after FcεRI cross-linking in RBL-2H3 cells occurs through raf1 and MEK, leading to the activation of downstream signals such as cytosolic PLA_2 ($cPLA_2$) [45] and the production of tumour necrosis factor-α (TNF-α) [43]. Aggregation of FcεRI leads to conversion of Ras, a small molecular weight guanine nucleotide binding protein (G-protein) to the active, GTP-bound state [21,22]. The entire MAP kinase pathway is inhibited by guanosine 5-(2-thiodiphosphate) (GDPβS) [45], suggesting that Ras may link FcεRI to the cytosolic MAP kinases. Activation of Ras occurs through tyrosine phosphorylation of the intermediate protein Shc which forms a complex with the adaptor protein, Grb2 and the guanine nucleotide exchange factor, Sos. This pathway appears to be downstream of Syk as evidenced by the enhancement of its signalling with over-expression of Syk and blockade by the expression of dominant-negative Syk [21,22] and negatively regulated by p38 MAP kinase [43].

JNK is activated after aggregation of FcεRI in a mouse mast cell line, MC/9 cells [44]. The activation of JNK occurs through activation of MEK kinase 1 (MEKK1), which activates JNK activator JNK kinase (JNKK). Significant inhibition of this pathway, but not the p42 MAPK pathway by wortmannin suggests that phosphatidylinositol 3-kinase (PI3K) may be involved in the MEKK1, JNKK and JNK pathway [44].

Phosphatidylinositol-3-kinase (PI3K)

Cross-linking of FcεRI activates PI3K [46]. PI3K consists of an 85 kDa adaptor subunit containing one SH3 and two SH2 domains and a 110 kDa catalytic subunit that binds to a region between the two SH2 domains of the 85 kDa protein [47]. Original studies of this complex noted its association with growth factor receptors and ability to phosphorylate the 3 position of the inositol ring of phosphatidylinositol (PI), phosphatidylinositol 4-phosphate (PIP) and PIP2 [reviewed in Reference 47]. It has been shown that p110 may also serve as a serine/threonine kinase that may phosphorylate other substrates in addition to p85 and itself [48]. PI3K is potently inhibited by the fungal toxin wortmannin [46,49] and the effects of this compound on RBL-2H3 cells demonstrate that PI3K is either downstream or independent of activation of Lyn, Syk or MAP kinase while it appears to play a role in secretion, membrane ruffling and fluid pinocytosis [49]. The findings that wortmannin inhibits the production of IP_3 with no significant effect on the Ca^{2+} response [49] may suggest that the SK pathway is not affected by wortmannin or that a Ca^{2+} influx mechanism, which is independent of PI3K and store depletion, is activated.

Other effectors

Protein kinase C

Mast cells contain Ca^{2+}-dependent (α, β_1, β_2) and Ca^{2+}-independent (δ,ε) forms of PKC as well as the atypical ζ isoform [50]. Ca^{2+}-dependent and independent PKCs may be translocated to membrane and activated in cells by DAG formed transiently from PI following activation of PLCγ or in a more sustained fashion from PC subsequent to activation of PC-specific PLD [51]. Both phospholipases are activated by clustering of FcεRI [50]. PKC activation is believed to regulate PI hydrolysis and to mediate downstream functions in mast cells including exocytosis and proto-oncogene expression. PKC-δ is phosphorylated on tyrosine as a consequence of FcεRI activation in RBL cells [52].

Phospholipase D

In stimulated cells, phospholipase D (PLD) cleaves phosphatidylcholine (PC) to produce phosphatidic acid and choline but also catalyses exchange reactions with primary alcohols to form novel phosphatidylalcohols [53]. These novel lipids are readily isolated from their endogenous counterparts and assessment of their formation is believed to constitute a specific assay for PLD. Phosphatidylethanol is formed in antigen-stimulated RBL-2H3 cells [50] and indirect evidence suggests that tyrosine phosphorylation may be involved [54]. Recent studies that implicate small molecular weight G-proteins in its activation demonstrate that PLD may have diverse roles in cell signalling [55,56]. Two different enzymes have been cloned and sequenced. PLD_I is activated by PKC and small G-proteins including ADP-ribosylation factor (ARF), Rho, Rac and CDC42 [57] and has been implicated in regulated secretion [58]. PLD_2 may play a role in cytoskeletal reorganization [59].

FUNCTIONAL CONSEQUENCES

Antigen stimulation of mast cells through FcεRI leads to the various functional responses via the signal transduction pathways discussed above. Secretion of histamine by exocytotic discharge of granules is absolutely dependent on the activation of certain isozymes of protein kinase C and increase in $[Ca^{2+}]_i$ [60]. PKC is activated by DAG generated via PLC and PLD and PLD may not be essential for secretion when PLC is activated as well [61]. Phosphorylation of the A form of myosin heavy chain by PKC, since B form does not exist in RBL-2H3 cells [62], and myosin light chain by Ca^{2+}-calmodulin-dependent myosin light chain kinase appears to be necessary for secretion [63]. Recent findings show that secretion is not regulated by MAP kinase [43]. MAP kinase appears to regulate activation of cytosolic phospholipase A_2 to release arachidonic acid and its metabolites. It also regulates the production of a cytokine, TNF-α, whereas subsequent secretion of TNF-α is regulated by PKC and $[Ca^{2+}]_i$ [43]. TNF-α plays an important role in inducing delayed inflammatory response to hypersensitivity immune reaction.

With respect to regulation of gene expression, clustering of FcεRI results in the induction of nuclear factor of activated T cells (NF-AT) [64] via Ras [65]. NF-AT regulates IL-5 gene expression [66]. MAP kinase pathways are also involved in gene expression. Activated MAP kinase not only phosphorylates cytosolic proteins such as $cPLA_2$ and pp90 ribosomal S6 kinase (Rsk) but also translocates to the nucleus where it can phosphorylate Elk-1, a ternary complex factor, which leads to increased transcription of c-fos mRNA [reviewed in Reference 67]. Rsk also translocates to the nucleus and phosphorylates many transcription factors including c-fos [68]. Other members of MAP kinase superfamily, JNK and p38 MAP kinase also translocate to the nucleus. The former activates c-jun and the latter, ATF-2 [69] and MAP kinase-activated protein kinase 2 [70].

Finally, it should be noted that FcεRI-mediated signalling may be negatively regulated by the SH2 domain-containing inositol polyphosphate 5-phosphatase (SHIP). SHIP is recruited by the inhibitory motif of FcγRIIB, which leads to the inhibition of Ca^{2+} influx when co-ligated with FcεRI or BCR in mast cells or B cells, respectively [71]. Recently, it has been shown that SHIP is tyrosine phosphorylated and recruited by the ITAM of β subunit of FcεRI. An interesting scenario is that the γ subunit associates with positively signalling Syk whereas the β subunit associate negatively signalling SHIP [72].

CONCLUSION

Recent advances in the field of FcεRI-mediated cell signalling now provide a general understanding of signal transduction pathways and how these pathways relate to functional responses. The indications are that PLCγ, p42 MAP kinase and probably sphingosine kinase pathways are downstream of Syk kinase. $cPLA_2$ is down stream of MAP kinase. Activation of PKC and an increase in $[Ca^{2+}]_i$ are sufficient and necessary signals for secretion. We have little understanding, however, of pathways initiated through PI3K and the relationship of these pathways to those initiated through Syk. Past studies have also focused on the mechanisms of activation of the various kinases and phospholipases discussed here, but it is apparent from studies with phosphatase inhibitors that phosphatases play an equally important role in regulation of the stimulatory pathways in mast cell. An understanding of the role of phosphatases and of the communications between different pathways is necessary for full comprehension of how extracellular signals are translated into all physiological responses via FcεRI.

Acknowledgements

We thank Dr Michael A. Beaven for a critical review of the manuscript.

References

1. Beaven MA, Baumgartner RA. Downstream signals initiated in mast cells by FcεRI and other receptors. Curr Opin Immunol. 1996;8:766–72.
2. Kinet J-P. Ig E receptor (FcεRI) and signal transduction. Eur Respir J. 1996;22:116s–18s.
3. Metzger M, Alcaraz G, Hohman R, Kinet J-P, Pribluda V, Quarto R. The receptor with high affinity for immunoglobulin E. Ann Rev Immunol. 1986;4:419–70.
4. Ravetch JV, Kinet J-P. Fc Receptors. Ann Rev Immunol. 1991;9:457–92.
5. Metzger H, Kinet J-P, Perez -Montfort R, Rivnay B, Wank SS. A tetrameric model for the structure of the mast cell receptor with high affinity for IgE. In: Yamamura Y, Tade T, eds, Progress in Immunology V. Florida: Academic Press; 1983:480–501.
6. Blank U, Ra C, Miller L, Metzger H, Kinet J-P, Complete structure and expression in transfected cells of high affinity IgE receptor. Nature. 1989;337:107–09.
7. Adamczewski M, Kinet J-P. The high-affinity receptor for immunoglobulin E. Chem Immunol. 1994;59:173–90.

8. Maurer D, Fiebiger E, Reininger B et al. Expression of functional high affinity immunoglobulin E receptors (FcεRI) on monocytes of atopic individuals. J Exp Med. 1994;179:745–50.
9. Reth M. Antigen receptor tail clue. Nature. 1989;338:383–84.
10. Paolini R, Numerof R, Kinet J-P. Kinase activation through the high-affinity receptor for immunoglobulin E. Immunomethods. 1994;4:35–40.
11. Samelson LE, Klausner RD. Tyrosine kinases and tyrosine-based activation motifs. J Biol Chem. 1992;267:24913–16.
12. Cambier JC. New nomenclature for the Reth motif or ARH1/ITAM/ARAM/YXXL). Immunol Today. 1995;16:110.
13. Kimura T, Kihara H, Bhattacharyya S, Sakamoto H, Appella E, Siraganian RP. Downstream signalling molecules bind to different phosphorylated immunoreceptor tyrosine-based activation motif peptides of the high affinity IgE receptor. J Biol Chem. 1996;271: 27962–8.
14. Jouvin M-H, Numerof RP, Kinet J-P. Signal transduction through the conserved motifs of the high affinity IgE receptor FcεRI. Semin Immunol. 1995;7:29–35.
15. Alber G, Miller L, Jelsema CL, Blank NV, Metzger H. Structure-function relationships in the mast cell high affinity receptor for IgE. J Biol Chem. 1991;266:22613–20.
16. Letourner F, Klausner R. T-cell and basophil activation through the cytoplasmic tail of T-cell receptor ζ family proteins. Proc Natl Acad Sci USA. 1991;88:8905–9.
17. Paolini R, Numerof R, Kinet J-P. Phosphorylation/dephosphorylation of high-affinity IgE receptors: a mechanism for coupling/uncoupling α large signalling complex. Proc Natl Acad Sci USA. 1992;89:10733–7.
18. Cambier JC. Antigen and Fc receptor signalling: the awesome power of the immunoreceptor tyrosine-based activation motif. J Immunol. 1995;155:3281–5.
19. Lin S, Cicala C, Scharenberg AM, Kinet J-P. The FcεRIβ subunit functions as an amplifier of FcεRIγ mediated cell activation signals. Cell. 1996;85:985–95.
20. El-Hillal O, Kurosaki T, Yamamura H, Kinet J-P, Scharenberg AM. Syk kinase activation by a src kinsae-initiated activation loop phosphorylation chain reaction. Proc Natl Acad Sci USA. 1997;94:1919 24.
21. Hirasawa N, Scharenberg A, Yamamura H, Beaven MA, Kinet J-P. A requirement for Syk in the activation of the microtubule-associated protein kinase/phospholipase A_2 pathway by FcεRI is not shared by a G protein-coupled receptor. J Biol Chem. 1995;270:10960 7.
22. Jabril-Cuenod B, Zhang C, Scharenberg AM et al. Syk-dependent phosphorylation of Shc: a potential link between FcεRI and the Ras/mitogen-activated protein kinase signalling pathway through Sos and Grb2. J Biol Chem. 1996;271: 16268 72.
23. Cambier JC, Johnson SA. Differential binding activity of ARH1/ITAM motifs. Immunol Lett. 1995;44:77–80.
24. Clapham DE. Calcium signalling. Cell.1995;80:259–68.
25. Berridge MJ. Inositol trisphosphate and calcium signalling. Nature. 1993;361:315–25.
26. Choi OH, Lee JH, Kassessinoff T, Cunha-Melo JR, Jones SVP, Beaven MA. Carbachol and antigen mobilize calcium by similar mechanisms in a transfected mast cell line (RBL-2H3 cells) that expresses m1 muscarinic receptors. J Immunol. 1993;151:5586–95.
27. Choi OH, Kim JH, Kinet J-P. Calcium mobilization via the sphingosine kinase in signalling by the FcεRI antigen receptor. Nature. 1996;380:634–6.
28. Ghosh TK, Bian J, Gill DL. Shpingosine 1-phosphate generated in the endoplasmic reticulum membrane activates release of stored calcium. J Biol Chem. 1994;269:22628–35.
29. Zhang H, Desai NN, Olivera A, Seki, T, Brooker G, Spiegel S. Sphingosine-1-phosphate, a novel lipid, involved in cellular proliferation. J Cell Biol. 1991;114:155–67.
30. Kindman LA, Kim S, McDonald TV, Gardner P. Characterization of a novel intracellular sphingolipid-gated Ca^{2+}-permeable channel from rat basophilic leukemia cells. J Biol Chem. 1994;269:13088–91.
31. Mao C, Kim SH, Almenoff JS, Rudner XL, Kearney DM, Kindman LA. Molecular cloning and characterization of SCaMPER, a sphingolipid Ca^{2+}-permeable channel from rat basohilic leukemia cells. Proc Natl Acad Sci USA. 1996;93:1993–6.
32. Beaven MA. Calcium signalling: sphingosine kinase versus phospholipase C? Curr Biol. 1996;6:798–801.
33. Scharenberg AM, Lin S, Cuenod B, Yamamura H, Kinet J-P. Reconstitution of interactions between tyrosine kinases and high affinity IgE receptor which are controlled by receptor clustering. Embo J. 1995;14:3385–94.

34. Zhang J, Berenstein EH, Evans RL, Siraganian RP. Transfection of Syk protein tyrosine kinase reconstitutes high affinity IgE receptor-mediated degranulation in a Syk-negative variant of rat basophilic leukemia RBL-2H3 cells. J Exp Med. 1996;184:71–80.
35. Putney JW Jr. A model for receptor-regulated calcium entry. Cell Calcium. 1986;7:1–12.
36. Putney JW Jr. Capacitative calcium entry revisited. Cell Calcium. 1990;11:611–24.
37. Berridge MJ. Capacitative calcium entry. Biochem J. 1995;312:1–11.
38. Hoth M, Penner R. Depletion of intracellular calcium stores activates a calcium current in mast cells. Nature. 1992;355:353–6.
39. Parekh AB, Penner R. Depletion-activated calcium current is inhibited by protein kinase in RBL-2H3 cells Proc Natl Acad Sci USA. 1995;92:7907–11.
40. Randriamampita C, Tsien RY. Emptying of intracellular Ca^{2+} stores releases a novel small messenger that stimulates Ca^{2+} influx. Nature. 1993;364:809–14.
41. Lee RJ, Oliver JM. Roles for Ca^{2+} stores release and two Ca^{2+} influx pathways in the FcεRI-activated Ca^{2+} responses of RBL-2H3 cells. Mol Biol Cell. 1995;6:825–39.
42. Santini F, Beaven MA. Tyrosine phosphorylation of a mitogen-activated protein kinase-like protein occurs at a late step in exocytosis. J Biol Chem. 1993:268:22716–22.
43. Zhang C, Baumgartner, RA, Yamada, K, Beaven MA. Mitogen-activated protein (MAP) kinase regulates production of tumor necrosis factor-α and release of arachidonic acid in mast cells. J Biol Chem. 1997;272:13397–402.
44. Ishizuka T, Oshiba A, Sakata N, Terada N, Johnson GL, Gelfand EW. Aggregation of the FcεRI on mast cells stimulates c-Jun amino-terminal kinase activity. A response inhibited by wortmannin. J Biol Chem. 1996;271:12762–6.
45. Hirasawa N, Santini F, Beaven MA. Activation of the mitogen-activated protein kinase/cytosolic phospholipase A_2 pathway in a rat mast cell line. Indications of different pathways for release of arachidonic acid and secretory granules. J Immunol. 1995;154:5391–402.
46. Yano H, Agatsuma T, Nakanishi S et al. Biochemical and pharmacological studies with KT7692 and LY294002 on the role of phosphatidylinositol 3-kinase in FcεRI-mediated signal transduction. Biochem J. 1995;312:145–50.
47. Kapeller R, Cantley LC. Phosphatidylinositol 3-kinase. BioEssays. 1994;16:565–76.
48. Dhand R, Hiles I, Panayotou G et al. PI 3-kinase is a dual enzyme: autoregulation by an intrinsic protein-serine kinase activity. EMBO J. 1994;13:522–33.
49. Barker SA, Caldwell KK, Hall A et al. Wortmannin blocks lipid and protein kinase activities associated with PI 3-kinase and inhibits a subset of responses induced by FcεRI cross-linking. Mol Biol Cell .1995;6:1145–58.
50. Rivera J, Beaven MA. Regulation of secretion from secretory cells by PKC. In: Parker PJ, Decker LV, eds, Protein Kinase C. Austin: R.G. Landes Company; 1996:131–64.
51. Nishizuka Y. Protein kinase C and lipid signalling for sustained cellular responses. FASEB J. 1995;9:484–96.
52. Haleem-Smith H, Chang E-Y, Szallasi Z, Blumberg PM, River J. Tyrosine phosphorylation of protein kinase C-δ in response to the activation of the high-affinity receptor for immunoglobulin E modifies its substrate recognition. Proc Natl Acad Sci USA. 1995;92: 9112–16.
53. Holbrook PG, Pannell LK, Murata Y, Daly JW. Molecular species analysis of a product of phospholipase D activation. Phosphatidylethanol is formed from phosphatidylcholine in phorbol ester- and bradykinin-stimulated PC12 cells. J Biol Chem. 1992;267:16834–40.
54. Lin P, Fung W-J, Li S, Chen T, Repetto B, Huang K-S, Gilfillan AM. Temporal regulation of the IgE-dependent 1,2-diacylglycerol production by tyrosine kinase activation in rat (RBL 2H3) mast-cell line. Biochem J. 1994;299:109–14.
55. Cockcroft S, Thomas GMH, Fensome A et al. Phospholipase D: a downstream effector of ARF in granulocytes. Science. 1994;263:523–6.
56. Brown HA, Gutowski S, Moomaw CR, Slaughter C, Sternweiss P. ADP-ribosylation factor, a small GTP-dependent regulatory protein, stimulates phospholipase D activity. Cell. 1993;75:1137–44.
57. Hammond SM, Jenco JM, Nakashima S et al. Characterization of two alternatively spliced forms of phospholipase D1. J Biol Chem. 1997;272:3860–8.
58. Ktistakis NT, Brown HA, Walters MG, Sternweiss PC, Rothe MG. Evidence that phospholipase D mediates ARF-dependent formation of golgi coated vesicles. J Cell Biol. 1996;134:295–306.

59. Colley WC, Sung T-C, Roll R et al. Phospholipase D2, a distinct phospholipase D isoform with novel regulatory properties that provokes cytoskeletal reorganization. Current Biol. 1997;7:191–201.
60. Ozawa K, Szallasi Z, Kazanietz MG et al. Ca^{2+}-Dependent and Ca^{2+}-independent isozymes of protein kinase C mediates exocytosis in antigen-stimulated rat basophilic RBL-2H3 cells: reconstitution of secretory responses with Ca^{2+} and purified isozymes in washed permeabilized cells. J Biol Chem. 1993;268:1749–56.
61. Nakamura Y, Nakashima S, Kumada T, Ojio K, Miyata H, Nozawa Y. Brefeldin A inhibits antigen- or calcium ionophore-mediated but not PMA-induced phospholipase D activation in rat basophilic leukemia (RBL-2H3) cells. Immunobiology. 1996;195:231–42.
62. Choi OH, Park C-S, Itoh K, Adelstein RS, Beaven MA. Cloning of the cDNA encoding rat myosin heavy chain-A and evidence for the absence of myosin heavy chain-B in cultured mast (RBL-2H3) cells. J Muscle Res Cell Motil. 1996;17:69–77.
63. Choi OH, Adelstein RS, Beaven MA. Secretion from rat basophilic RBL-2H3 cells is associated with phosphorylation of myosin light chains by myosin light chain kinase as well as phosphorylation by protein kinase C. J Biol Chem. 1994;269:536–41.
64. Hutchinson LE, McCloskey MA. FcεRI-mediated induction of nuclear factor of activated T-cells. J Biol Chem. 1995;270:16333–8.
65. Prieschl EE, Pendl GG, Harrer NE, Baumruker T. $P21^{ras}$ links FcεRI to NF-AT family member in mast cells. The AP3-like factor in this cell type is an NF-AT family member. J Immunol. 1995;155:4963–70.
66. Prieschl EE, Goulleux V, Walker C, Harrer NE, Baumruker T. A nuclear factor of activated T cell-like transcription factor in mast cells is involved in IL-5 gene regulation after IgE plus antigen stimulation. J Immunol. 1995;154:6112–19.
67. Waskiewicz AJ, Cooper JA. Mitogen and stress response pathways: MAP kinase cascades and phosphatase regulation in mammals and yeast. Curr Opin Cell Biol. 1995;7:798–805.
68. Chen RH, Abate C, Blenis J. Phosphorylation of c-Fos transrepression domain by mitogen-activated protein kinase and 90-kDa ribosomal S6 kinase. Proc Natl Acad Sci USA. 1993;90: 10952–6.
69. Gupta S, Campbell D, Derijard B, Davis RJ. Transcription factor ATF2 regulation by the JNK signal transduction pathway. Science. 1995;267:389–93.
70. Rouse J, Cohen P, Trigon S et al. A novel kinase cascade triggered by chemical stress and heat shock which stimulates MAP kinase-activated protein kinase-2 and phosphorylation of the small heat shock protein. Cell. 1994; 78:1039–49.
71. Ono M, Bolland S, Tempst P, Ravetch JV. Role of the inositol phosphatase SHIP in negative regulation of the immune system by the receptor FcγRIIB. Nature. 1996;383:263–6.
72. Kimura T, Sakamoto H, Appella E, Siraganian RP. The negatively signalling molecule SH2 domain-containing inositolpolyphosphate 5-phosphatase (SHIP) binds to the tyrosine-phosphorylated β subunit of the high affinity IgE receptor. J Biol Chem. 1997;272:13991–6.

9
Phagocyte Fc receptors for IgG

S. E. McKENZIE, Z. K. INDIK and A. D. SCHREIBER

INTRODUCTION

This chapter focuses on the Fc receptors for IgG (FcγR) on phagocytic cells, e.g. polymorphonuclear neutrophils (PMNs), monocytes and macrophages. Eosinophils and basophils/mast cells which also have phagocytic properties are not discussed. PMNs and monocytes circulate in the bloodstream and are derived from pluripotent haematopoietic stem cells in the bone marrow. Macrophages are derived from blood-borne monocytes and together with PMNs are found in tissues. The ingestion of antibody-coated microorganisms and cells is one of the important immunological functions of PMNs, monocytes and macrophages. The identity of the Fcγ receptors on these phagocytes, the ligand binding and signal transducing subunits, the patterns and control of expression and the immediate and downstream signals in phagocyte activation will be discussed.

The Fc receptors for IgG on phagocytes belong to 3 classes: FcγRI (CD64), FcγRII (CD32) and FcγRIII (CD16), distinguishable by size and structure, gene origin and reactivity with anti-FcγR monoclonal antibodies. The cellular pattern of expression, the structure, and the subunits for the 3 classes of FcγR on phagocytes are summarized in Figure 9.1. Of the FcγRI genes, the FcγRIA gene is responsible for the expression of the known FcγRI protein. FcγRIIIA is expressed on monocytes/macrophages and FcγRIIIB is expressed on PMNs. The three FcγRII genes encode FcγRIIA, FcγRIIB and FcγRIIC. FcγRI and FcγRIIIA physically associate with γ chain homodimers which constitute the signal transducing subunit for the ligand-binding subunits. In contrast, FcγRIIA mediates signals in the absence of associated subunits through sequences within its own cytoplasmic domain (see below).

J.G.J. van de Winkel and P.M. Hogarth (eds.), The Immunoglobulin Receptors and their Physiological and Pathological Roles in Immunity. 95–107.

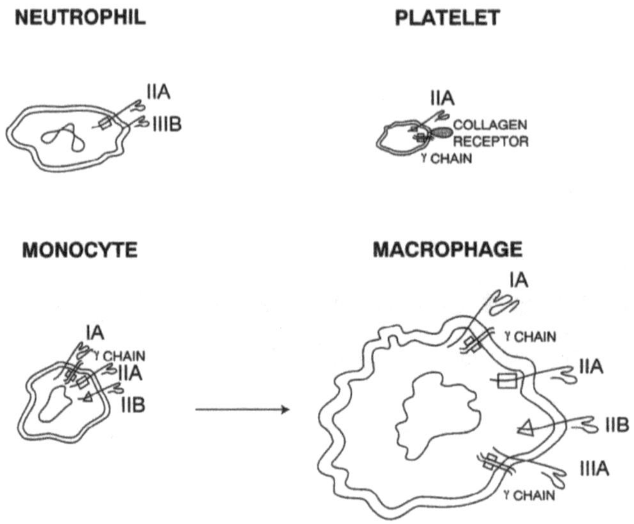

Figure 9.1. Cellular pattern of expression, structure and subunits for the three classes of Fcγ receptors on phagocytes. □ = ITAM, immunoreceptor tyrosine activation motif; △ = ITIM, immunoreceptor tyrosine inhibition motif; IA = FcγRI; IIA = FcγRIIA; IIB = FcγRIIB; IIIA = FcγRIIIA and IIIB = FcγRIIIB

Soluble receptors have been detected for FcγRIIA, FcγRIIIB and FcγRIIIA. Soluble FcγRIIA species are produced primarily by alternative splicing but also occur by proteolytic cleavage of the membrane form. Soluble FcγRIIIB is released upon cleavage of the GPI-linked membrane form and constitutes the majority of FcγRIII in serum, while soluble FcγRIIIA is released by proteolysis. The physiologic and pathologic roles of these receptors are under investigation. Clearly, soluble Fc receptors may affect the functions mediated by the cell membrane FcγR. This chapter focuses on the membrane FcγR and their cellular functions (see Chapter 28; this volume, for soluble FcR).

Since mice are increasingly being used in transgenic and knockout models of FcR biology, it is important to recognize the differences between mouse and man. Mice express FcγRI, FcγRII and FcγRIII proteins. However, there are no murine equivalents of human FcγRIIA, FcγRIIC and FcγRIIIB. The divergence between FcγRII and FcγRIII genes in mouse and man has important functional consequences for phagocytes. Our model for the evolution of human FcγRII is depicted in Figure 9.2 and is based on new information since the development of earlier models [1,2]. We believe that FcγRIIB is the primordial gene and gave rise to FcγRIIC by way of an insertion of approximately 4 kb between exons C2 and C3 of FcγRIIB. This insertion resulted in the inclusion of a new cytoplasmic exon containing the ITAM (immunoreceptor tyrosine activation motif) sequence and relocated the FcγRIIB ITIM (immunoreceptor tyrosine inhibitory motif) sequence to the 3′-untranslated region. Next,

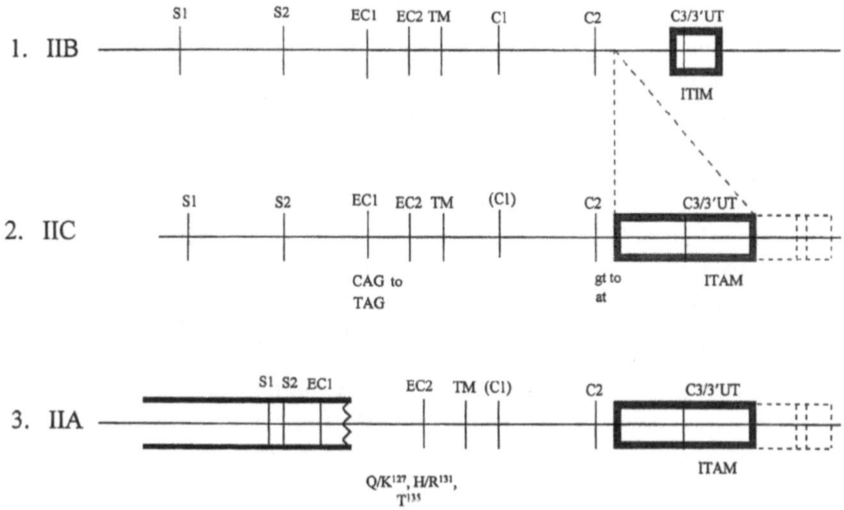

Figure 9.2 Model for the evolution of the human FcγRII class, from FcγRIIB to FcγRIIC to FcγRIIA. Exons are shown as vertical lines; S = signal peptide, EC = extracellular domain, TM = transmembrane domain, C = cytoplasmic domain, UT = untranslated region; additional abbreviations as in Figure 9.1. The evolutionary changes in FcγRIIC and FcγRIIA are indicated at the respective genes

FcγRIIC gave rise to FcγRIIA by replacement of the 5′-flanking region, which in FcγRIIA diverges from FcγRIIB and FcγRIIC as well as from the recently identified FcγRIIIA and FcγRIIIB 5′-flanking regions [3–6]. We have observed that FcγRIIB and FcγRIIC are 99% homologous through 2.5 kb of 5′-flanking region and completely distinct from FcγRIIA (own unpublished observations). Interestingly, there is considerable homology between the human FcγRIIA and mouse FcγRIIIA 5′-flanking and 5′-untranslated regions (unpublished observations, GenBank accession numbers X68090 and U11854). Further changes recently identified in both FcγRIIC and FcγRIIA support our conclusion that FcγRIIA is the most recent evolutionary form. The FcγRIIC gene is frequently either deleted entirely [7,8] or possesses pseudogene mutations (CAG to TAG in EC1, gt-at at C2 splice junction [1,2; unpublished observations] which eliminate its expression. No consequences have been shown to arise from absence of FcγRIIC expression. The FcγRIIA gene has undergone additional changes which cluster in a 14 amino acid stretch of the distal EC2 exon encoding amino acid positions 127, 135 and most notably 131 (see below) [9]. In summary, these changes endow FcγRIIA with properties particularly relevant to phagocyte function, notably the unique ability among FcγRs to bind the human IgG_2 subclass and the ability to transmit a phagocytic signal through its own cytoplasmic domain in the absence of associated subunits.

CONTROL OF RECEPTOR EXPRESSION

For PMNs, the array of Fc receptors which are expressed depends on the state of cellular activation. Resting PMNs express FcγRII and FcγRIIIB, with approximately 10 times more FcγRIIIB than FcγRII. Most of the FcγRII on the PMN surface is the FcγRIIA isoform. FcγRI is expressed during neutrophil development, but in the absence of specific stimulation, FcγRI is turned off and absent from the cell surface by the time the mature PMN is released to the circulation. IFN-γ, G-CSF and other mediators of inflammation have been reported to induce PMN expression of FcγRI [10]. The actions of these mediators on PMN FcγRI expression have been observed both *in vitro* and *in vivo*. The mechanism of FcγRI inducibility resides in specific DNA sequences in the FcγRI gene promoter referred to as the gamma response region (GRR). The interaction of this region with specific DNA-binding proteins brings about increased expression [11,12]. Unlike FcγRI, FcγRIIIB levels diminish with stimulation, such as following incubation with FMLP, due to both receptor internalization and shedding of the receptor. Thus, the ligand binding and cellular activation properties of PMN FcγRs vary with the state of cellular activation.

Monocytes in the circulation express FcγRI and FcγRII. Most of the FcγRII is the FcγRIIA isoform, but RNA and indirect protein evidence also support the presence of FcγRIIB expression in monocytes [13]. FcγRI and FcγRII expression can be increased by IFN-γ and decreased by glucocorticoids [14]. FcγRIIIA is also expressed upon differentiation of monocytes into macrophages *in vitro*. *In vivo*, M-CSF therapy has also been associated with the circulation of FcγRIIIA positive monocytes [15]. *In vitro*, the expression of FcγRIIIA is enhanced by TGF-β, M-CSF and IL-10 and decreased by IL-4 [16,17]. The FcγRIIIA promoter includes the GRR sequence, but it remains to be established if transcriptional activation via this region is responsible for increased FcγRIIIA expression [4–6]. Studies with transgenic mice expressing FcγRIIIA and FcγRIIIB have begun to elucidate cis-acting sequences important for the differential expression of these highly homologous receptor genes [18].

The FcγR γ chain is expressed by PMNs, monocytes and macrophages. While the γ chain promoter has been characterized [19], the factors regulating myeloid cell expression have not been identified. In transgenic mice expressing human FcγRI, there is diminished murine FcγRIII expression [10], due perhaps to competition for FcR γ chains. In human monocytes/ macrophages, FcγRI levels do not decrease as FcγRIIIA is expressed. Thus, IFN-γ and M-CSF may increase γ chain expression as well as FcγRIIIA expression in monocytes/ macrophages. Similarly, in the normal differentiation of monocytes into macrophages, γ chain expression as well as FcγRIIIA expression may be increased. Although co-expression with the γ chain is not required for efficient expression of FcγRI in a cell culture model system [20,21], crossbreeding of FcγRI transgenic mice with γ chain knockout mice has demonstrated a requirement for the γ chain for high level FcγRI surface expression *in vivo* [22].

THE PHAGOCYTE CELLULAR FUNCTIONS TRIGGERED BY RECEPTOR ENGAGEMENT

When FcγRs on PMNs, monocytes and macrophages are crosslinked, a wide range of cellular functions are activated. These include phagocytosis of IgG-coated bacteria and other microorganisms; phagocytosis of IgG-coated red blood cells, platelets, leukocytes or tumour cells; ingestion of non-cellular immune complexes (IC); antigen presentation; release of inflammatory mediators including reactive oxygen species; antibody-dependent cellular cytotoxicity (ADCC); and altered gene expression.

FcγRI is occupied by monomeric IgG at physiological concentrations, while FcγRII and FcγRIII possess low affinity for monomeric IgG. For each class of receptor, crosslinking is essential as the first step in triggering cellular responses. Because phagocytes express multiple classes of FcγRs simultaneously, it has required some effort to dissect which receptor(s) contributes to which cellular functions. Approaches include studies of the receptors in isolation [in transfectants or on platelets (FcγRIIA only)], use of ligands which preferentially engage one class of receptor, and transgenic and knockout mouse models. For PMNs, it has been established that both FcγRI/γ and FcγRIIA can trigger phagocytosis, endocytosis of IC and release of inflammatory mediators. FcγRIIIB cannot perform these functions in isolation but can interact with FcγRIIA and/or CR3 (CD11b/18), resulting in transduction of signals for performance of several functions [23–27]. The high number and rapid lateral mobility of FcγRIIIB molecules on the PMN surface contribute to its function. During the immune response *in vivo*, it is likely that multiple classes of receptors are often engaged simultaneously [28]. In one exception, $hIgG_2$ binds efficiently only to FcγRIIA, notably the H^{131} isoform [29,30]. Furthermore, even in the presence of complement activation and the potential simultaneous engagement of complement receptors, the FcγRs on PMNs are likely important for phagocytosis of opsonized bacteria [31–33]. It had been previously observed that FcγRs and complement receptors interact synergistically in the binding of IgG-coated cells [34]. More recently, CR3 was observed to mediate the phagocytosis of IgG-coated cells in association with a phagocytosis-deficient mutant form of FcγRIIA. Thus, CR3 can interact functionally not only with FcγRIIIB but also with FcγRIIA [35].

On monocytes/macrophages, FcγRI/γ, FcγRIIA and FcγRIIIA/γ can each mediate phagocytosis, endocytosis of IC, release of inflammatory mediators, ADCC and antigen presentation. In mice transgenic for human FcγRI, phagocytosis and antigen-presenting cell (APC) functions can be performed by the human receptor, as elucidated by bispecific antibody studies [10]. In γ chain knockout (KO) mice, no phagocytosis or ADDC takes place despite binding of immune complexes [36,37]. In addition, macrophages of FcγRIII knockout mice have been demonstrated to lack the ability to mediate phagocytosis of IgG_1-coated cells [38].

Since the surface expression of FcγRIIIA and FcγRI is impaired in the absence of the γ chain in these mice, studies in humans genetically or pharmacologically lacking γ chain function are of potential importance. For example, in human monocytes/macrophages, γ chain depletion inhibits phagocytosis of IgG coated cells [39]. Mice transgenic for human FcγRIIA have also been created and are under investigation (own unpublished observations). One area of current interest is concerned with whether the different FcγR classes use different signalling pathways to effect similar functions. As described below, the proximal signals appear similar, with select differences between FcγRIIA on one hand and FcγRI/γ and FcγRIIIA/γ on the other [40–43] and perhaps with subtle differences in downstream signalling molecules.

STRUCTURE–FUNCTION RELATIONSHIPS IN Fcγ RECEPTOR SIGNALLING FOR PHAGOCYTOSIS

The first step in cellular activation is ligand binding. FcγRI has three Ig-like extracellular (EC) domains, while FcγRII and FcγRIII have two. All of the FcγR EC domains are glycosylated. Although the role of glycosylation in ligand binding has not been established, deglycosylated receptors are frequently recognized by ligand-blocking anti-FcγR mAbs, consistent with a negligible role for glycosylation in ligand binding [44]. Amino acid polymorphisms in the EC domains influence ligand binding and thus phagocytosis. The best studied examples are FcγRIIIB NA1/NA2 and FcγRIIA H/R^{131} (see Chapter 27), but others are emerging [45]. We recently described an FcγRIIA K^{127} isoform with novel ligand binding properties (binding of both $mIgG_1$ and $hIgG_2$) which induces phagocytosis [9]. Elucidation of the FcγR EC crystal structure should help clarify the role of amino acid variation in ligand binding (see Chapter 3). Interestingly, formation of the FcγR/γ chain complex can lead to higher affinity for monomeric IgG, analogous to a number of heterodimeric cytokine receptor systems [46].

The second step in cellular activation is receptor clustering, followed within seconds by changes in protein tyrosine phosphorylation. Molecules phosphorylated on tyrosine include FcγRIIA and the FcγR γ subunit as well as non-receptor protein tyrosine kinases (PTK) of the Src and Syk families [42,43,47–49]. We initially observed that FcγRIIA is tyrosine phosphorylated, can mediate a phagocytic signal in transfected COS-1 cells in the absence of other Fc receptors and that tyrosine containing sequences of the FcγRIIA cytoplasmic domain are required for phagocytic signalling [20,48]. The structural basis for the phosphorylation of FcγRIIA and the γ chain resides in the ITAM [42,48–51]. The γ chain has the classic ITAM configuration YXXL (X_7) YXXL, while FcγRIIA has a unique configuration, YXXL (X_{12}) YXXL, and an additional tyrosine (also present in the human but not in the murine γ chain) seven amino acids upstream [42]. According to current models, the initial event following receptor crosslinking is most likely activation of a non-receptor protein tyrosine kinase (PTK) coincident with the action of a membrane-associated phosphatase(s). Phosphatase removes a phosphate from a regulatory tyrosine residue in the PTK and allows its activation. The initial non-receptor PTKs, most likely members of the Src family, phosphorylate the tyrosines of the ITAM. These tyrosine-phosphorylated residues then interact with SH2 domains of a number of proteins, including Syk kinase, which is required for phagocytosis [39]. Several lines of evidence, including, most convincingly, Syk knockouts produced by antisense technology [39], demonstrate that Syk itself most likely is not responsible for phosphorylation of the γ chain or FcγRIIA ITAM regions. The Syk family PTK is activated coincident with FcγRIIA or γ chain tyrosine phosphorylation. While some Syk may be receptor-associated in resting cells, Syk association and activation is increased by tyrosine phosphorylation of the ITAMs [52]. Although it had been believed that both members of the Syk family (Syk and ZAP-70) were similar functionally, we observed distinct differences between Syk and ZAP-70 with respect to phagocytic signalling [49]. The specific ITAM amino acid sequence (γ chain vs FcγRIIA), the pattern of tyrosine phosphorylation and the identity of the available Src PTKs most likely determine why Syk is more efficient than ZAP-70 in phagocytosis [49,53]. It is of note that Syk rather than ZAP-70 is expressed in most phagocytic cells.

Several approaches have allowed dissection of the molecular events involved in phagocyte activation by the Fcγ receptors. These include transfection of wild type, chimeric and mutant receptors into heterologous cells and also immunoprecipitation, tyrosine phosphorylation and tyrosine kinase assays applied to both endogenous cells and transfected heterologous cells. In the COS-1 cell model system, studies have clearly established that the tyrosine residues in the γ chain and FcγRIIA ITAM motifs are necessary for the phosphorylation of the cytoplasmic domain of the receptor, for engagement of Syk kinase, and for Fcγ receptor mediated phagocytosis [40–42,49,50,52,54,55]. We have observed differences between FcγRIIA and the γ chain in the initial signal transduction step. For FcγRI/γ and FcγRIIIA/γ, deletion of the γ chain ITAM or substitution of phenylalanine for tyrosine within the ITAM is associated with elimination of phagocytosis and receptor tyrosine phosphorylation [40,41,49,54–56]. In contrast, substitution of a single YXXL tyrosine in the FcγRIIA ITAM reduces but does not eliminate receptor tyrosine phosphorylation, FcγRIIA association with Syk kinase and phagocytosis mediated by FcγRIIA [40,42,52]. The third cytoplasmic tyrosine appears most important for FcγRIIA binding to Syk kinase and for FcγRIIA tyrosine phosphorylation [52]. In addition, for FcγRIIA but not for FcγRI/human γ or FcγRIIIA/human γ, the cytoplasmic tyrosine 7 amino acids upstream of the ITAM plays a role in receptor tyrosine phosphorylation, binding to Syk kinase and in phagocytic signalling [40,42,52].

Introduction of human Syk kinase into COS-1 cells significantly enhances the phagocytic signal mediated by FcγRIIA and produces an even greater increase in phagocytosis mediated by FcγRI/γ and FcγRIII/γ [55]. On the other hand, Syk antisense oligonucleotides introduced into monocytes abrogate Syk expression, leading to inhibition of the ingestion of IgG coated cells mediated by FcγRIIA, FcγRI/γ and FcγRIIIA/γ [39]. Several investigators have also used phosphopeptides to induce tyrosine phosphorylation of Syk kinase and to interfere with the association of Syk and the phosphorylated ITAMs [57,58]. Taken together, these studies have helped to establish the functional interaction between Syk kinase and the Fc receptor ITAM sequences.

A new issue in phagocyte Fcγ receptor mediated cellular activation is the possibility of inhibitory signals mediated by the FcγRIIB molecule. It has been clearly established in B-cells that FcγRIIB serves as an inhibitor of B-cell receptor (BCR) mediated cell activation. The cellular activation signals mediated by crosslinking of the BCR are abrogated by co-crosslinking the BCR and FcγRIIB [59–62]. Inhibition of B-cell receptor mediated signalling by FcγRIIB is dependent upon the presence of a motif containing a single tyrosine, the ITIM, in the cytoplasmic tail of FcγRIIB [63–65]. In the COS-1 cell system, co-crosslinking of FcγRIIB and FcγRIIA inhibits FcγRIIA mediated phagocytosis by decreasing FcγRIIA tyrosine phosphorylation [65]. Inhibition of phagocytosis is observed to a lesser extent when FcγRIIB is co-crosslinked with FcγRI/γ or FcγRIIIA/γ. Thus, the same ligand, IgG, in complexing with both FcγRIIA and FcγRIIB, may induce both a positive signal for phagocytosis and a negative signal to regulate phagocytosis. Therefore, the proportion of FcγRIIB present on the monocyte/macrophage or neutrophil lineages compared to FcγRIIA, is probably a factor in the regulation of FcγRIIA-mediated phagocytosis. Furthermore, for FcγRIIA and the BCR, the pathways for regulation of signalling by FcγRIIB appear to be different, since inhibition of BCR tyrosine phosphorylation is not observed following FcγRIIB crosslinking.

The platelet is a particularly interesting model cell for the study of FcγR signalling since FcγRIIA is the only FcγR expressed in platelets [13,47,66]. Platelet activation by crosslinking FcγRIIA has enabled dissection of the interactions of the ITAM cytoplasmic tail, Src family tyrosine kinases and Syk kinase [47,66–68]. Platelets/megakaryocytes have been employed to demonstrate properties applicable also to phagocytes. For example, FcγRIIA and the γ chain expressed in the same cells (platelets) can transduce signals by distinct functional pathways. In the platelet, the γ chain is associated with a collagen receptor but not with FcγRIIA [68]. Collagen receptor crosslinking results in γ chain tyrosine phosphorylation but not in FcγRIIA phosphorylation. Conversely, FcγRIIA crosslinking in platelets is not associated with γ chain tyrosine phosphorylation. These observations are in keeping with recent observations in a monocytic cell line in which tyrosine phosphorylation was localized to the receptor complex being stimulated [69].

SIGNALS DOWNSTREAM OF CROSSLINKED RECEPTORS AND NON-RECEPTOR PROTEIN TYROSINE KINASES

Much work is now being focused on defining the downstream molecular events following crosslinking of the Fcγ receptors. In addition to the receptor cytoplasmic domains and the Src and Syk PTKs, several other molecules are rapidly and transiently phosphorylated on tyrosine following Fcγ receptor crosslinking [43]. One set of molecules is likely to be involved in changes in intracellular calcium concentration. PLC-γ and PI-3 kinase are specific early targets for activation following Fcγ receptor crosslinking. Phosphoinositide turnover and the cleavage of PIP_3 results in both the activation of intracellular and plasma membrane calcium channels and in the activation of protein kinase C. In addition, we have observed that PI-3 kinase is required for mediation of phagocytosis by each of the phagocytic Fcγ receptors and that PI-3 kinase and Syk kinase are associated *in vivo* [56].

A second set of molecules is probably involved in the reorganization of the actin cytoskeleton and in events such as locomotion and the phagocytic engulfment of antibody coated cells and bacteria. In addition to the focal accumulation of filamentous actin accompanying FcγR-mediated phagocytosis in macrophages [70], there are a number of cellular targets involved with actin cytoskeleton reorganization. Phosphorylated targets following Fcγ receptor crosslinking include p58/paxillin and p125/focal adhesion kinase (Fak) [43]. The precise mechanisms by which phosphorylation of these actin/cytoskeleton associated molecules leads to the reorganization seen in cellular locomotion and the development of the phagosome is not well defined. A role for small GTPases in actin reorganization has also been identified. In murine macrophages and transfected COS-1 cells, inactivation of the small GTPase Rho results in the impairment of F-actin accumulation around phagocytic cups in concert with inhibition of phagocytosis [71].

A third set of molecules probably effects nuclear events such as *de novo* gene transcription and modification of gene transcription. Several investigators have noted relationships between phosphorylation of the ITAM subunits, activation of Syk kinase and association of adaptor protein molecules such as Grb-2, SOS and Shc [72]. These molecules participate in a cascade that can ultimately lead to Ras and Raf interaction and activation of GTP-associated molecules. Another target of tyrosine kinase activation is the p62/GAP-associated protein [17]. Ras-associated responses may ultimately lead to activation of the respiratory burst and secretion of granular contents. In addition, Ras activation through Raf may be associated with triggering of MAP kinase related pathways which modify transcription factor activity and translocation into the nucleus. Finally, in addition to the activation of the Ras/Raf pathway and the ultimate activation of MAP kinase with its effects on gene transcription, there is activation by tyrosine phosphorylation of the Vav protooncogene in phagocytic cells following Fcγ receptor-mediated crosslinking [17,43]. Vav is important for a number of intracellular functions and our data suggest that Syk and Vav are associated in macrophages following FcγR crosslinking [43].

Thus, there are a number of parallel pathways which trigger distinct cellular processes following initial Fcγ receptor crosslinking, ITAM phosphorylation and Syk activation. It is anticipated that current approaches utilizing phosphopeptides, pharmacologic inhibitors and selected chimeric and mutant molecules will elucidate more specifically the molecular interactions involved in downstream signalling events in phagocytic cells.

SUMMARY AND FUTURE PROSPECTS

The major Fcγ receptors of the neutrophil, monocyte and macrophage have been identified and the cellular programs which they mediate have begun to be dissected at the molecular level. Advances in the biochemisty of signal transduction pathways as well as genetic manipulation of the cellular programs *in vivo* and *in vitro* will likely elucidate the biology of the Fcγ receptors in phagocytic cells.

References

1. Qiu WQ, de Bruin D, Brownstein BH, Pearse R, Ravetch JV. Organization of the human and mouse low affinity FcγRII genes: duplication and recombination. Science. 1990;248:732–7.
2. Warmerdam PA, Nabben NM, van de Graef SA, van de Winkel JG, Capel PJ. The human low affinity Immunoglobulin G Fc receptor IIC gene is a result of an unequal crossover event. J Biol Chem. 1993;268:7346–9.
3. McKenzie SE, Keller MA, Cassel DL, Rappaport EF, Schwartz E, Surrey, S. Characterization of the 5′-flanking transcriptional regulatory region of the human Fcγ receptor gene, FcγRIIA. Mol Immunol. 1992;29:1165–74.
4. Gessner JE, Grussenmeyer T, Kolanus W, Schmidt RE. The human low affinity immunoglobulin G Fc receptor III-A and III-B genes. Molecular characterization of the promoter regions. J Biol Chem. 1995;270:1350–61.
5. Gessner JE, Grussenmeyer T, Schmidt RE. Differentially regulated expression of human IgG Fc receptor class III genes. Immunobiology. 1995;193:341–55.
6. Feinman R, Qiu WQ, Pearse RN et al. PU.1 and an HLH familiy member contribute to the myeloid-specific transcription of the FcγRIIIA promoter. EMBO J. 1994;13:3852–60.
7. Reilly AF, Surrey S, Rappaport EF, Schwartz E, McKenzie SE. Variation in human *FCGR2C* gene copy number. Immunogenetics. 1994;40:456.
8. de Haas M, Kleijer M, van Zwieten R, Roos D, von dem Borne, AE. Neutrophil FcγRIIIb deficiency, nature and clinical consequences: a study of 21 individuals from 14 families. Blood.1995:2403–13.
9. Norris CF, Pricop L, Millard SS et al. A naturally occurring mutation in FcγRIIA: A Q to K^{127} change confers unique IgG binding properties to the R^{131} allelic form of the receptor. Blood. 1998;91:656–62.
10. Heijnen IA, van Vugt MJ, Fanger NA et al. Antigen targeting to myeloid-specific human FcγRI/CD64 triggers enhanced antibody responses in transgenic mice. J Clin Invest. 1996; 97:331–8.
11. Wilson KC, Finbloom DS. Interferon-γ rapidly induces in human monocytes a DNA-binding factor that recognizes the gamma response region within the promoter of the gene for the high-affinity Fc gamma receptor. Proc Natl Acad Sci USA. 1992;89:11964–8.
12. Pearse RN, Feinman R, Shuai K, Darnell JE Jr, Ravetch JV. Interferon γ-induced transcription of the high-affinity Fc receptor for IgG requires assembly of a complex that includes the 91-kDa subunit of transcription factor ISGF3. Proc Natl Acad Sci USA. 1993; 90:4314–18.
13. Cassel DL, Keller MA, Surrey S et al. Differential expression of FcγRIIA, FcγRIIB and FcγRIIC in hematopoietic cells: analysis of transcripts. Mol Immunol. 1993;30:451–60.
14. Comber PG, Lentz V, Schreiber AD. Modulation of the transcriptional rate of Fcγ receptor mRNA in human mononuclear phagocytes. Cell Immunol. 1992;145:324–38.
15. Munn DH, Bree AG, Beall AC et al. Recombinant human M-CSF in nonhuman primates-selective expansion of a CD16(+) monocyte subset with phenotypic similarity to primate natural killer cells. Blood. 1996;88:1215–24.
16. Calzada-Wack JC, Frankenberger M, Ziegler-Heitbrock HW. Interleukin-10 drives human monocytes to CD16 positive macrophages. J Inflamm. 1996;46:78–85.
17. Darby C, Geahlen RL, Schreiber AD. Stimulation of macrophage Fc gamma RIIIA activates the receptor-associated protein tyrosine kinase Syk and induces phosphorylation of multiple

proteins including p95Vav and p62/GAP-associated protein. J Immunol. 1994;152:5429–5437.
18. Li M, Wirthmueller U, Ravetch JV. Reconstitution of human FcγRIII cell type specificity in transgenic mice. J Exp Med. 1996;183:1259–63.
19. Kuster H, Thompson H, Kinet JP. Characterization and expression of the gene for the human Fc receptor γ subunit. J Biol Chem. 1990;265:6448–52.
20. Indik Z, Kelly C, Chien P. Levinson AI, Schreiber AD. Human FcγRII, in the absence of other Fcγ receptors, mediates a phagocytic signal. J Clin Invest. 1991;88:1766–71.
21. Indik ZK, Hunter S, Huang MM et al. The high affinity Fc gamma receptor (CD64) induces phagocytosis in the absence of its cytoplasmic domain: the gamma subunit of Fc gamma RIIIA imparts phagocytic function to Fc gamma RI. Exp Hematol. 1994;22:599–606.
22. van Vugt MJ, Heijnen AF, Capel PJ et al. FcR gamma-chain is essential for both surface expression and function of human Fc gamma RI (CD64) in vivo. Blood. 1996;87:3593–9.
23. Salmon JE, Millard SS, Brogle NL, Kimberly RP. Fcγ receptor IIIb enhances Fcγ receptor IIa function in an oxidant-dependent and allele-sensitive manner. J Clin Invest 1995;95: 2877–85.
24. Edberg JC, Kimberly RP. Modulation of Fcγ and complement receptor function by the glycosyl-phosphatidyl inositol-anchored form of FcγRIII. J Immunol. 1994;152:5826–35.
25. Krauss JC, Poo H, Xue W, Mayo-Bond L, Todd III, RF, Petty HR. Reconstitution of antibody-dependent phagocytosis in fibroblasts expressing Fcγ receptor IIIB and the complement receptor type 3. J Immunol. 1994;153:1769–77.
26. Kindzelskii AL, Laska ZO, Todd RF, Petty HR. Urokinase-type plasminogen activator receptor reversibly dissociates from complement receptor type 3 ($\alpha M/\beta_2$; CD11b/CD18) during neutrophil polarization. J Immunol. 1996;156:297–309.
27. Green JM, Schreiber AD, Brown EJ. Role for a glycan phosphoinositol anchor in Fcγ receptor synergy. J Cell Biol. 1997;139:1209–17.
28. Vossebeld PH, Kessler J, von dem Borne AD, Roos D, Verhoeven AJ. Heterotypic FcγR clusters evoke a synergistic Ca^{2+} response in human neutrophils. J Biol Chem. 1995;270: 10671–9.
29. Parren PH, Warmerdam PA, Boeije LC et al. On the interaction of IgG subclasses with the low affinity FcγRIIA (CD32) on human monocytes, neutrophils and platelets. Analysis of a functional polymorphism to human IgG_2. J Clin Invest. 1992;90:1537–46.
30. Reilly AF, Norris CF, Surrey S et al. Genetic diversity human Fc receptor II for Immunoglobulin G: Fcγ receptor IIA ligand-binding polymorphism. Clin Diag Lab Immunol. 1994;1:640–4.
31. Hall MA, Hickman ME, Baker CJ, Edwards MS. Complement and antibody in neutrophil-mediated killing of type V group B streptococcus. J Infect Dis. 1994;170:88–93.
32. Noya FJ, Baker CJ, Edwards MS. Neutrophil Fc receptor participation in phagocytosis of type III group B streptococci. Infect Immun. 1993;61:1415–20.
33. Sanders LA, Feldman RG, Voorhorst-Ogink MM et al. Human immunoglobulin G (IgG) Fc receptor IIA (CD32) polymorphism and IgG_2-mediated bacterial phagocytosis by neutrophils. Infect Immun. 1995;63:73–81.
34. Schreiber AD, Parsons J, McDermott P, Cooper RA. Effect of corticosteroids on the human monocyte and complement receptors. J Clin Invest. 1975;56:1189–97.
35. Worth RC, Mayobond L, van de Winkel JGJ, Todd RF, Petty HR. CR3 ($\alpha M\beta_2$ CD11b/CD18) restores IgG-dependent phagocytosis in transfectants expressing a phagocytosis-defective FcγRIIA (CD32) tail-minus mutant. J Immunol. 1996;157:5660–5.
36. Takai T, Li M, Sylvestre D, Clynes R, Ravetch JV. FcR γ-chain deletion results in pleiotrophic effector cell defects. Cell. 1994;76:519–29.
37. Clynes R, Ravetch JV. Cytotoxic antibodies trigger inflammation through Fc receptors. Immunity. 1995;3:21–6.
38. Hazenbos WLW, Gessner JE, Hofhuis FMA et al. Impaired IgG-dependent anaphylaxis and Arthus reaction in FcγRIII (CD16) deficient mice. Immunity. 1996;5:181–8.
39. Matsuda M, Park JG, Wang DC, Hunter S, Chien P, Schreiber AD. Abrogation of the FcγRIIA-mediated phagocytic signal by stem-loop Syk antisense oligonucleotides. Mol Biol Cell. 1996;7:1095–106.
40. Cauley DC, Indik ZK, Schreiber AD. Cytoplasmic tyrosine requirements for phagocytic signalling differ between the human γ chain and FcγRIIA. Submitted.

41. Park JG, Murray RK, Chien P, Darby C, Schreiber AD. Conserved cytoplasmic tyrosine residues of the γ subunit are required for a phagocytic signal mediated by FcγRIIIA. J Clin Invest. 1993;92:2073–9.
42. Mitchell MA, Huang M-M, Indik ZK, Chien P, Pan XQ, Schreiber AD. Substitutions and deletions in the cytoplasmic domain of the phagocytic receptor FcγRIIA. Effect on receptor tyrosine phosphorylation and phagocytosis. Blood. 1994;84:1753–9.
43. Pan XQ, Indik ZK, Schreiber AD. Protein trosine phosphorylation following activation of monocyte/macrophge Fcγ receptors. J Immunol. 1994;152:3231.
44. Rappaport EF, Cassel DL, Walterhouse DO et al. A soluble form of the human Fcγ receptor, Fc_{γ}RIIA: cloning, transcript analysis and detection. Exp Hematol. 1993;21:689–96.
45. de Haas M, Koene HR, Kleijer M et al. A triallelic Fcγ receptor type IIIA polymorphism influences the binding of human IgG by NK cell FcγRIIIa. J Immunol. 1996;156:3948–55.
46. Miller KL, Duchemin AM, Anderson CL. A novel role for the Fc receptor gamma subunit: enhancement of FcγR ligand affinity. J Exp Med. 1996;183:2227–33.
47. Huang, MM, Indik Z, Brass LF, Hoxie JA, Schreiber AD, Brugge JS. Activation of FcγRII induces tyrosine phosphorylation of multiple proteins including FcγRII. J Biol Chem. 1992; 267:5467–73.
48. Schreiber AD. Fcγ receptor signal transduction. American Association of Immunologists (FASEB), April 5,1992, Anaheim, CA, USA.
49. Park JG, Schreiber AD. Determinants of the phagocytic signal mediated by the type IIIA Fc gamma receptor, Fc gamma RIIIA: sequence requirements and interaction with protein-tyrosine kinases. Proc Natl Acad Sci USA. 1995;92:7381–5.
50. Indik ZK, Pan X-Q, Huang, M-M, McKenzie SE, Levinson AI, Schreiber AD. Insertion of cytoplasmic tyrosine sequences into the nonphagocytic receptor FcγRIIB establishes phagocytic function. Blood. 1994;83:2072–80.
51. Tuijnman WB, Capel PJA, van de Winkel JGJ. Human Low affinity IgG receptor FcγRIIA (CD32) introduced into mouse fibroblasts mediates phagocytosis of sensitized erythrocytes. Blood. 1992;79:1651–6.
52. Pan X-Q, Indik ZK, Kim MK, Chien P, Schreiber AD. Cytoplasmic tyrosine requirements for FcγRIIA-Syk association and tyrosine phosphorylation. J Allergy Clin Immunol. 1997:S485.
53. Taylor N, Jahn T, Smith S et al. Differential activation of the tyrosine kinases ZAP-70 and Syk after FcγRI stimulation. Blood. 1997;89:388–96.
54. Park JG, Isaacs RE, Chien P, Schreiber AD. In the absence of other Fc receptors, FcγRIIIA transmits a phagocytic signal that requires the cytoplasmic domain of its γ subunit. J Clin Invest. 1993;92:11967–73.
55. Indik ZK, Park JG, Pan XQ, Schreiber AD. Induction of phagocytosis by a protein tyrosine kinase. Blood. 1996;85:1175–80.
56. Indik ZK, Park JG, Hunter S, Schreiber AD. The molecular dissection of Fcγ receptor mediated phagocytosis. Blood. 1995;86:4389–99.
57. Chacko GW, Brandt JT, Coggeshall KM, Anderson CL. Phosphoinositide 3-kinase and p72syk noncovalently associate with the low affinity Fc gamma receptor on human platelets through an immunoreceptor tyrosine-based activation motif. Reconstitution with synthetic phosphopeptides. J Biol Chem. 1996;271:10775–81.
58. Shiue L, Zoller MJ, Brugge JS. Syk is activated by phosphotyrosine-containing peptides representing the tyrosine-based activation motifs of the high affinity receptor for IgE. J Biol Chem. 1995;270:10498–502.
59. DeFranco AL, Law AL. Tyrosine Phosphatases and the antibody response. Science. 1995; 268:263–4.
60. D'Ambrosio D, Hippen KL, Minskoff SA et al. Recruitment and activation of PTP1C in negative regulation of antigen receptor signalling by FcγRIIB1. Science. 1995;268:263–4.
61. Daeron M, Latour S, Malbec O et al. The same tyrosine-based inhibition motif, in the intra-cytoplasmic domain of FcγRIIB, regulates negatively BCR-, TCR- and FcR-dependent cell activation. Immunity. 1995;3:635–46.
62. Ono M, Bolland S, Tempst P, Ravetch JV. Role of the inositol phosphatase SHIP in negative regulation of the immune system by the receptor Fc(γ)RIIB. Nature. 1996;383:263–6.
63. Muta T, Kurosaki T, Misulovin Z, Sanchez M, Nussenzweig MC, Ravetch JV. A 13-amino-acid motif in the cytoplasmic domain of FcγRIIB modulates B-cell receptor signalling. Nature. 1994;369:340.

64. Takai T, Ono M, Hikida M, Ohmori H, Ravetch JV. Augmented humoral and anaphylactic responses in FcγRII-deficient mice. Nature. 1996;379:346–9.
65. Hunter S, Indik ZK, Kim MK, Park JG, Schreiber AD. Regulation of Fcγ receptor mediated phagocytosis by a non-phagocytic Fcγ receptor. Blood. 1998;91:1762–8.
66. Chacko GW, Duchemin AM, Coggeshall KM, Osborne JM, Brandt JT, Anderson CL. Clustering of the platelet Fcγ receptor induces noncovalent association with the tyrosine kinase p72syk. J Biol Chem. 1994;269:32435–40.
67. Yanaga F, Asselin J, Schieven GL, Watson SP. Phenylarsine oxide inhibits tyrosine phosphorylation of phospholipase Cγ2 in human platelets and phospholipase Cγ1 in NIH-3T3 fibroblasts. FEBS Lett. 1995;368:377–80.
68. Yanaga F, Poole A, Asselin J et al. Syk interacts with tyrosine-phosphorylated proteins in human platelets activated by collagen and cross-linking of the FcγRIIA receptor. Biochem J. 1995;311:471–8.
69. Pfefferkorn LC, Swink SL. Intracluster restriction of Fc receptor γ chain tyrosine phosphorylation subverted by a protein tyrosine phosphatase inhibitor. J Biol Chem. 1996;271:11099–105.
70. Greenberg S, El Khoury J, Di Vigilio F, Kaplan EM, Silverstein, SC. Ca(2+)-independent F actin assembly and disassembly during Fc receptor-mediated phagocytosis in mouse macrophages. J Cell Biol. 1991;113:757–67.
71. Hackett DJ, Rotstein OD, Schreiber AD, Grunstein S. Rho is required for the initiation of calcium signalling and phagocytosis by Fcγ receptors in macrophages. J Exp Med. 1997;186: 955–66.
72. Park RK, Liu Y, Durden DL. A role for Shc, Grb2 and Raf-1 in FcγRI signal relay. J Biol Chem. 1996; 271:13342–8.

10
Human IgA Fc receptors (FcαR)

H. C. MORTON, M. van EGMOND and J. G. J. van de WINKEL

INTRODUCTON

Cellular receptors specific for the Fc region of immunoglobulins (FcR) are expressed by all cells of the immune system. Specific FcR enable immunoglobulins to trigger a diverse array of cellular effector functions such as phagocytosis, ADCC, and cytokine release [1]. Human B cells produce five distinct classes of immunoglobulin, IgA, IgD, IgE, IgG and IgM, with the vast majority ($\sim$80%) committed to IgA production. In fact, more IgA is produced per day (66 mg/kg/day) than all other isotypes combined [2]. IgA is found at substantial concentrations in serum and in mucosal secretions. In humans two sub-classes of IgA, IgA_1 and IgA_2, exist. Serum IgA is (mostly) monomeric and predominantly (90%) IgA_1, whereas secretions contain a greater proportion of IgA_2. In mucosal secretions, where it is the predominant isotype, IgA occurs almost exclusively as a dimeric complex containing two additional polypeptides, J chain and secretory component, which are involved in polymerization and the transport of polymeric IgA into mucosal fluids [3–5].

It is now well documented that IgA-opsonized particles provoke potent cellular effector functions through interaction with specific cell-bound IgA receptors (FcαR) resulting in their clearance and/or destruction. Only one human FcαR has been characterized in terms of its molecular and biochemical structure, and cell expression pattern [5]. This molecule, termed FcαRI or CD89, is expressed exclusively by cells of the myeloid lineage and will form the basis of this review. Although the existence of FcαR on human lymphoid cells has been reported, detailed experimental evidence is still lacking. Similarly, FcαR of other species have not yet been characterized. Several other receptors which bind IgA, namely, the polymeric immunoglobulin receptor and the asialoglycoprotein receptor have been extensively reviewed elsewhere and will not be discussed here [4–6].

J.G.J. van de Winkel and P.M. Hogarth (eds.), The Immunoglobulin Receptors and their Physiological and Pathological Roles in Immunity. 109–117.

FcαRI/CD89

Genetics

Uniquely amongst human FcR genes, the gene for FcαRI maps to chromosome 19, at position 19q13.4 [7]. Most human, mouse, and rat FcR genes mapped so far are located on chromosome 1 [1]. The FcαRI gene spans ~12 kb and is composed of 5 exons [8] (Figure 10.1). The first exon (S1) encodes the 5′ untranslated region (UTR), including an ATG initiation codon, and the majority of the leader peptide. Exon 2 (S2) is 36 bp long and includes the rest of the leader peptide including the predicted signal peptidase cleavage site. Such mini-exons are characteristic of FcR genes from all species so far examined. However, in all genes (except FcαRI) the S2 mini-exon is only 21 bp in size [1]. The third and fourth exons (EC1 and EC2) each code for a single extracellular Ig-like domain. The last exon (TM/C) encodes a short extracellular segment, the predicted transmembrane helix, and a 41 amino acid cytoplasmic tail. The longest isolated FcαRI cDNA is 1.6 kb in length and is composed of a 39 bp 5′-UTR, an 861 bp open reading frame, and a 711 bp 3′-UTR. The 3′-UTR is very AT-rich, includes an Alu-sequence, but lacks a polyadenylation signal [9]. Recently, several other FcαRI cDNAs corresponding to alternatively spliced products of the FcαRI gene have been detected by RT-PCR techniques [10–13] (Figure 10.1). It is still unclear, however, whether these transcripts give rise to functional proteins and (if so) what their biological relevance may be.

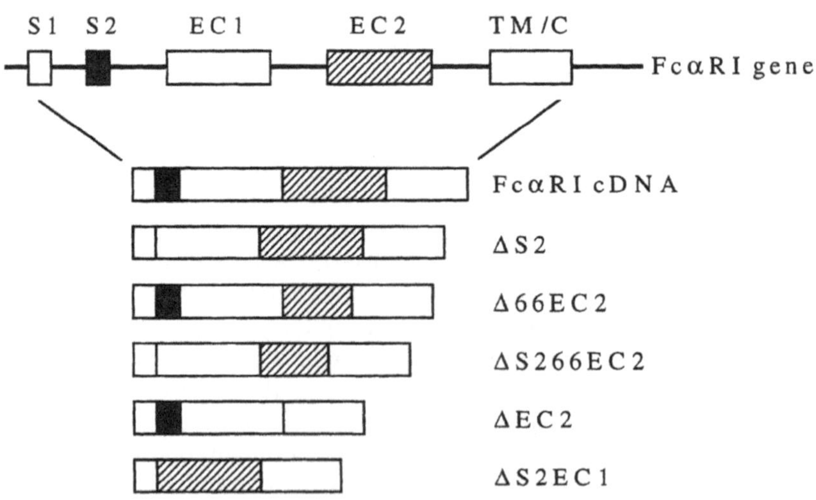

Figure 10.1 Diagram of the FcαRI gene, cDNA and alternatively spliced transcripts

Protein structure

The 1.6 kb FcαRI cDNA encodes a 287 amino acid protein [9]. The 21 amino acid hydrophobic leader sequence located at the 5′ end is removed during processing to form the mature 266 amino acid transmembrane protein. FcαRI consists of two Ig-like domains followed by a stretch of hydrophobic amino acids (which represent the predicted transmembrane domain) and a short cytoplasmic tail. The protein core of FcαRI has a predicted molecular mass of 30 kDa with differential glycosylation of the six potential N-linked sites, and the probability of additional *O*-glycosylation contributing to the observed size of the mature receptor (55–100 kDa). It has recently been shown, that FcαRI (like many other FcR) associates with the FcR γ chain homodimer, which is responsible for signal transduction via FcαRI [5,14–16]. Association between FcαRI and FcR γ chain is dependent upon charged residues located within their transmembrane domains. [15] (Figure 10.2).

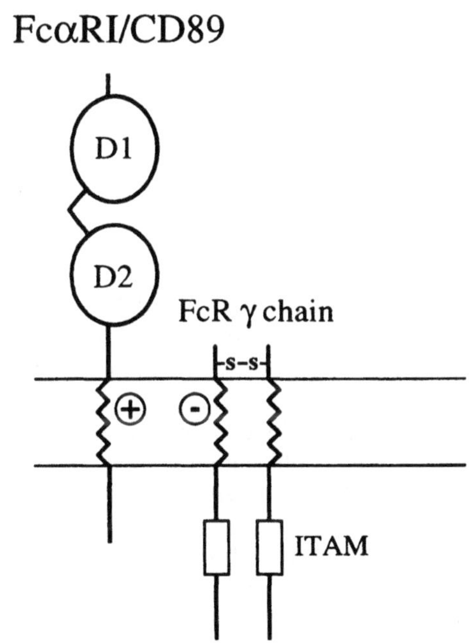

Figure 10.2 Structure of the FcαRI and FcRγ chain signalling complex

Cell distribution, modulation of expression and ligand binding

Expression of FcαRI is restricted to cells of the myeloid lineage, specifically neutrophils, monocytes, macrophages, eosinophils, and corresponding cell lines [9,17–20] (Figure 10.3). FcαRI purified from neutrophils, monocytes, and U937

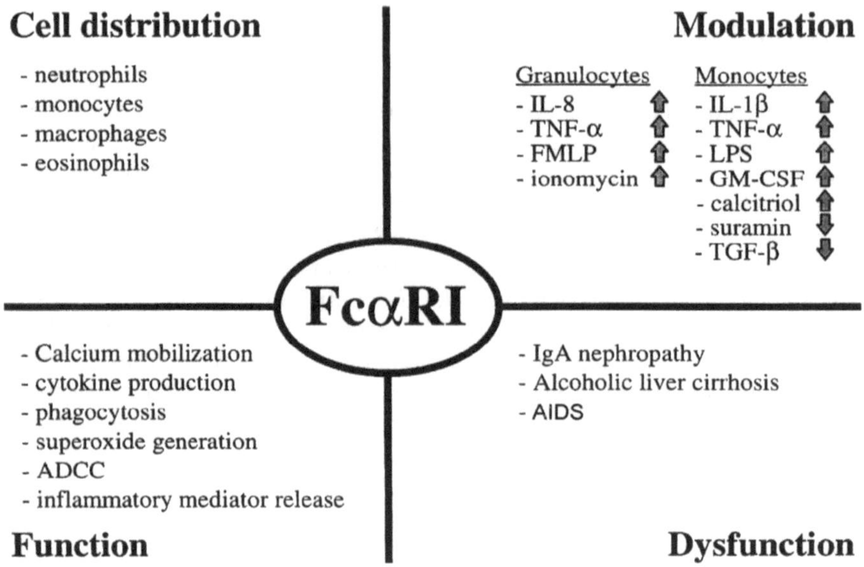

Figure 10.3 Summary of FcαRI cellular expression pattern, modulation, function and dysfunction

cells has a molecular mass of 55–75 kDa, whereas eosinophil FcαR is larger (70–100 kDa) due to more extensive glycosylaton [17,19,20]. Several FcαRI-specific monoclonal antibodies exist (A3, A59, A62, A77, and My43), only one of which (My43) inhibits binding of IgA. Lymphoid cells do not express FcαRI either at the protein or mRNA level [9,21,22].

FcαRI expression can be modulated by a number of cytokines and other factors [5] (Figure 10.3). Neutrophils upregulate FcαRI expresion in response to FMLP, IL-8, and TNF-α [23–25]. FcαRI upregulation by neutrophils remains unaffected by protein synthesis inhibitors, suggesting the existence of an intracellular pool of FcαRI [23]. In both neutrophils and eosinophils FcαRI levels appear to be modulated by a Ca^{2+}-dependent signalling process [19,23]. FcαRI on monocytes, and monocyte-like cell lines can be up-regulated by calcitriol, LPS, TNF-α, GM-CSF and IL-1β, and downregulated by suramin and TGF-β [26–29]. Current evidence also suggests the possibility that signalling pathways regulating FcαRI expression differ between granulocytes and monocytes [19].

The affinity of FcαRI for IgA is estimated to be $\sim 5 \times 10^{7}$ M^{-1}, which, considering the normal serum concentration of IgA, suggests that FcαRI is saturated *in vivo* [3,30]. FcαRI binds monomeric IgA_1 and IgA_2 and secretory IgA with similar affinity, suggesting binding not to be obstructed by secretory component or J-chain [3].

Function (and dysfunction)

Current evidence, based on transfectant model systems, suggests FcαRI-mediated signalling to depend upon association with the FcR γ chain. The FcR γ chain is a homodimeric molecule containing ITAM activatory motifs within its cytoplasmic domain. These tyrosine containing sequences are responsible for the interaction with intracellular kinases and, thus, initiation of the intracellular signalling cascade [5,31]. FcR γ chain has been shown critical for triggering FcαRI-mediated phosphorylation of cellular proteins, increases in intracellular free Ca^{2+}, cytokine production, phagocytosis, and antigen presentation [5] (Figure 10.3). Little is known of the intracellular events following FcαRI ligation, but the kinase $p72^{syk}$ proposedly plays a role [32]. Recently it has been suggested that (a proportion of) normal blood phagocytes express FcαRI in the absence of FcR γ chain [16]. Expression of FcαRI (ligand binding subunit) alone may explain why some investigators fail to detect efficient triggering by IgA [33–35].

Stimulation of neutrophils with either aggregated serum or secretory IgA triggers a respiratory burst that can be primed by FMLP, while monomeric and dimeric IgA, as well as mAb My43, can also induce a similar response if a cross-linking second antibody is used [22,30,36–39]. Despite the fact that monocytes express lower numbers of FcαRI relative to FcγR, stimulation of monocytes with serum or secretory IgA leads to the production of superoxide levels that are comparable to those generated via IgG [40]. Similarly, IgA_2-immune complexes trigger neutrophil activation more effectively than IgG-containing complexes [41]. In the case of eosinophils, IgA-complexes trigger eosinophil degranulation and release of eosinophil-derived neurotoxin [42]. The observation that the most potent degranulatory signal for eosinophils is provided by secretory IgA is complicated, however, by the observation that eosinophils possess a specific secretory component receptor in addition to FcαRI [5,43].

FcαR are efficient trigger molecules for ADCC in various *in vitro* systems. Phagocytes have been stimulated to kill target cells such as bacteria, *Schistosoma mansoni* schistosomula, and erythrocytes in an IgA-dependent manner [44–46]. Recently, the therapeutic potential of FcαRI has been demonstrated by the use of bispecific antibodies (BsAb) comprising anti-FcαRI mAb A77 F(ab′) X anti-tumour antigen mAb F(ab′) which were shown to trigger ADCC and phagocytosis of tumour cells by FcαRI-positive effector cells [5].

Neutrophils can easily phagocytose IgA-opsonized bacteria and yeast particles [36,47]. However, priming of neutrophils with either GM-CSF or IL-8 resulting in increased IgA binding was necessary for an efficient phagocytosis of IgA-coated latex beads [24,48,49]. Stimulation of monocytes with IL-1, TNF-α, GM-CSF, and LPS, enhances FcαRI expression and augments IgA-mediated phagocytosis [50]. Similarly, HL-60 cells differentiated using calcitriol, upregulate FcαRI expression and aquire phagocytic ability for IgA-coated erythrocytes [18]. In at least one experimental system, utilizing IgA (and IgG) opsonized influenza virus particles, however, the authors report specific antibody to actually inhibit binding/phagocytosis by blocking monocyte recognition of the viral haemagglutinin molecule [51].

IgA can also trigger the release of various inflammatory mediators and cytokines through interaction with FcαRI. Monocytes secrete IL-1β, TNF-α, IL-6, leukotrienes C_4, B_4 and prostaglandin E_2 following triggering of FcαRI using either IgA immune complexes, or CD89 mAb crosslinked with a second antibody [52–54]. IIA1.6 B cells transfected with FcαRI/FcR γ chain secrete IL-2 following stimulation with IgA or anti-FcαRI mAb [15].

Together with the asialoglycoprotein receptor, FcαRI has been proposed to play an important role in the removal of potentially harmful IgA-IC from the bloodstream (via endocytosis) while avoiding a full blown immune reaction [8,55,56]. Recently, defective FcαRI-mediated endocytosis has been proposed to contribute to the pathogenesis of diseases such as IgA nephropathy (IgAN), alcoholic liver cirrhosis (ALC), and AIDS. These are all characterized by high serum polymeric IgA concentrations, and increased levels of IgA-IC [55–57]. Defective clearance of IgA-IC leads to their deposition in the kidney where they may trigger chronic tissue damage [58].

CONCLUSIONS

The human FcαRI gene is clearly distinct from the other human FcR genes due to its chromosome location and the size of the S2 exon. However, in common with many FcR, FcαRI is able to associate with the FcR γ chain through a unique (for human FcR) charge-based mechanism. Functionally, FcαRI has been shown to represent a powerful trigger molecule expressed by myeloid effector cells, and as such may have therapeutic potential. A true understanding of the role of FcαRI in health and disease, however, awaits more extensive characterization by *in vitro* and *in vivo* model systems.

References

1. Ravetch JV, Kinet J-P. Fc receptors. Annu Rev Immunol. 1991;9:457–92.
2. Mestecky J, McGhee JR. Immunoglobulin A (IgA): Molecular and cellular interactions involved in IgA biosynthesis and immune response. Adv Immunol. 1987;40:153–245.
3. Kerr MA. The structure and function of human IgA. Biochem J. 1990;271:285–96.
4. Mostov KE. Transepithelial transport of immunoglobulins. Ann Rev Immunol. 1994;12:63–84.
5. Morton HC, van Egmond M, van de Winkel JGJ. Structure and function of human IgA Fc receptors. Crit Rev Immunol. 1996;16:423–40.
6. Stockert RJ. The Asialoglycoprotein receptor: relationships between structure, function, and expression. Physiol Rev. 1995;75:591–609.
7. Kremer EJ, Kalatzis V, Baker E, Callen DF, Sutherland GR, Maliszewski CR. The gene for the human IgA Fc receptor maps to 19q13.4. Hum Genet. 1992;89:107–8.
8. De Wit TPM, Morton HC, Capel PJA, van de Winkel JGJ. Structure of the gene for the human myeloid IgA Fc receptor (CD89). J Immunol. 1995;155:1203–9.
9. Maliszewski CR, March CJ, Schoenborn MA, Gimpel S, Shen L. Expression cloning of a human Fc receptor for IgA. J Exp Med. 1990;172:1665–72.
10. Morton HC, Schiel AE, Janssen SWJ, van de Winkel JGJ. Alternatively spliced forms of the human myeloid Fcα receptor (CD89) in neutrophils. Immunogenetics. 1996;43:246–7.

11. Patry C, Sibille Y, Lehuen A, Monteiro RC. Identification of Fcα receptor (CD89) isoforms generated by alternative splicing that are differentially expressed between blood monocytes and alveolar macrophages. J Immunol. 1996;156:4442–8.
12. Pleass RJ, Andrews PD, Kerr MA, Woof JM. Alternative splicing of the human IgA Fc receptor CD89 in neutrophils and eosinophils. Biochem J. 1996;318:771–7.
13. Reterink TJF, Verweij CL, van Es LA, Daha MR. Alternative splicing of IgA Fc receptor (CD89) transcripts. Gene. 1996;175:279–80.
14. Pfefferkorn LC, Yeaman GR. Association of IgA-Fc receptors (FcαR) with FcγRI 2 subunits in U937 cells. Aggregation induces the tyrosine phosphorylation of γ2. J Immunol. 1994;153: 3228–36.
15. Morton HC, van den Herik-Oudijk IE, Vossebeld P et al. Functional association between the human myeloid IgA Fc receptor (CD89) and FcR γ chain. Molecular basis for CD89/FcR γ chain association. J Biol Chem. 1995;270:29781–7.
16. Saito K, Suzuki K, Matsuda H, Okimura K, Ra C. Physical association of Fc receptor γ homodimer with IgA receptor. J Allergy Clin Immunol. 1995;96:1152–60.
17. Albrechtsen M, Yeaman GR, Kerr MA. Characterization of the IgA receptor from human polymorphonuclear leucocytes. Immunology. 1988;64:201–5.
18. Maliszewski CR, Shen L, Fanger MW. Expression of receptors for IgA on human monocytes and calcitriol-treated HL-60 cells. J Immunol. 135;3878–81.
19. Monteiro RC, Hostoffer RW, Cooper MD, Bonner JR, Gartland GL, Kubagawa H. Definition of immunoglobulin A receptors on eosinophils and their enhanced expression in allergic individuals. J Clin Invest. 1993;92:1681–5.
20. Monteiro RC, Kubagawa H, Cooper MD. Cellular distribution, regulation, and biochemical nature of an Fcα receptor in humans. J Exp Med. 1990;171:597–613.
21. Monteiro RC, Cooper MD, Kubagawa H. Molecular heterogeneity of Fcα receptors detected by receptor-specific monoclonal antibodies. J Immunol. 1992;148:1764–70.
22. Shen L, Lasser R, Fanger MW. My 43 a monoclonal antibody that reacts with human myeloid cells inhibits monocyte IgA binding and triggers function. J Immunol. 1989;143: 4117–22.
23. Hostoffer RW, Krukovets I, Berger M. Increased FcαR expression and IgA-mediated function on neutrophils induced by chemoattractants. J Immunol. 1993;150:4532–40.
24. Nikolova EB, Russell MW. Dual function of human IgA antibodies: inhibition of phagocytosis in circulating neutrophils and enhancement of responses in IL-8-stimulated cells. J Leukoc Biol. 1995;57:875–82.
25. Hostoffer RW, Krukovets I, Berger M. Enhancement by tumor necrosis factor-alpha of Fc alpha receptor expression and IgA-mediated superoxide generation and killing of *Pseudomonas aeruginosa* by polymorphonuclear leukocytes. J Infect Dis. 1994;170:82–7.
26. Shen L, Collins JE, Schoenborn MA, Maliszewski CR. Lipopolysaccharide and cytokine augmentation of human monocyte IgA receptor expression and function. J Immunol. 1994; 152:4080–6.
27. Gessl A, Willheim M, Spittler A, Agis H, Krugluger W, Boltz Nitulescu G. Influence of tumour necrosis factor-α on the expression of Fc IgG and IgA receptors, and other markers by cultured human blood monocytes and U937 cells. Scand J Immunol. 1994;39:151–6.
28. Boltz Nitulescu G, Willheim M, Spittler A, Leutmezer F, Tempfer C,Winkler S. Modulation of IgA, IgE, and IgG Fc receptor expression on human mononuclear phagocytes by 1,25-dihydroxyvitamin D_3 and cytokines. J Leukoc Biol. 1995;58:256–62.
29. Reterink TJF, Leverht EWN, Klar-Mohamad N, van Es LA, Daha MR. Transforming growth factor-beta (TGFβ) down-regulates IgA Fc-receptor (CD89) expression on human monocytes. Clin Exp Immunol. 1996;103:161–6.
30. Stewart WW, Mazengera RL, Shen L, Kerr MA. Unaggregated serum IgA binds to neutrophil FcαR at physiological concentrations and is endocytosed but cross-linking is necessary to elicit a respiratory burst. J Leukoc Biol. 1994;56:481–87.
31. Cambier JC. Antigen and Fc receptor signalling: The awesome power of the immunoreceptor tyrosine based activation motif (ITAM). J Immunol. 1995;155:3281–5.
32. Ueland JM, Shen L, Fanger M. The tyrosine kinase syk is coupled to FcαR. FASEB J. 1995;9: A804.
33. Kemp AS, Cripps AW, Brown S. Suppression of leukocyte chemokinesis and chemotaxis by human IgA. Clin Exp Immunol. 1980;40:388–95.

34. Van Epps DE, Brown SL. Inhibition of formylmethionyl-leucyl-phenylalanine-stimulated neutrophil chemiluminescence by human immunoglobulin A paraproteins. Infect Immun. 1981;34:864–70.
35. Saito K, Kato C, Katsuragi H, Komatsuzaki A. IgA-mediated inhibition of human leucocyte function by interference with FcR γ and C3b receptors. Immunology. 1991;74:99–106.
36. Gorter A, Hiemstra PS, Leijh PCJ et al. IgA- and secretory IgA-opsonized *S. aureus* induce a respiratory burst and phagocytosis by polymorphonuclear leucocytes. Immunology. 1987;61: 303–9.
37. Stewart WW, Kerr MA. The specificity of the human neutrophil IgA receptor (FcαR) determined by measurement of chemiluminescence induced by serum or secretory IgA_1 or IgA_2. Immunology. 1990;71:328–34.
38. Shen L. A monoclonal antibody specific for immunoglobulin A receptor triggers polymorphonuclear neutrophil superoxide release. J Leukoc Biol. 1992;51:373–8.
39. MacKenzie SJ, Kerr MA. IgM monoclonal antibodies recognizing FcαR but not FcγRIII trigger a respiratory burst in neutrophils although both trigger an increase in intracellular calcium levels and degranulation. Biochem J. 1995;306:519–23.
40. Shen L, Collins J. Monocyte superoxide secretion triggered by human IgA. Immunology. 1989;68:491–6.
41. Zhang W, Voice J, Lachmann PJ. A systematic study of neutrophil degranulation and respiratory burst *in vitro* by defined immune complexes. Clin Exp Immunol. 1995;101:507–14.
42. Abu-Ghazaleh RI, Fujisawa T, Mestecky J, Kyle RA, Gleich GM. IgA-induced eosinophil degranulation. J Immunol. 1989;142:2393–2400.
43. Lamjhioued B, Gounni AS, Gruart V, Pierce A, Capron A, Capron M. Human eosinophils express a receptor for secretory component. Role in secretory IgA-dependent activation. Eur J Immunol. 1995;25:117–25.
44. Lowell GH, Smith LF, Griffiss JM, Brandt BL. IgA-dependent, monocyte-mediated antibacterial activity. J Exp Med. 1980;152:452–7.
45. Dunne DW, Richardson BA, Jones FM, Clark M, Thorne KJI, Butterworth AE. The use of mouse human chimaeric antibodies to investigate the roles of different antibody isotypes, including IgA2, in the killing of *Schistosoma mansoni* schistosomula by eosinophils. Parasite Immunol. 1993;15:181–5.
46. Clark DA, Dessypris EN, Jenkins Jr DE, Krantz SB. Acquired immune hemolytic anemia associated with IgA erythrocyte coating: Investigtion of hemolytic mechanisms. Blood. 1984; 64:1000–5.
47. Yeaman GR, Kerr MA. Opsonization of yeast by human serum IgA anti-mannan antibodies and phagocytosis by human polymorphonuclear leukocyes. Clin Exp Immunol. 1987;68:200–8.
48. Weisbart RH, Kacena A, Schuh A, Golde DW. GM-CSF induces human neutrophil IgA-mediated phagocytosis by an IgA Fc receptor activation mechanism. Nature.1988;332:647–8.
49. Burnett D, Chamba A, Stockley RA, Murphy TF, Hill SL. Effect of recombinant GM-CSF and IgA opsonization on neutrophil phagocytosis of latex beads coated with P6 outer membrane protein from *Haemophilus influenzae*. Thorax. 1993;48:638–42.
50. Shen L, Collins JE, Schoenborn MA, Maliszewski CR. Lipopolysaccharide and cytokine augmentation of human monocyte IgA receptor expression and function. J Immunol. 1994; 152:4080–6.
51. Scott CB, Ratcliffe DR, Cramer EB. Human monocytes are unable to bind to or phagocytose IgA and IgG immune complexes formed with influenza virus *in vitro*. J Immunol. 1996;157: 351–9.
52. Levy Polat G, Laufer J, Fabian I, Passwell JH. Cross-linking of monocyte plasma membrane Fcα, Fcγ or mannose receptors induces TNF production. Immunology. 1993;80:287–92.
53. Ferreri NR, Howland WC, Spiegelberg HL. Release of leukotrienes C_4 and B_4 and prostaglandin E_2 from human monocytes stimulated with agregated IgG, IgA and IgE. J Immunol. 1986;136:4188–93.
54. Patry C, Herbelin A, Lehuen A, Bach JF, Monteiro RC. Fcα receptors mediate release of tumour necrosis factor-α and interleukin-6 by human monocytes following receptor aggregation. Immunology. 1995;86:1–5.
55. Silvain C, Patry C, Launay P, Lehuen A, Monteiro RC. Altered expression of monocyte IgA Fc receptors is associated with defective endocytosis in patients with alcoholic cirrhosis: potential role for IFN-γ. J Immunol. 1995;155:1606–18.

56. Monteiro RC, Grossetete B, Nguyen AT, Jungers P, Lehuen A. Dysfunctions of Fcα receptors by blood phagocytic cells in IgA nephropathy. Contrib Nephrol. 1995;111:116–22.
57. Grossetete B, Viard JP, Lehuen A, Bach JF, Monteiro RC. Impaired Fcα receptor expression is linked to increased immunoglobulin A levels and disease progression in HIV-1-infected patients. AIDS. 1995;9:229–34.
58. Galla JH. IgA nephropathy. Kidney Int. 1995;47:377–87.

11
Platelet Fc receptors for IgG and IgE*

A. GREINACHER

Platelets are a major source for human Fc receptors. Together with their precursors, the megakaryocytes, they express membrane bound Fc receptors and produce soluble Fc receptors. Receptors for IgG, the Fcγ receptor, and IgE, the Fcε receptor, are expressed at the platelet surface, respectively.

THE PLATELET Fcγ RECEPTOR

Of the three types of Fcγ receptors (FcγRI, CD64; FcγRII, CD 32; FcγRIII, CD 16) only the FcγRII has been identified on platelets, and the FcγRIIA gene seems to be predominantly transcribed in the megakaryocyte lineage [1]. Thus the platelet is a model to study the FcγRIIa receptor without interference of FcγRIIb, FcγRIIc or FcγRI and FcγRIII. As platelets do not contain a nucleus, in contrast to mononuclear cells, up-regulation of FcγRIIa transcription is excluded during experiments using platelets.

Structure of FcγRIIa on platelets

Until now, there is no evidence that the structure of FcγRIIa on platelets differs from the structure of FcγRIIa receptors on other cells [2]. By means of binding studies with the monoclonal antibody (mAb) IV.3, the mean number of FcγRIIa receptors on the platelets was estimated to be approximately 1200–1400 per platelet [3,4]. However, using F(ab) fragments of IV.3, approximately double

*Literature up to November 1996 is included.

J.G.J. van de Winkel and P.M. Hogarth (eds.), The Immunoglobulin Receptors and their Physiological and Pathological Roles in Immunity. 119–134.

this number of FcγRIIa per platelet was found [5]. If one concludes from this that a single mAb-IV.3–IgG can bind bivalently to two FcγRIIa molecules, then either FcγRIIa molecules are inserted very flexibly in the platelet membrane, or they are primarily organized into clusters.

There are large inter-donor variations in the number of FcγRIIa molecules per platelet. The numbers of receptors vary up to three-fold, even when measured with the same method [3,6–8]. Furthermore, the number of copies can increase transiently in acute disease states [8]. Kiss *et al.* [9] showed in a megakaryocytic cell line that c-kit ligand led to a significant and specific increase in FcγRIIa mRNA and enhanced FcγRIIa expression. As other cytokines such as IL3, IL6, IL11, and GM-CSF had only neglegible effects, c-kit ligand seems to enhance FcγRIIa transcription specifically.

As in other cell types, the platelet FcγRIIa has two polymorphic sites, one at position 27 and the other at position 131. From these only the $Arg^{131}His$ polymorphism has functional relevance [10,11]. This is discussed in detail in chapter 23 of this book. The $Arg^{131}His$ polymorphism is not correlated with the observed inter-donor variations of platelet FcγRIIa receptor expression [12].

There is indirect evidence that function and probably also expression of platelet FcγRIIa depend on physical properties and platelet activation as binding of immune complexes is increased at low ionic strength [5] and FcγRIIa expression is enhanced following stimulation of platelets with thrombin or ADP [13].

Soluble FcγRIIa

Rappaport *et al.* [14] cloned a platelet FcγRIIa mRNA lacking the transmembrane exon of FcγRIIa, produced by alternative splicing of the FcγRIIa transcript. In contrast to neurophils and monocytes, in which the transmembrane $Fc\gamma RIIa_1$ mRNA predominates, $Fc\gamma RIIa_1$ mRNA and the $Fc\gamma RIIa_2$ mRNA transcripts which lack the transmembrane exon are present in platelets and megakaryocytes in nearly equal amounts [1,15]. The mAb IV.3, directed against the extracellular region of FcγRIIa precipitates a 35 kDa protein, which was suggested to be the product of the truncated $Fc\gamma RIIa_2$ mRNA [14]. Gachet *et al.* [16] demonstrated that $Fc\gamma RIIa_2$ is rapidly released by thrombin-activated platelets, indicating that $Fc\gamma RIIa_2$ is stored in the platelet granules. Whether soluble platelet $Fc\gamma RIIa_2$ is an endogenous platelet protein (which is likely) or whether it is absorbed from the plasma, as fibrinogen, is unresolved. There is very little information on the biological function of soluble platelet $Fc\gamma RIIa_2$. *In vitro* $Fc\gamma RIIa_2$ can inhibit FcγRIIa-dependent platelet activation by monoclonal antibodies (mAbs) at a concentration of 12 μg/ml. As 10^9 platelets/ml can produce 1 to 2 ng/ml $Fc\gamma RIIa_2$, a 5000-fold increase of $Fc\gamma RIIa_2$ local concentration would be required to inhibit FcγRIIa dependent platelet activation in vivo [16], if one were to extrapolate these results to the *in vivo* situation.

Since the number of platelets is so great the membrane-bound platelet FcγRIIa represents approximately 60% of total cell membrane bound circulating FcγRIIa [17]. Considering that mRNA transcripts for $Fc\gamma RIIa_1$ and $Fc\gamma RIIa_2$ are found in comparable numbers in megakaryocytes and platelets, soluble and cell bound platelet FcγRIIa might be the major Fc-receptor pool in the circulation.

Anti-platelet-glycoprotein monoclonal antibodies as tools for FcγRIIa investigation

MAbs have been the major tool for characterization of blood cell glycoproteins (GP) and some mAbs have platelet activating capacity [18,19]. At the end of the 1980s it became obvious that several mechanisms contribute to the platelet activating/aggregating effect:

(1) Platelet activation by clustering of the antigens by the mAb [20].

(2) Few mAbs of the IgG class and many mAbs of the IgM class cause complement dependent platelet lysis. Examples are anti-CD9 mAbs (MM257, BA-2), and anti-FcγRII mAbs (NNKY 3-2, NNKY 4-7) [12, 21–23].

(3) Some anti-GPIIb/IIIa mAbs induce a conformational change of the fibrinogen receptor enabling the receptor to bind fibrinogen, which results in platelet aggregation without major platelet activation. (D3GP3 and D336 [24–26]; LIBS-3, [27]).

(4) Activation of platelets via the platelet FcγRIIa [28]. This is usually concluded from the following experimental evidence:

- The intact antibodies activate platelets as measured by the release of granule contents (demonstration of mAb-induced platelet aggregation or agglutination is insufficient).

- $F(ab')_2$ or Fab-fragments of the antibody do not cause platelet activation but inhibit the activating effect of the intact antibody.

- Fc-fragments of human- or mouse-IgG or the mAb IV.3 inhibit the activating effect of the intact antibody.

In a recent review Rubinstein *et al.* [29] summarized the platelet responses to different monoclonal antibodies (Table 11.1). Although Table 11.1 does not give a complete description of all platelet activating mAbs, it is obvious that the platelet activating capacity can often be inhibited by the mAb IV.3. However, inhibition of platelet activation by mAb IV.3 does not prove that signal transduction is mediated via the FcγRIIa. The interaction of the $F(ab')_2$ binding part of the antibody with its antigen might also contribute to the observed

Table 11.1 FcγRIIa-dependent platelet responses to monoclonal antibodies [modified from Reference 29]. The numbers of mAbs and their characteristics are given in the table

Antigen	*Copies per platelet*	IgG_1	IgG_{2a}	IgG_{2b}	IgG_3	*Inhibit activation*	*Not tested*	*Does not inhibit/ inconclusive data*
CD 9	40–60 000	7	2	1	1	6	2	2
GPIIb/IIIa-complex	30 000–70 000	12		1		10	3	
CD 36	12 000	3				3		
Others: β_2m, FcγRIIa, CD 69, unknown	600–15 000	9	2			7	2	2
	600–70 000	31	4	2	1	26	7	4

effects. For example, it has been shown for the intact anti CD9 mAb 50H.19 that its platelet activating effect could be inhibited by the mAb VI.3. However, clustering of the CD9 antigen by $F(ab')_2$ fragments immobilized to a polystyrene surface also cause platelet activation, indicating that FcγRIIa in the first experimental setting acts only as a passive, immobilizing surface [20,30] for the mAb but not as the signal-transducing receptor.

Most of the mAbs causing FcγRIIa dependent platelet activation belong to the IgG_1 class, this may reflect that FcγRIIa has the highest affinity for murine IgG_1. Many FcγRIIa activating mAbs have anti-CD9 or anti-GPIIb/IIIa specificity. The reason for this is currently unclear. The corresponding antigens are present in a high copy number at the platelet surface. Although this might enable complex formation of the mAb with the FcγRIIa it cannot be the major reason. MAbs directed towards other high-copy platelet antigens have not been reported to cause FcγRIIa dependent platelet activation i.e. mAbs with GPIb/IX complex or PECAM specificity. The corresponding antigens are found at the platelet surface with ~30 000 and ~40 000 copies, respectively. On the other hand, some mAbs, such as Jun-1 [31], directed toward an unknown antigen with less than 1000 copies per platelet also show typical FcγRIIa-dependent platelet activation.

Interplatelet and intraplatelet activation of FcγRIIa by mAbs

The question of how FcγRIIa-dependent platelet activation occurs is still controversial. One hypothesis is that clustering of FcγRIIa with other platelet receptors enhances platelet activation, as has been suggested for HLA class I molecules [32]. An experimentally supported model is FcγRIIa oligomerization as the primary activation trigger. Fab fragments of an anti-FcγRIIa mAb, crosslinked by goat anti-mouse $F(ab')_2$-fragments cause FcγRIIa dependent platelet activation [28,33]. This model is probably adequate for mAbs directed toward antigens which are either clustered at the platelet surface, or can cluster in response to mAb binding [2]. As the number of copies of the major involved GPs is much higher than the number of FcγRIIa copies per platelet, this model of oligomerization also favours that the platelet FcγRIIa is mobile at the platelet surface enabling clustering of the receptor. However, this model does not differentiate between interplatelet and intraplatelet interaction of mAbs and FcγRIIa.

To distinguish between these mechanisms similar experimental approaches have been used by different groups [31,34,35]. Two samples of platelets were used. In the first sample, platelets were pre-treated with the mAb IV.3 (or its Fab fragments) to block FcγRIIa and incubated with an intact FcγRIIa-activating mAb directed against a platelet surface antigen (CD9, GPIIb/IIIa). These platelets were not activated by this treatment. The second sample of platelets was obtained either from patients with a hereditary deficiency for the corresponding antigen, or the corresponding antigen was blocked by $F(ab')_2$ fragments of the mAb, providing inhibition of binding of the platelet-activating intact mAb. Mixing of both platelet samples resulted in platelet activation, clearly demonstrating interplatelet activation by the mAbs.

More complicated and not satisfactorily resolved is the experimental approach to distinguish inter- from intra-platelet activation. To inhibit interplatelet contact, platelets were either diluted in buffer or in dextran, or embedded in agarose [31]. The mAbs were then added to the solution, or exposed to the platelets by diffusion through the agar. The latter experimental setting most convincingly inhibited interplatelet interaction and the authors concluded that the mAbs JS-1 (anti-multimerin) and Jun-1 (directed towards an unknown platelet antigen), and the anti-GPIIb/IIIa mAb RAJ-1 are able to induce intraplatelet activation via the FcγRIIa. This was also postulated by de Reys *et al.* [22] using a single platelet counting technique. However, other mAbs with GPIIb/IIIa-specificity (UR-1, 6C9) [34] or CD9-specificity (AG1, MAb-7) [34,35] showed exclusively interplatelet activation. Thus both inter- and intraplatelet FcγRIIa-dependent activation by mAbs seems to be possible and the comparison of platelet responses to different mAbs is a tool to gain more informations on the respective roles of the antigen and the FcγRIIa, as suggested by Slupsky *et al.* [36].

However, not all mAbs cause platelet activation after binding to FcγRIIa. Some antibodies are able to interact with (bind to?) the FcγRIIa without activating the platelet but inhibiting the activating effect of another mAb, indicating that probably also the sterical organization of the antigen–antibody complexes is important for FcγRIIa-activation by mAbs [37]. As a consequence one should interpret results of experiments using more than one mAb with caution in regard to the FcγRIIa function.

MAbs and FcγRIIa His^{131}Arg polymorphism

As in other cells, the affinity of the FcγRIIa Arg^{131} variant to murine IgG_1 mAbs is higher than that of the FcγRIIa His^{131} variant (Arg^{131}/Arg^{131} > Arg^{131}/His^{131} > His^{131}/His^{131}) [2,12,37,38]. These differences in affinity between the two variants are not dependent on the specificity of the mAbs [12,39]. Currently this phenomenon is used for selecting 'sensitive' FcγRIIa-Arg^{131}–platelets for laboratory testing of FcγRIIa-activating human antibodies [40,41].

FcγRIIa and other platelet glycoproteins

Some mAbs directed to GPIIb/IIIa (i.e. D3GP3, OE5, AP-3) inhibit binding of mAb IV.3 to platelets [35,37,42] as well as platelet aggregation by heat-aggregated IgG. MAbs with anti-GPIb/IX specificity (AP1, 6 D1), which have been reported to interact with FcγRIIa, however, only inhibit platelet aggregation by heat aggregated IgG but do not block IV.3 binding [42]. In addition, anti-GPIb/IX mAbs inhibit platelet FcγRIIa activation by antibodies from patients with heparin-induced thrombocytopenia [43].

This indicates a topographical association of GPIIb/IIIa and FcγRIIa [37], and a functional association of GPIb/IX and FcγRIIa which is probably linked via the platelet cytoskeleton [42]. These associations with GPIIb/IIIa or GPIb/IX are not crucial for platelet activation via the FcγRIIa, as patients with a hereditary deficiency of either GPIIb/IIIa or GPIb/IX can be activated via the FcγRIIa-like normal platelets [44–46].

Platelet activation by collagen induces tyrosine phosphorylation of the Fc receptor γ-chain, indicating that collagen induced signal transduction involves a signalling pathway closely related to signalling by immune receptors. As FcγRIIa activation does not involve phosphorylation of the Fc receptor γ-chain it seems not to be directly associated with the still unidentified collagen receptor involved [47].

Platelet FcγRIIa signal transduction

Incubation of platelets with aggregated IgG in a plasma free medium results in a strong platelet activation response, which cannot be inhibited by acetylsalicylic acid and thrombin antagonists [49]. This response is enhanced in a low ionic strength medium [41]. The net charge of IgG immune complexes constitutes a critical property that conditions the ability of immune complexes to trigger platelet activation [50]. Platelets are not activated by single IgG molecules – indeed the high concentrations of monomeric IgG in plasma seem to inhibit immune complex-mediated platelet activation [49]. It is currently thought that the generation of a FcγRIIa activation signal occurs by clustering of the receptor. However, it is not clear whether this means aggregation, stabilisation, immobilization of FcγRIIa, or stabile or transient interaction with other receptors.

Crosslinking of FcγRIIa either by immune complexes or by mAbs, leads to calcium increase [51] and phosphatidylinositide metabolism. This can only partially be inhibited by the cyclooxygenase inhibitors acetylsalicylic acid, indomethacin or the phospholipase inhibitor dibromoacetophenone [33]). Tyrosine phosphorylation of the integral immune receptor tyrosine-based activation motif (ITAM) of the receptor [28,52,53], and of phospholipase Cγ2 [54] are important in FcγRIIa activation. During activation, the cytoplasmic tyrosine kinase $p72^{syk}$ [55] and the phosphoinositide 3-kinase (a serine, threonine and lipid kinase) bind to FcγRIIa following tyrosine phosphorylation [56]. In addition, $pp125^{FAK}$ is tyrosine-phosphorylated. While $pp125^{FAK}$phosphorylation appears to be dependent on prior protein kinase C activation [46], activation of $p72^{syk}$ does not depend on protein kinase C [46] or phospholipase C but seems to be mediated by its src-homology (SH)2 domains [57]. Following platelet FcγRIIa stimulation at least two further tyrosine phoshorylated proteins of 38 kDa and 63 kDa associate with the receptor [58]. In this way FcγRIIa may be linked to two different signal transduction pathways. However, the functional significance of this recent observation is unknown.

Although still rather speculative one may draw a model of platelet FcγRIIa signal transduction: clustering of FcγRIIa leads to tyrosine phosphorylation of ITAM, also called antigen recognition activation motif (ARAM). The phosphorylated region binds proteins containing SH_2 and SH_3 domains as $p72^{Syk}$ and $pp125^{FAK}$. $p72^{Syk}$ itself might then act as an adapter to recruit phosphoinositol 3 kinase to activated FcγRIIa.

Interaction of platelet FcγRIIa with human antibodies

Most of the recent data relating to the platelet FcγRIIa were deduced from experiments using murine mAbs. To understand the biological relevance of the receptor, experiments with human antibodies are more important, although little data currently exists. In a recent review, Anderson *et al.* [17] speculated on the possible pathophysiological relevance of the platelet FcγRIIa:

(1) homing functions which allow platelets to be attracted to inflammatory sites where primary haemostatic mechanisms are needed;

(2) serves as an auxiliary platelet activating device to augment platelet activation at inflammatory sites;

(3) serves to clear immune-complexes from the circulation [48].

In contrast to these biological functions, interaction of platelet FcγRIIa with platelet allo- or autoantibodies seems to be only of minor clinical relevance.

HLA-antibodies and platelet FcγRIIa

One of the most common clinical problems in polytransfused patients are HLA antibodies causing platelet transfusion refractoriness. Heinrich *et al.* [59] demonstrated platelet activation by HLA antibodies. Crosslinking of HLA class I molecules by the mAb w6.32 induces FcγRIIa-dependent platelet activation [60] and human anti-HLA class I antibodies cause platelet activation nearly as potently as does thrombin [61] if complete antibodies are used. However, this activating effect was dependent on an additional plasma factor, shown by Brandt *et al.* to be complement [62]. The platelet-activating effect of the HLA-antibodies in their experimental setting could not be inhibited by the mAb IV.3. This suggests that human anti-HLA class I antibodies induced platelet activation *in vitro* does not depend on the FcγRIIa but is complement mediated.

Platelet allo- and autoantibodies and FcγRIIa

The majority of human platelet autoantibodies have anti-GPIb/IX or anti-GPIIb/IIIa specificity. Antibodies against the latter antigen might, in analogy to mAbs, be able to induce FcγRIIa-dependent platelet activation. However, we have so far failed to demonstrate such an effect (unpublished observation). There is a single report of a patient who developed first an autoantibody with anti-CD9 specificity, inducing aggregation of normal platelets. This patient later developed a second autoantibody, which inhibited FcγRIIa-dependent platelet aggregation induced by an anti-CD9 mAb and also inhibited IV.3 binding [63]. Whether the second antibody had FcγRIIa specificity, as suggested by the authors, should be reconsidered taking into account that autoantibodies with anti-GPIIb/IIIa specificity are frequent, and anti-GPIIb/IIIa mAbs are able to interfere with the FcγRIIa by inhibiting binding of other mAbs without FcγRIIa activation [37].

By far the clinically most important pathophysiological condition in which platelet FcγRIIa has been demonstrated to play a substantial role is heparin-induced thrombocytopenia (HIT).

Heparin-induced thrombocytopenia and FcγRIIa

HIT is caused by immunoglobulins, usually IgG, which activate platelets in the presence of pharmacological concentrations of heparin [64–68]. The major target antigen is a macromolecular complex of heparin (or other polysulphated polysaccharides) and platelet factor 4 (PF4) that binds to the platelet surface [68–74]. Platelet activation by HIT-antibodies is mediated via the platelet FcγRIIa in *in vitro* experiments [44,76]. It is dependent on intact IgG and can be inhibited by the mAb IV.3 and by Fc-fragments of human IgG. *In vivo*, patients with acute HIT, present with platelet activation [77], develop thrombocytopenia and paradoxical thrombotic vessel occlusions. The current concept of pathogenesis in HIT is that the clinical complications are related to multiple events [69,78]:

(1) Multimolecular complexes of PF4 and heparin are formed [79].

(2) Some patients develop antibodies against these complexes.

(3) The antibodies activate platelets via the FcγRIIa, resulting in PF4 release and enhanced antigen formation [44,76].

(4) Platelet activation causes generation of platelet derived microparticles and thrombin [80], both of which have procoagulant activity.

(5) HIT-antibodies also bind to PF4–heparan, sulphate complexes at the endothelial cell surface, resulting in endothelial cell activation [73,74,81].

In HIT patients FcγRIIa density of the platelets is enhanced [8]; consistent with this, platelets from HIT patients showed increased aggregation reactivity to aggregated IgG or HIT antibodies [8]. There might be a functional difference between acutely up-regulated platelet FcγRIIa expression and a constitutively higher number of platelet FcγRIIa, as FcγRIIa density on platelets of normal donors does not correlate with their reactivity to HIT antibodies [67]. In one study HIT antibodies primarily belonged to the IgG_2 subclass [82]; however, we were unable to confirm this observation. Several groups found an association of FcγRIIa $Arg^{131}His$ allotype and susceptibility of platelets toward HIT antibody activation and clinical complications. The His/His FcγRIIa genotype, with high affinity to IgG_2, was found to be predominant in HIT patients by three groups [40,83,84]. However, this was not confirmed by others [85,86]. Thus, although the involvement of FcγRIIa in HIT is widely accepted, the impact of the $His^{131}Arg$ FcγRIIa allotype is still a matter of debate.

Antistreptokinase antibodies and FcγRIIa

Beside HIT, streptokinase (SK)-induced antibodies are also able to activate platelets via the FcγRIIa *in vitro* [87]. SK is produced by β-haemolytic streptococci and anti-SK-antibodies may appear following bacterial infection or therapeutic administration of SK during thrombolysis. It has been speculated that these antibodies might have deleterious effects in patients undergoing lysis for acute myocardial infarct, if platelets activated by SK–anti-SK–antibody complexes cause further vessel occlusions [41,87].

Antiphospholipid syndrome and FcγRIIa

A pathological condition with striking similarities to HIT is the antiphospholipid syndrome (APS). The APS is clinically associated with thrombosis, recurrent fetal loss and thrombocytopenia [88–90]. Most antiphospholipid antibodies are directed towards a complex of insolubilized cardiolipin with β_2-glycoprotein I [91]. Despite the similarities of APS and HIT antigens, the role of

FcγRIIa in APS is unclear. Platelet activation by subthreshold concentrations of platelet agonists can be enhanced in an FcγRIIa dependent fashion by mAbs directed to β_2-glycoprotein I [92]. However, human antiphospholipid-antibodies showed no effect on platelet aggregation and secretion in another study [93]. An interesting hypothesis of Arnout [90] is that in a 'double hit' injury antiphospholipid antibodies cause clinical problems by enhancing another very weak platelet activating stimuli.

Platelet FcγRIIa as a therapeutic target

To inhibit platelet FcγRIIa in patients currently only high dose i.v. IgG can be used *in vivo.* In HIT patients high dose i.v. IgG augments a rapid increase in platelet counts [94,95]. We have shown that inhibition of FcγRIIa either by the mAb IV.3 or, by Fc-fragments of human IgG, or by high concentrations of 7s IgG preparations can suppress platelet activation by HIT antibodies. An interesting finding of the study was the large differences of the capability of various i.v. IgG preparations to inhibit FcγRIIa dependent platelet activation [70]. These differences should be taken into account if one analyses clinical studies with high dose i.v. 7s IgG.

THE PLATELET IgE RECEPTOR

Platelets express the low affinity receptor for IgE FcεRII (CD23) [96] with approximately 600 binding sites per platelet [97,98]. However, FcεRII expression is usually restricted to 10–20 % of normal human platelets but increases to up to 50% of platelets in allergic patients [98]. This increase in FcεRII expression is caused by reduced receptor degradation following IgE binding and increased expression [99]. The structure of the FcεRII is reviewed in another chapter of this book. In contrast to the FcγRIIa, FcεRII expression seems to be linked to the expression of GPIIb/IIIa, as patients with a hereditary defect of this receptor do not express FcεRII [100]. In IgE antibody-dependent cytotoxicity reactions, platelets are able to damage parasite targets such as *Schistosoma mansoni* larvae and filaria [101,102]. FcεRII activation following IgE crosslinking leads to release of cytotoxic agents [103]. IgE-stimulated platelets produce oxygen radicals, which seems to be restricted to FcεRII stimulation by IgE [105,107]. The monoclonal antibody, BB10, which specifically inhibits FcεRII-dependent platelet activation, allowed further investigation of FcεRII function. Incubation of platelets from allergic or atopic donors with anti-IgE leads to platelet activation, most likely by crosslinking of the FcεRII [106,107]. Platelets from patients with house mite allergy are activated via the FcεRII if coincubated *in vitro* with the antigen or an antigen mimicking protein [108].

This response can be specifically diminished by treatment of the patient e.g. with disodium cromoglycate [109,110]. IgE-mediated platelet activation results in serotonin release [111], which in animal experiments seems to contribute to contact sensitivity responses [112]. These observations may be of relevance for treatment of allergic or atopic patients. *In vivo* it has been shown that following treatment with disodium cromoglycate platelet survival time increases in patients with stable asthma [113].

IgE–FcεRII interactions not only enhance platelet activation, high concentrations of IgE seem to inhibit platelet activation by arachidonic acid [114]. This might be related to the postulated link between GPIIb/IIIa and FcεRII. Of relevance for the clinical immuno-haematologist might be the observation of Wilhelm *et al.* [115] that specific IgE antibody levels are enhanced in patients developing febrile non-haemolytic transfusion reactions following platelet transfusions. Although there is good evidence for IgE-platelet–FcεRII interaction with biological relevance in parasite defence, data on the relevance of platelet FcεRII in other clinical situations are still too sparse to allow further conclusions.

Acknowledgements

This review is dedicated to Professor Dr Christian Mueller-Eckhardt, who started the work on platelet Fc receptors 30 years ago.

This work has been supported by the Deutsche Forschungsgemeinschaft GR 1096/2-2.

References

1. Cassel DL, Keller MA, Surrey S et al. Differential expression of Fc gamma RIIA, Fc gamma RIIB and Fc gamma RIIC in hematopoietic cells:analysis of transcripts. Mol Immunol. 1993;30:451 60.
2. Gosselin EJ, Brown MF, Anderson CL, Zipf TF, Guyre PM. The monoclonal antibody 41H16 detects the Leu 4 responder form of human Fc gamma RII. J Immunol. 1990;144 (5):1817-22.
3. Karas SP, Rosse WF, Kurlander RJ. Characterization of the IgG-Fc receptor on human platelets. Blood. 1982;60:1277 82.
4. Rosenfeld SI, Anderson CL. Fc receptor of human platelets. In: Kunicki TJ, Georg JN, eds. Platelet Immunobiology, Molecular and Clinical Aspects. Lippincott Company: Philadelphia; 1989:337 53.
5. King M, McDermott P, Schreiber AD. Characterization of the Fc gamma receptor on human platelets. Cell Immunol. 1990;128(2):462-79.
6. Looney RJ, Anderson CL, Ryan DH, Rosenfeld SI. Structural polymorphism of the human platelet Fcγ receptor. J Immunol. 1988;141:2680–3.
7. Rosenfeld SI, Ryan DH, Looney RI, Anderson CL, Abraham GN, Leddy JP. Human Fcγ receptor:stable inter-donor variation in quantitative expression on platelets correlates with functional responses. J Immunol. 1987;138:2869–73.
8. Chong BH, Pilgrim RL, Cooley MA, Chesterman CN. Increased expression of platelet IgG Fc receptors in immune-induced thrombocytopenia. Blood. 1993;81:988–93.
9. Kiss C, Surrey S, Schreiber AD, Schwartz E, McKenzie SE. Human c-kit ligand (stem cell factor) induces platelet Fc receptor expression in megakaryoblastic cells. Exp Hematol. 1996; 24:1232–7.
10. Tate BJ, Witort E, McKenzie IFC, Hogarth PM. Expression of the high responder/non-responder human FcγRII. Analysis by PCR and transfection into FcR-COS cells. Immunol Cell Biol. 1992;70:79–87.

11. Warmerdam PAM, van de Winkel JGJ, Vlug A, Westerdaal NAC, Capel PJA. A single amino acid in the second Ig-like domain of the human Fcγ receptor II is critical for human IgG_2 binding. J Immunol. 1991;147:1338–43.
12. Tomiyama Y, Kunicki TJ, Zipf TF, Ford SB, Aster RH. Response of human platelets to activating monoclonal antibodies: importance of Fc gamma RII (CD32) phenotype and level of expression. Blood. 1992;80:2261–8.
13. McCrae KR, Shattil SJ, Cines DB. Platelet activation induces increased Fc gamma receptor expression. J Immunol. 1990;144:3920–7.
14. Rappaport EF, Cassel DL, Walterhouse DO et al. A soluble form of the human Fc receptor Fc gamma RIIA:cloning, transcript analysis and detection. Exp Hematol. 1993;21:689–96.
15. Markovic B,Wu Z, Chesterman CN, Chong BH. Quantitation of soluble and membrane-bound Fc gamma RIIA (CD32A) mRNA in platelets and megakaryoblastic cell line (Meg-01). Br J Haematol. 1995;91:37–42.
16. Gachet C, Astier A, de la Salle H et al. Release of Fc gamma RIIa2 by activated platelets and inhibition of anti-CD9-mediated platelet aggregation by recombinant Fc gamma RIIa2. Blood. 1995;85:698–704.
17. Anderson CL, Chacko GW, Osborne JM, Brandt JT. The Fc receptor for immunoglogulin G (FcγRII) on Human platelets. Semin Thromb Hemost. 1995;21:1–9.
18. Horton MA, Hogg N. Platelet antigens: new and previously defined clusters. In: McMichael AJ, ed., Leucocyte Typing III. White Cell Differentiation Antigens. Oxford: University Press Oxford; 1987:733–46.
19. von dem Borne AEGK, Modderman PW, Admiraal LG, Nieuwenhuis HK. Joint report of the platelet section. Platelet antibodies, the overall results. In: Knapp W, ed., Leucocyte Typing IV. White Cell Differentiation Antigens. Oxford: University Press Oxford; 1989:951–66.
20. Griffith L, Slupsky J, Seehafer J, Boshkov L, Shaw AR. Platelet activation by immobilized monoclonal antibody:evidence for a CD9 proximal signal. Blood. 1991;78(7):1753–9.
21. Carroll RC, Rubinstein E, Worthington RE, Boucheix C. Extensive C1q-complement initiated lysis of human platelets by IgG subclass murine monoclonal antibodies to the CD9 antigen. Thromb Res. 1990;59:831–9.
22. De Reys S, Blom C, Lepoudre B et al. Human platelet aggregation by murine monoclonal antiplatelet antibodies is subtype-dependent. Blood. 1993;81:1792–800.
23. Nomura S, Yamaguchi K, Kido H et al. New monoclonal anti-human Fc gamma receptor II antibodies induce platelet aggregation. Clin Exp Immunol. 1991;86:179–84.
24. Gulino D, Ryckewaert JJ, Andrieux A, Rabiet MJ, Marguerie G. Identification of a monoclonal antibody against platelet GPIIb that interacts with a calcium-binding site and induces aggregation. J Biol Chem. 1990;265:9575–81.
25. Kouns WC, Jennings LK. Activation-independent exposure of the GPIIb-IIIa fibrinogen receptor. Thromb Res. 1991;63:343–54.
26. Rubinstein E, Kouns WC, Jennings LK, Boucheix C, Carroll RC. Interaction of two GPIIb/IIIa monoclonal antibodies with platelet Fc receptor (FcγRII). Br J Haematol. 1991;78:80–6.
27. Frelinger AL, Du X, Plow EF, Ginsberg MH. Monoclonal antibodies to ligand-occupied conformers of integrin alpha IIb beta3 (glycoprotein IIb-IIIa) alter receptor affinity, specificity, and function. J Biol Chem. 1991;266:17106–11.
28. Osborne JM, Brandt JT, Chacko GW, Anderson CL. CD 32 dependence of antibody-mediated platelet aggregation. In: Schlossmann SF, Boumsell L, Wally Gilks et al., eds, Leucocyte Typing V. White Cell Differentiation Antigens. Oxford: University Press Oxford; 1993:1251–5.
29. Rubinstein E, Boucheix C, Worthington RE, Carroll RC. Anti-platelet antibody interactions with Fcγ receptor. Semin Thromb Haemost. 1995;21:10–22.
30. Slupsky JR, Seehafer JG, Tang SC, Masellis-Smith A, Shaw AR. Evidence that monoclonal antibodies against CD9 antigen induce specific association between CD9 and the platelet glycoprotein IIb-IIIa complex. J Biol Chem. 1989;264:12289–93.
31. Horsewood P, Hayward CP, Warkentin TE, Kelton JG. Investigation of the mechanisms of monoclonal antibody-induced platelet activation. Blood. 1991;78:1019–26.
32. Rubinstein E, Urso I, Boucheix C, Carroll RC. Platelet activation by cross-linking HLA class I molecules and Fc receptor. Blood. 1992;79:2901–8.
33. Anderson GP, Anderson CL Signal transduction by the platelet Fc receptor. Blood. 1990; 76(6):1165–72.

34. Anderson GP, van de Winkel JG, Anderson CL. Anti-GPIIb/IIIa (CD41) monoclonal antibody-induced platelet activation requires Fc receptor-dependent cell-cell interaction. Br J Haematol. 1991;79:75–8.
35. Rubinstein E, Boucheix C, Urso I, Carroll RC. Fc gamma receptor-mediated interplatelet activation by a monoclonal antibody against beta 2 microglobulin. J Immunol. 1991;147: 3040–6.
36. Slupsky JR, Cawley JC, Griffith LS, Shaw ARE, Zuzel M. Role of FcγRII in platelet activation by monoclonal antibodies. J Biol Chem. 1992;148:3148–94.
37. Bachelot C, Saffroy R, Gandrille S, Aiach M, Rendu F. Role of FcγRIIa gene polymorphism in human platelet activation by monoclonal antibodies. Thromb Haemost. 1995;74:1557–63.
38. Parren PW, Warmerdam PA, Boeije LC et al. On the interaction of IgG subclasses with the low affinity Fc gamma RIIa (CD32) on human monocytes, neutrophils, and platelets. Analysis of a functional polymorphism to human IgG_2. J Clin Invest. 1992;90:1537–46.
39. Roberts JJ, Rodgers SE, Drury J, Ashman LK, Lloyd JV. Platelet activation induced by a murine monoclonal antibody directed against a novel tetra-span antigen. Br J Haematol. 1995;89:853–60.
40. Burgess JK, Lindeman R, Chesterman CN, Chong BH. Single amino acid mutation of Fc gamma receptor is associated with the development of heparin-induced thrombocytopenia. Br J Haematol. 1995;91:761–6.
41. Lecompte T, Stieltjes N, Shao-Kai L, Morel MC, Kaplan C, Meyer Samama M. Heparin- and streptokinase-dependent platelet-activating immunoglobulin G: mechanism and diagnosis. Semin Thromb Haemost 1995;21:95–105.
42. Kang J, Cabral C, Kushner L, Salzman EW. Membrane glycoproteins and platelet cytoskeleton in immune complex-induced platelet activation. Blood. 1993;81:1505–12.
43. Adelmann B, Sobel M, Fujimura Y, Ruggieri ZM, Zimmerman TS. Heparin-associated thrombocytopenia: observations on the mechanism of platelet aggregation. J Lab Clin Med. 1989;113:204–10.
44. Kelton JG, Sheridan D, Santos A et al. Heparin-induced thrombocytopenia. Blood. 1988; 72:925–30.
45. Greinacher A, Liebenhoff U, Kiefel V, Presek P, Mueller-Eckhardt C. Heparin-associated thrombocytopenia: the effects of various intravenous IgG preparations on antibody mediated platelet activation – a possible new indication for high dose i.v. IgG. Thromb Haemost. 1994; 71:641–5.
46. Haimovich B, Regan C, DiFazio L et al. FcγRII receptor triggers $pp125^{FAK}$ phosphorylation in platelets. Biol Chem. 1996;271:16332–37.
47. Gibbins J, Asselin J, Farnadals R, Barnes M, Law C-L, Watson SP. Tyrosine phosphorylation of the Fc receptor γ-chain in collagen-stimulated platelets. J Biol Chem. 1996;271:18095–9.
48. Kelton JG, Horsewood P. Antibody-mediated platelet activation. J Lab Clin Med. 1993;121: 382–4.
49. Mueller Eckhardt C, Lüscher EF. Immune reactions of human blood platelets. I. A comparative study on the effects on platelets of heterologous antiplatelet antiserum, antigen-antibody complexes, aggregated gammaglobulin and thrombin. Thromb Diath Haemorrh. 1968;20:155–67.
50. Schattner M, Lazzari M, Trevani AS et al. Activation of human platelets by immune complexes prepared with cationized human IgG. Blood. 1993;82:3045–51.
51. Kuroda K, Ozaki Y, Qi R et al. FcγII receptor-mediated platelet activation induced by anti-CD9 monoclonal antibody opens Ca^{2+} Channels which are distinct from those associated with Ca^{2+} store depletion. J Immunol. 1995;155:4427–36.
52. Blake RA, Asselin J, Walker T, Watson SP. Fc gamma receptor II stimulated formation of inositol phosphates in human platelets is blocked by tyrosine kinase inhibitors and associated with tyrosine phosphorylation of the receptor. FEBS Lett. 1994;342:15–8.
53. Huang MM, Indik Z, Brass LF, Hoxie JA, Schreiber AD, Brugge JS. Activation of Fc gamma RII induces tyrosine phosphorylation of multiple proteins including Fc gamma RII. J Biol Chem. 1992;267:5467–73.
54. Blake RA, Schieven GL, Watson SP. Collagen stimulates tyrosine phosphorylation of phospholipase C-gamma 2 but not phospholipase C-gamma 1 in human platelets. FEBS Lett. 1994;353:212–6.
55. Chacko GW, Duchemin AM, Coggeshall KM, Osborne JM, Brandt JT, Anderson CL. Clustering of the platelet Fc gamma receptor induces noncovalent association with the tyrosine kinase p72syk. J Biol Chem. 1994;269:32435–40.

56. Chacko GW, Brandt JT, Coggeshall KM, Anderson CL. Phosphoinositide 3-kinase and $p72^{syk}$ noncovalently associate with the low affinity Fc gamma receptor on human platelets through an immunoreceptor tyrosine-based activation motif. Reconstitution with synthetic phosphopeptides. J Biol Chem. 1996;271:10775–81.
57. Yanaga F, Poole A, Asselin J et al. Syk interacts with tyrosine-phosphorylated proteins in human platelets activated by collagen and cross-linking of the Fc gamma-IIA receptor. Biochem J. 1995;311:471–8.
58. Robinson A, Gibbins J, Rodriguez-Linares B et al. Characterization of Grb2-binding proteins in human platelets activated by Fc gamma RIIA cross-linking. Blood. 1996;88: 522–30.
59. Heinrich D, Stephinger U, Mueller-Eckhardt Ch. Specific interaction of HLA antibodies (eluates) with washed platelets. Br J Haematol. 1977;35:441–52.
60. Rubinstein E, Urso I, Boucheix C, Carroll RC. Platelet activation by cross-linking HLA class I molecules and Fc receptor. Blood. 1992;79:2901–8.
61. Christie DJ, Leja DN, Carlson DL, Swinehart CD. Characterization of platelet activation induced by HLA antibodies associated with alloimmune thrombocytopenia [published erratum appears in J Lab Clin Med. 1993;122:227] J Lab Clin Med. 1993;121:437–43.
62. Brandt JT, Julius CJ, Osborne JM, Anderson CL. The mechanism of platelet aggregation induced by HLA-related antibodies. Thromb Haemost. 1996;76:774–9.
63. Yanabu M, Nomura S, Fukuroi T et al. Platelet activation induced by an antiplatelet autoantibody against CD9 antigen and its inhibition by another autoantibody in immune thrombocytopenic purpura. Br J Haematol. 1993;84(4):694–701.
64. Babcock RB, Dumper CW, Scharfman WB. Heparin-induced immune thrombocytopenia. N Engl J Med. 1976;295:237–41.
65. Chong BH, Pitney WR, Castaldi PA. Heparin-induced thrombocytopenia:association of thrombotic complications with heparin-dependent IgG antibody that induces thromboxane synthesis and platelet aggregation. Lancet. 1982;2:1246 9.
66. Sheridan D, Carter C, Kelton JG. A diagnostic test for heparin-induced thrombocytopenia. Blood. 1986;67:27–30.
67. Warkentin TE, Hayward CPM, Smith CA, Kelly PM, Kelton JG. Determinants of donor platelet variability when testing for heparin-induced thrombocytopenia. J Lab Clin Med. 1992;120:371–9.
68. Greinacher A, Michels I, Kiefel V, Mueller-Eckhardt C. A rapid and sensitive test for diagnosing heparin-associated thrombocytopenia. Thromb Haemost. 1991;66:734 6.
69. Greinacher A. Antigen generation in heparin-associated thrombocytopenia: the nonimmunologic type and the immunologic type are closely linked in their pathogenesis. Semin Thromb Hemost. 1995;21:106–16.
70. Greinacher A, Michels I, Liebenhoff U, Presek P, Mueller-Eckhardt C. Heparin-associated thrombocytopenia:immune complexes are attached to the platelet membrane by the negative charge of highly sulphated oligosaccharides. Br J Haematol. 1993;84:711–6.
71. Amiral J, Bridey F, Dreyfus M, Vissac AM, Fressinaud E, Wolf M, Meyer D. Platelet factor 4 complexed to heparin is the target for antibodies generated in heparin-induced thrombocytopenia (letter). Thromb Hemost. 1992;68 95–6.
72. Visentin GP, Ford SE, Scott JP, Aster RH. Antibodies from patients with heparin-induced thrombocytopenia/thrombosis are specific for platelet factor 4 complexed with heparin or bound to endothelial cells. J Clin Invest. 1994;93:81 8.
73. Kelton JG, Smith JW, Warkentin TE, Hayward CPM, Denomme GA, Hordewood P. Immunoglobulin G from patients with heparin-induced thrombocytopenia binds to a complex of heparin and platelet factor 4. Blood. 1994;83:3232–9.
74. Greinacher A, Pötzsch B, Amiral J, Dummel V, Eichner A, Mueller-Eckhardt C. Heparin-associated thrombocytopenia:isolation of the antibody and characterization of a multimolecular PF4-heparin complex as the major antigen. Thromb Haemost. 1994;71:247–51.
75. Amiral J, Bridey F, Wolf M et al. Antibodies to macromolecular platelet factor 4-heparin complexes in heparin-induced thrombocytopenia: a study of 44 cases. Thromb Haemost. 1995;73:21–8.
76. Chong BH, Fawaz I, Chesterman CN, Berndt MC. Heparin-induced thrombocytopenia:mechanism of interaction of the heparin-dependent antibody with platelets. Br J Haematol. 1989;73:235–40.
77. Chong BH, Murray B, Berndt MC, Dunlop LC, Brighton T, Chesterman CN. Plasma P-selectin is increased in thrombotic consumptive platelet disorders. Blood. 1994;83:1535–41.

78. Aster RH. Heparin-induced thrombocytopenia and thrombosis [editorial;comments]. N Engl J Med. 1995;332:1374–6.
79. Greinacher A, Alban S, Dummel V, Franz G, Mueller-Eckhardt C. Characterization of the strucural requirements for a carbohydrate based anticoagulant with a reduced risk of inducing the immonological type of heparin-associated thrombocytopenia. Thromb Haemost. 1995;74:886–92.
80. Warkentin TE. Heparin-induced thrombocytopenia: IgG-mediated platelet activation, platelet microparticle generation, and altered procoagulant/anticoagulant balance in the pathogenesis of thrombosis and venous limb gangrene complicating heparin-induced thrombocytopenia. Transf Med Rev. 1996;10:249–58.
81. Cines DB, Tomaski A, Tannenbaum S. Immune endothelial-cell injury in heparin-associated thrombocytopenia. N Engl J Med. 1987;316:581–9.
82. Arepally, G. Poncz M, McKenzie SE, Cines D. Characterization of antibody subclass specificity and antigenic determinants in heparin-associated thrombocytopenia. Blood. 1994;84:Abstract 740.
83. Brandt JT, Isenhart CE, Osborne JM, Ahmed A, Anderson CL. On the role of platelet Fc gamma RIIa phenotype in heparin-induced thrombocytopenia. Thromb Haemost. 1995; 74:1564–72.
84. Warkentin TE, Sheppard JI, Denomme GA, Kelton JG. FcγIIa receptor genotype heterogenieity and platelet activation by heparin-induced thrombocytopenia IgG (HIT-IgG): A possible explanation for the predominance of the 'low responder' (His^{131}) FcγIIa receptor genotype in HIT patients. Blood. 1995;86(Suppl 1):537a.
85. Greinacher A, Baurichter G, Olbrich K et al. Association of human platelet FcγRIIa (CD32) genotype with heparin-induced thrombocytopenia (HIT). Ann Hematol. 1996;72:349.
86. Arepally G, McKenzie SE, Poncz M, Cines DB. FcγRIIa polymorphism, subclass-specific IgG anti-heparin/PF4 antibodies and clinical course in patients with heparin-induced thrombocytopenia and thrombosis. Blood. 1996;88(1):Abstract 1111.
87. Lebrazi J, Helft G, Abdelouahed M et al. Human anti-streptokinase antibodies induce platelet aggregation in an Fc receptor (CD32) dependent manner. Thromb Haemost. 1995; 74:938–42.
88. Roubey RAS. Autoantibodies to phospholipid-binding plasma proteins: a new view of lupus anticoagulants and other 'antiphospholipid' autoantibodies. Blood. 1994;84:2854–67.
89. Galli M, Finazzi G, Barbui T. Thrombocytopenia in the antiphospholipid syndrome. Br J Haematol. 1996;93:1–5.
90. Arnout J. The pathogenesis of the antiphospholipid syndrome: a hypothesis based on parallelisms with heparin-induced thrombocytopenia. Thromb Haemost. 1996;75:536–41.
91. Galli M, Comfurius P, Maassen C et al. Anticardiolipin antibodies (ACA) directed not to cardiolipin but to a plasma protein cofactor. Lancet. 1990;335:1544–7.
92. Arvieux J, Roussel B, Pouzol P, Colomb MG. Platelet activating properties of murine monoclonal antibodies to beta 2-glycoprotein I. Thromb Haemost. 1993;70:336–41.
93. Shi W, Chong BH, Chesterman CN. Beta 2-glycoprotein I is a requirement for anticardiolipin antibodies binding to activated platelets:differences with lupus anticoagulants. Blood. 1993;81:1255–62.
94. Frame JN, Mulvey KP, Phares JC, Anderson MJ. Correction of severe heparin-associated thrombocytopenia with intravenous immunoglobulin. Am Intern Med. 1989;111:946–7.
95. Grau E, Linares M, Olaso A, Ruvira J, Sauchris J. Heparin-induced thrombocytopenia-response to intravenous immunoglobulins in vivo and in vitro. Am J Hematol. 1992;39:312–3.
96. Dessaint JP, Tsicopoulos A, Pancré V et al. Platelets, IgE and allergy. In: Kaplan-Gouet C, Schlegel N, Salomon C, McGregor J, eds, Platelet Immunology: Fundamental and Clinical Aspects. Colloque INSERM: John Libbey Eurotext Ltd. 1991;206:187–94.
97. Joseph M, Capron A, Ameisen JC et al. The receptor for IgE on blood platelets. Eur J Immunol. 1986;16:306–12.
98. Cines DB, van der Keyl H, Levinson AI. In vitro binding of an IgE protein to human platelets. J Immunol. 1986;136:3433–40.
99. Pancré V, Joseph M, Capron A et al. Recombinant human immune interferon induces increased IgE receptor expression on human platelets. Eur J Immunol. 1988;18:829–32.
100. Ameisen JC, Joseph M, Cean JP et al. A role for glycoprotein IIb-IIIa complexe in the binding of IgE to human platelets and platelet IgE-dependent cytotoxic functions. Br J Haematol. 1986;64:21–32.

101. Joseph M, Auriault C, Capron A, Vorng H, Viens P. A new function for platelets: IgE-dependent killing of schistosomes. Nature. 1983;303:810–2.
102. Haque A, Cuna W, Bonnel B, Capron A, Joseph M. Platelet-mediated killing of larvae from different filarial species in the presence of Dipetalonema viteae stimulated IgE antibodies. Parasite Immunol. 1985;7:517–26.
103. Capron A, Dessaint JP. Immunologic aspects of schistosomiasis. Ann Rev Med. 1992;43: 209–18.
104. Cesborn JY, Capron A, Vargaftig BB et al. Platelets mediate the action of diethylcarbamazine on microfilariae. Nature. 1987;325:533–6.
105. Vanhee D, Joseph M, Vorng H, Tonnel AB. A colorimetric assay to evaluate the immune reactivity of blood platelets based on the reduction of a tetrazolium salt. J Immunol Meth. 1993;159:253–9.
106. Masini E, Di-Bello MG, Raspanti S, Sacchi TB, Maggi E, Mannaione PF. Platelet aggregation and histamine release by immunological stimuli. Immunopharmacology. 1994; 28:19–29.
107. Cardot E, Pestel J, Callebaut I et al. Specific activation of platelets from patients allergic to dermatophagoides pteronyssinus by synthetic peptides derived from the allergen Der p I. Int Arch Allergy Immunol. 1992;98:127–34.
108. Jeannin P, Pestel J, Bossus M, Lassalle P, Tartar A, Tonnel AB. Comparative analysis of biological activities of Der p I-derived peptides on Fc epsilon receptor-bearing cells form Dermatophagoides pteronyssinus-sensitive patients. Clin Exp Immunol. 1993;92:133–8.
109. Tsicopoulos A, Tonnel AB, Wallaert B et al. Decrease of IgE-dependent platelet activation in Hymenoptera hypersensitivity after specific rush desensitization. Clin Exp Immunol. 1988; 71:433–8.
110. Rogala B, Gumprecht J, Gawlik R, Strojek K. Platelet aggregation in IgE-mediated allergy with elevated soluble Fc epsilon RII/CD23 level. J Investig Allergol Clin Immunol. 1995;5: 161–6.
111. Askenase PW, Geba GP, Levin J et al. A role for platelet release of serotonin in the initiation of contact sensitivity. Int Arch Allergy Immunol. 1995;107:145–7.
112. Bermejo N, Gueant JL, Bata E, Gerard P, Moneret-Vautrin DA, Laxenaire MC. Platelet serotonin is a mediator potentially involved in anaphylactic reaction to neuromuscular blocking drugs. Br J Anaesth. 1993;70:322–5.
113. Tunon de Lara JM, Rio P, Marthan R, Vuillemin L, Ducassou D, Taytard A. The effect of sodium cromoglycate on platelets:an in vivo and in vitro approach. J Allergy Clin Immunol. 1992;89:994–1000.
114. Rio P, Tunon de Lara JM, Marthan R, Taytard A. IgE binding inhibits arachidonic acid-induced chemiluminescence of human platelets. Clin Exp Allergy. 1992;22:91–7.
115. Wilhelm D, Kluter H, Kloche M, Kirchner H. Impact of allergy screening for blood donors: relationship to nonhemolytic transfusion reactions. Vox Sang. 1995;69:217–21.

12
Polymorphisms of FcγRIIIa on NK cells and macrophages

H. R. KOENE, A. E. G. Kr. VON DEM BORNE, D. ROOS and M. de HAAS

INTRODUCTION

Natural killer (NK) cells are CD3-negative lymphocytes that play an important role in the innate, MHC-nonrestricted immunity against tumour cells and intracellular pathogens. Complete deficiency of NK cells is associated with recurrent viral and bacterial infections [1]. NK cells are able to mediate spontaneous and antibody-dependent cellular cytotoxicity and secrete a range of cytokines upon activation [2]. The mechanisms by which NK cells discriminate between self and non-self are beginning to be revealed, and involve an MHC class I-mediated inhibitory signal via killer cell inhibitory receptors [KIRs; reviewed in Reference 3].

FcγRIIIa is the only FcγR that is expressed on the plasma membrane of NK cells. Expression of FcγRIIIa is dependent on association with homo- or heterodimers of the TCR-ξ chain or the FcεRI-γ chain [4]. These chains contain immune receptor tyrosine-based activation motifs (ITAMs) and are the major transducers of signals from this receptor into the cell, although the cytoplasmic tail of FcγRIIIa also seems to be involved in signalling [5]. Successive truncations of the cytoplasmic tail proved that the cytoplasmic amino acids are not essential for expression of the receptor, but that the four most membrane-proximal amino acids play a crucial role in signalling by FcγRIIIa.

A soluble form of NK cell-derived FcγRIIIa is present in plasma from healthy individuals, albeit in lower concentrations compared with that of neutrophil-derived sFcγRIIIb [6]. The concentration of sFcγRIIIa is elevated in plasma from patients with diseases in which NK cells are involved, e.g. NK cell lymphocytosis and rheumatoid arthritis [6].

J.G.J. van de Winkel and P.M. Hogarth (eds.), The Immunoglobulin Receptors and their Physiological and Pathological Roles in Immunity. 135–140.

POLYMORPHISMS OF FcγRIIIa

Polymorphisms of members of the FcγR family are receiving increasing attention because they may serve as heritable risk factors for a variety of diseases (see Chapters 22–24). Two genetic polymorphisms have been described for FcγRIIIa [7,8]. In the membrane-proximal, IgG-binding domain of FcγRIIIa, a valine (V) or a phenylalanine (F) can be present at position 158 [7]. A triallelic polymorphism has been discovered for amino-acid position 48, in which leucine (L), arginine (R) or histidine (H) can be present [8]. We functionally analysed the two polymorphisms and found that the FcγRIIIa-158V/F polymorphism affected the IgG binding capacity of the receptor, independently of the 48L/R/H phenotype [9]. The FcγRIIIa-158V isoform has a higher affinity for IgG_1, IgG_3 and possibly IgG_4 than does the FcγRIIIa-158F isoform. Moreover, freshly isolated homozygous FcγRIIIa-158V-positive NK cells carry more cytophilic (i.e. monomeric) IgG than do homozygous 158F-positive NK cells. Because of the strong genetic linkage between the FcγRIIIA-158V/F and the -48L/R/H polymorphisms [9], we previously attributed high IgG binding to the presence of a histidine or arginine at amino-acid position 48 [8]. All FcγRIIIa-48H and -48R-positive individuals also carried one or two FcγRIIIA-158V genes, resulting in strong binding of IgG to their NK cells.

FcγRIIIa has a higher affinity for IgG than does the PI-linked FcγRIIIb, which is expressed constitutively only by neutrophils [10,11]. Extrapolating data from FcγRIIa mutants to FcγRIII, the higher IgG binding affinity of FcγRIIIa, compared with FcγRIIIb, was suggested to be the result of the presence of the phenylalanine at amino acid position 158 [12]. Our findings are in contrast with this hypothesis, because the FcγRIIIa-158V displays the highest affinity for IgG and the lower-affinity neutrophil FcγRIIIb also carries a valine at position 158. Other factors are, therefore, probably responsible for the affinity differences between the two receptors. Recently, experiments with exchange mutants of FcγRIIIa and FcγRIIIb have shown that association of FcγRIIIa with the γ-chain increases the affinity of the receptor for its ligand [13], although the differences in IgG binding affinity between FcγRIIIa and FcγRIIIb cannot be completely attributed to the differences in membrane anchoring. A PI-linked form of FcγRIIIa was shown to have a slightly higher affinity for IgG compared with (wildtype) PI-linked FcγRIIIb. However, the IgG binding affinity of wildtype, γ chain-associated FcγRIIIa was much higher than that of the mutated PI-linked form of FcγRIIIa [13]. Although the 158V/F-genotype of the FcγRIIIA construct was not reported, these experiments suggest that only a smalll part of the affinity differences result from amino acid differences between the extracellular parts of FcγRIIIa and FcγRIIIb and that the differences in ligand binding affinity are largely caused by the type of in membrane anchoring of FcγRIIIa and FcγRIIIb.

FcγRIIIa POLYMORPHISMS AND DISEASE

FcγRIIIa and viral infections

The possible clinical consequences of the FcγRIIIa polymorphisms have been under investigation. Two studies suggest a role of FcγRIIIa polymorphisms in the pathogenesis of viral infections. Jawahar *et al.* described a patient with recurrent herpes simplex and other infections, with a decreased number of circulating NK cells, that were not reactive with CD16 mAb B73.1 [14]. This patient was found to have a homozygous FcγRIIIA-48H-positive genotype. Independently, we identified two unrelated patients suffering from severe chickenpox and other viral infections who had the same FcγRIIIa phenotype [8]. These two patients were also typed for the FcγRIIIA-158V/F polymorphism and were both found to be homozygous 158V-positive [9]. Because the amino-acid polymorphism at position 158 influences the IgG binding capacity of NK cell FcγRIIIa, in contrast to the amino-acid polymorphism at position 48 [9], it is possible that a causal relationship exists between the disease and the FcγRIIIa-158 allotype. This would imply that a higher affinity for IgG interferes with NK cell function. It has been suggested that the amount of cytophilic IgG inversely correlates with the level of natural killer activity of NK cells [15–17]. The finding that FcγRIIIa-158V-positive NK cells carry more cytophilic IgG than do 158F-positive NK cells might therefore imply that the NK activity of the former is relatively low, resulting in a decreased anti-viral response. Furthermore, crosslinking of FcγRIIIa by IgG or CD16 mAbs induces activated NK cells to undergo apoptosis *in vitro* [18–21]. It remains to be established whether the differences in IgG binding affinity of the FcγRIIIa-158V and -158F isoforms influence the life span or the function of NK cells.

NK activity of stimulated NK cells is suppressed in HIV-positive individuals [22]. Patients with symptomatic HIV infection have decreased NK and LAK cell acitivity compared with asymptomatic HIV-infected patients. Because the FcγRIIIa-158V/F polymorphism might influence NK cell activity and life span as discussed above, we hypothesized that it might have an effect on susceptibility to or progression of HIV infection. We determined the FcγRIIIA-158V/F gene frequencies in individuals participating in the Amsterdam Cohort Studies on AIDS. The study population consisted of 140 individuals with a known date of HIV-1 seroconversion, 52 of whom developed clinical AIDS. Twenty-three participants were still asymptomatic after 9 years (long-term nonprogressors). We found no association between the FcγRIIIa-158V/F polymorphism and either the presence or the progression of HIV infection (unpublished observations). This suggests that the FcγRIIIa-158V/F polymorphism does not have a major influence on the immunity against HIV in adults. However, the contribution of NK cells to general immunity might be more important in children than in adults. Future studies will have to be performed to elucidate whether the FcγRIIIa-158V/F polymorphism is a risk factor for any kind of viral infections.

FcγRIIIa and systemic lupus erythematosus

FcγR polymorphisms seem to be involved in the pathogenesis of systemic lupus erythematosus (SLE) [23–26](see also Chapters 21 and 23). Macrophage expression of FcγR allotypes with relatively low affinity for IgG might result in decreased clearance of IC and subsequently tissue deposition and tissue damage. Harley *et al.* observed a genetic linkage between SLE and the FcγR locus on chromosome 1, based on microsatellite analysis [25]. Furthermore, several studies reported an association between SLE and the FcγRIIa-131R allotype in Venezualan, Dutch-Caucasian and Afro-American patients [23,24,26]. In some of these studies, the association was only present [24] or was more pronounced [26] in SLE patients with nephritis. In another study, no association between SLE and the FcγRIIa-131R/H polymorphism was observed at all in several ethnically different patient groups [27]. To investigate the contribution of the FcγRIIIa-158V/F and other FcγR polymorphisms in this context, we typed a group of Caucasian SLE patients for all known FcγR polymorphisms. In this group, only the distribution of the FcγRIIIa-158V/F polymorphism was significantly different from that in healthy controls, the FcγRIIIA-158F allele being more frequently present (Koene *et al.*, Arthritis Rheum., in press). In contrast to some of the previous work, the distribution of the FcγRIIa-131 allotypes was similar to that in healthy controls. A subdivision of the patient group based on the presence or absence of nephritis did not alter these results.

POSSIBLE THERAPEUTIC IMPLICATIONS

The possible therapeutic consequences of the FcγR polymorphisms are of interest. Several studies have shown positive effects of targeting tumour cells to FcγRIII on effector cells, using bispecific monoclonal antibodies (mAbs) (see Chapter 25). Earlier, CD16 mAb 3G8 was used in the experimental treatment of idiopathic thrombocytopenic purpura [28]. Both the FcγRIIIa-48L/R/H as well as the FcγRIIIa-158V/F polymorphism independently influence the binding of several CD16 mAbs [8,9]. For example, the amino acid at position 158 of FcγRIIIa influAb 3G8 [11] (P.M. Guyre, personal communication). Recently, it was suggested that FcγRIIIa on NK cells is involved in the first-dose cytokine-release syndrome that is seen after administration of CAMPATH 1-H [29]. Further studies should be performed to determine whether the FcγRIIIa-158V/F polymorphism plays a role in these reactions. Overall, it is conceivable that allele-dependent differences in affinity for humanized and (or) CD16 mAbs need to be taken into account when designing FcγRIII-directed bispecific mAbs for immunotherapy.

CONCLUDING REMARKS

The FcγRIIIa-158V/F polymorphism influences the IgG binding capacitiy of

NK cell FcγRIIIa and constitutes a sensitive risk factor for SLE and, possibly, viral infections in children. Although these clinical data are suggestive, further experiments will have to be performed to determine whether the polymorphism influences the function and the life span of NK cells and (or) macrophages.

References

1. Biron CA, Byron KS, Sullivan JL. Severe herpesvirus infetions in an adolescent without natural killer cells. N Engl J Med. 1989;320:1731–3.
2. Bancroft GJ. The role of natural killer cells in innate restistance to infection. Curr Opin Immunol. 1993;5:503–10.
3. Lanier LL. Natural killer cells: from no receptors to too many. Immunity. 1997;6:371–8.
4. Wirthmueller U, Kurosaki T, Murakami MS, Ravetch JV. Signal transduction by FcγRIII (CD16) is mediated through the gamma chain. J Exp Med. 1992;175:1381–90.
5. Hou X, Dietrich J, Odum N, Geisler C. The cytoplasmic tail of FcγRIIIAα is involved in signalling by the low affinity receptor for immunoglobulin G. J Biol Chem. 1996;271:22851–62.
6. De Haas M, Kleijer M, Minchinton RM, Roos D, Von dem Borne AEGKr. Soluble FcγRIIIa is present in plasma and is derived from natural killer cells. J Immunol. 1994;152:900–7.
7. Ravetch JV, Perussia B. Alternative membrane forms of FcγRIII (CD16) on human NK cells and neutrophils: Cell-type specific expression of two genes which differ in single nucleotide substitutions. J Exp Med. 1989;170:481–97.
8. De Vries E, Koene HR, Vossen JM, Gratama JW, Vn dem Borne AEGKr Waaijer JLM, Haraldsson A, De Haas M, Van Tol MJD. Identification of an unusual Fcγ receptor IIIa on natural killers in a patient with recurrent infections. Blood. 1996;88:3022–7.
9. Koene HR, Kleijer M, Algra J, Roos D, Von dem Borne AEGKr, De Haas M. FcγRIIIa-158V/F polymorphism influences the binding of IgG by NK cell FcγRIIIa, independently of the FcγRIIIa-48L/R/H phenotype. Blood. 1997;90:1109–14.
10. Huizinga TWJ, Van Kemenade F, Koenderman L et al. The 40-kDa Fcγ receptor (FcγRII) on human neutrophils is essential for the IgG-induced respiratory burst and IgG-induced phagocytosis. J Immunol. 1989;142:2365–9.
11. Vance BA, Huizinga TWJ, Wardwell K, Guyre PM. Binding of monomeric human IgG defines an expression polymorphism of FcγRIII on large granular lymphocyte/natural killer cells. J Immunol. 1993;151:6429–39.
12. Hulett MD, Hogarth PM. Molecular basis of Fc receptor function. Adv Immunol. 1994;57: 1–127.
13. Miller KL, Duchemin A, Anderson CL. A novel role for the Fc receptor gamma subunit: enhancement fo FcγR ligand affinity. J Exp Med. 1996;183:2227–33.
14. Jawahar S, Moody C, Chan M, Finberg R, Geha R, Chatila T. Natural killer (NK) cell deficiency associated with an epitope-deficient Fcγ receptor type IIIA (CD16-II). Clin Exp Immunol. 1996;103:408–13.
15. Sulica A, Gherman M, Galatiuc C, Manciulea M, Herberman RB. Inhibition of human natural killer cell activity by cytophilic immunoglobulin G. J Immunol. 1982;128:1031–6.
16. Gherman M, Manciulea M, Bancu AC, Sulica A, Stanworth DR, Herberman RB. Regulation of human natural cytotoxicity by IgG. Characterization of the structural site on monomeric IgG responsible for inhibiting natural killer cell activity. Mol Immunol. 1987;24: 743–50.
17. Sulica A, Metes D, Gherman M, Whiteside TL, Herberman RB. Divergent effects of FcγRIIIA ligands on the functional activities of human natural killer cells *in vitro*. Eur J Immunol. 1996;26:1199–203.
18. Azzoni L, Anegon I, Calabretta B, Perussia B. Ligand binding to FcγR induces c-myc-dependent apoptosis in IL-2-stimulated NK cells. J Immunol. 1995;154:491–9.
19. Ortaldo JR, Mason AT, O'Shea JJ. Receptor-induced death in human natural killer cells: Involvement of CD16. J Exp Med. 1995;181:339–44.
20. Eischen CM, Schilling JD, Lynch DH, Krammer PH, Leibson PJ. Fc-receptor induced expression of Fas ligand on activated NK cells facilitates cell-mediated cytotoxicity and subsequent autocrine NK cell apoptosis. J Immunol. 1996;156:2693–9.

21. Ortaldo JR, Winkler-Pickett RT, Nagata S, Ware CF. Fas involvement in human NK cell apoptosis: lack of a requirement for CD16-mediated events. J Leukoc Biol. 1997;61:209–15.
22. Ullum H, Gotzsche PC, Victor J, Dickmeiss E, Skinhoj P, Klarlund Pedersen B. Defective natural immunity: an early manifestation of human immunodeficiency virus infection. J Exp Med. 1995;182:789–99.
23. Blasini AM, Stekman IL, Leon-Ponte M, Caldera D, Rodriguez MA. Increased proportion of responders to a murine anti-CD3 monoclonal antibody of the IgG_1 class in patients with systemic lupus erythematosus (SLE). Clin Exp Immunol. 1993;94:423–8.
24. Duits AJ, Bootsma H, Derksen RHWM et al. Skewed distribution of IgG Fcγ recptor IIa (CD32) polymorphism is associated with renal disease in systemic lupus erythematosus patients. Arthritis Rheum. 1995;39:1832–6.
25. Harley JB, Sheldon P, Neas B et al. Systemic lupus erythematosus: considerations for a genetic approach. J Invest Dermatol. 1994;103:144S–9S.
26. Salmon JE, Millard S, Schachter LA et al. FcγRIIA alleles are heritable risk factors for lupus nephritis in African Americans. J Clin Invest. 1996;97:1348–54.
27. Botto M, Theodoridis E, Thompson EM et al. FcγRIIa polymorphism in systemic lupus erythematosus (SLE): no association with disease. Clin Exp Immunol. 1996;104:264–8.
28. Clarkson SB, Bussel JB, Kimberly RP, Valinsky JE, Nachman RL, Unkeless JC. Treatment of refractory immune thrombocytopenic purpura with an anti-Fcγ-receptor antibody. N Engl J Med. 1986;314:1236–9.
29. Wing MG, Moreau T, Greenwood J et al. Mechanism of first-dose cytokine-release syndrome by CAMPATH 1-H: involvement of CD16 (FcγRIII) and CD11a/CD18 (LFA-1) on NK cells. J Clin Invest. 1996;98:2819–26.

13
Cooperation between IgG Fc receptors and complement receptors in host defence

E. J. BROWN

INTRODUCTION

Successful host protection against bacterial infectious diseases requires recognition and destruction of potential invading pathogen by monocytes and polymorphonuclear neutrophils (PMN), the so-called 'professional phagocytes'. Many pathogenic bacteria have evolved mechanisms to evade direct recognition by phagocytes. As a result, the most successful host strategy for phagocytic elimination of potentially virulent bacteria requires coating of the bacteria with serum components, which are in turn recognized by specific receptors on the professional phagocytes. The process of serum protein binding to potentially dangerous environmental agents, both infectious and noninfectious, is known as opsonization. It has long been known that the two most important opsonic activities in serum are antibody and complement. Patients with genetic deficiencies of either complement opsonization or antibody production have markedly increased susceptibility to bacterial infections, demonstrating that both antibody and complement are necessary for optimal host defence. Since the work of Ehlenberger and Nussenzweig in 1975 [1], it has been clear that antibody and complement cooperate in opsonization for phagocytosis. In the intervening two decades since their experiments, the molecular nature of antibody and complement receptors has been elucidated. This increased understanding has allowed many laboratories to begin to identify the molecular mechanisms involved in cooperation between these two important opsonins. The purpose of this chapter is to review current understanding of the mechanism, extent, and significance of this cooperation.

J.G.J. van de Winkel and P.M. Hogarth (eds.), The Immunoglobulin Receptors and their Physiological and Pathological Roles in Immunity. 141–149.

COMPLEMENT RECEPTORS

Understanding of cooperation between Fc and complement receptors begins with knowledge of the structure of these receptors. IgG Fc (FcγR) are described in detail elsewhere in this volume, so this chapter will confine itself to a description of complement receptors. The major complement opsonins are fragments of the C3 molecule. C3 is activated by cleavage of its α chain by enzymes generated through either of two pathways of complement activation [reviewed in Chapter 2]; this cleavage results in release of a 9 kDa fragment known as C3a from the 200 kDa C3 molecule. A labile thioester bond in the remaining 190 kDa C3b fragment is exposed, and this can participate in formation of either an ester or an amide bond with electron donor oxygen or nitrogen atoms in molecules on the complement activating surface. Thus, C3b is deposited covalently onto an invading pathogen. Covalently bound C3b can be further cleaved at two sites by the concerted action of a serum serine protease, called Factor I and its essential cofactor, Factor H. The Factor I cleavage releases another small fragment from the C3 α chain, and the remaining covalently bound 180 kDa protein is known as iC3b. iC3b can then be further cleaved to release all but a 45 kDa C3 remnant, known as C3d, from the cell surface. The opsonic complement receptors of phagocytes recognize C3b and iC3b.

The primary receptors for C3b is complement receptor 1 (CR1) [3,4]. CR1 is a type I transmembrane protein which is a member of the regulators of complement activation or RCA family. The basic structural unit of the proteins in the RCA family is the short consensus repeat (SCR). Each SCR contains approximately 60 amino acids and, notably four disulphide bonds. The highly disulphide bonded structure is thought to give the SCRs a very rigid structure. CR1 contains 16, 20, or 24 SCRs, arranged as longer tandem repeats comprising four SCRs, known as long homologous repeats (LHR). CR1 has two binding sites for C3b, one within the first two SCR of the most amino terminal LHR and the second in a similar position within the second LHR. CR1 has a cytoplasmic tail of about 60 amino acids. Other than its ability to bind to the actin cytoskeleton, the functions of the CR1 cytoplasmic tail are poorly understood.

There are two leukocyte receptors for iC3b, known as CR3 and CR4 [5,6]. Both are members of the integrin family of adhesion molecules (Figure 13.1). Integrins are heterodimers composed of an α chain and a β chain from separate gene families. Both α and β chains are type 1 membrane proteins, with cytoplasmic tails varying from 30–70 amino acids. Ligand binding by integrins requires divalent cations; Mg^{2+} is generally considered to be the physiological divalent cation required for ligand binding. CR3 and CR4 are members of a subfamily of integrins expressed only on leukocytes, known as the LeuCAM, CD18, or β_2 integrins. There are four CD18 integrins; all are expressed on phagocytes. CR3 is the predominant CD18 integrin expressed on PMN and peripheral blood monocytes. Expression of CR4 increases during monocyte

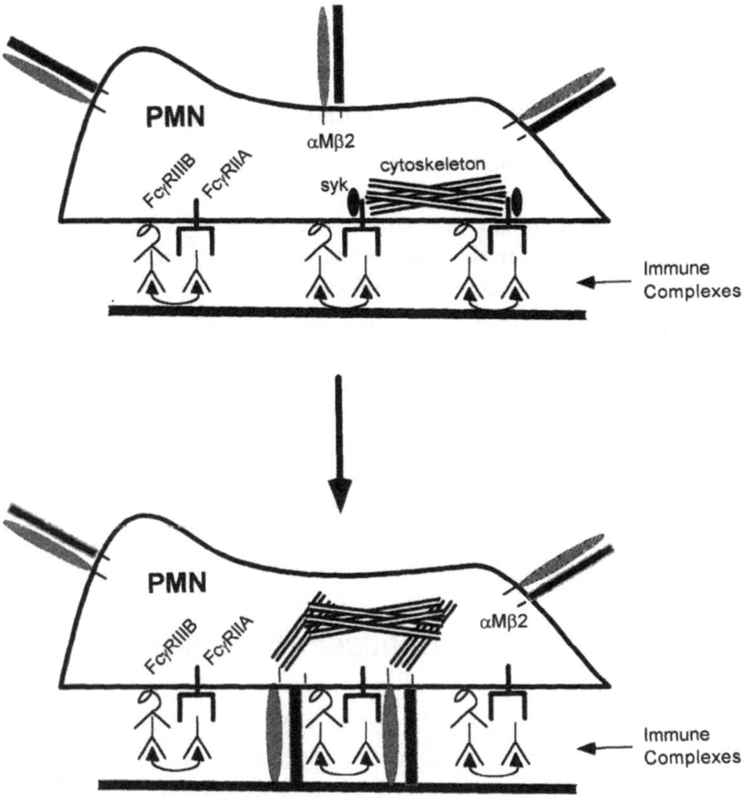

Figure 13.1 Proposed mechanism for involvement of CR3 in FcγR-dependent adhesion. (Top) Initial PMN adhesion is mediated by FcγRIIA and FcγRIIIB. Interaction with the actin cytoskeleton probably depends on activation of syk by FcγRIIA. (Bottom) With time,CR3 (αMβ2) associates with the adhesion sites, through interactions with FcγRs. Stable interaction with the cytoskeleton is mediated by CR3, not FcγRs. This interaction is required for phagocytosis and certain aspects of signal transduction associated with immune complex interaction with PMN

maturation into macrophages *in vitro* and *in vivo* [7]. A major feature of integrins including CD18 integrins is that their cytoplasmic tails are important sites for linking the plasma membrane to the actin cytoskeleton. This is a property of all integrin β chain cytoplasmic tails. The interaction of actin microfilaments with integrins is indirect, mediated via a number of linking proteins including talin, vinculin, and α-actinin. A second important feature of phagocyte integrins is that ligand binding is regulated by the state of cell activation. When PMN and monocytes are circulating, their integrins have very little ability to recognize ligand and mediate adhesion; when the cells are activated, such as at extravascular sites of infection, integrin-mediated adhesion is dramatically up-regulated [8,9]. While the details of how integrin avidity is regulated are unknown, the process clearly requires signal transduction, a process known as 'inside-out signalling'.

COOPERATION BETWEEN COMPLEMENT AND IgG

Freshly isolated PMN and monocytes can bind but not ingest phagocytic targets opsonized with C3 alone. Failure to phagocytose has been correlated with immobility of the complement receptors in the plane of the membrane [10,11]. The original explanation for this phenomenon was that the target-bound C3b and iC3b interacting with their receptors was a kind of molecular glue causing intercellular adhesion which increased the efficiency of presentation of IgG to its phagocyte receptors. The ability of complement to augment IgG function this way probably depends on the nature of the opsonized ligand. On most bacteria, the covalent bond with C3b occurs primarily with molecules exposed on the bacterial surface. However, when immune complexes containing only protein and antibody activate complement, much of the C3b is deposited on the antibody molecule itself, which can interfere with recognition of IgG by cellular Fcγ receptors. This may be the reason that immune complexes are actually cleared from the circulation more rapidly in the absence of complement [12].

COMPLEMENT-INDEPENDENT C3 RECEPTOR COOPERATION WITH Fcγ RECEPTORS

A potentially more interesting phenomenon is the requirement for complement receptors in Fcγ receptor function in the absence of complement activation. This phenomenon was first discovered when it was found that antibodies to CR3 blocked IgG-mediated ingestion by monocytes [13,14] and that phagocytes from patients with a genetic defect in the expression of β_2 integrins failed to ingest IgG opsonized targets normally. When transfected into NIH3T3 cells, FcγRIIIB, the glycan phosphatidylinositol-linked PMN FcγR, can mediate binding of IgG-coated particles, but this binding does not lead to phagocytosis [15]. However, cotransfection of CR3 renders these cells competent for IgG-dependent phagocytosis. Thus both in normal phagocytes and in transfection models, interaction between FcγR and CR3 seems to be required for optimal phagocytosis. This requirement for CR3 in FcγR functions extends to other consequences of interaction of IgG with PMN, including generation of the arachidonate metabolite LTB4 [16], and possibly activation of the respiratory burst [17] and increase in $[Ca^{2+}]_i$, although these last two are controversial [16]. This is a property of CR3 but not CR1; in fact, CR1-mediated phagocytosis requires CR3 as well, even for CR1 ligands which do not bind to CR3 [18]. Thus, CR3 has a role in phagocytosis which is independent of its ability to bind ligands on the phagocytic target.

The molecular mechanisms involved in this requirement for CR3 in phagocyte responses to FcγR ligation are not entirely understood. It is clear that a close physical relationship exists between the receptors. This was first shown that immune complexes (IC) could inhibit the binding of anti-CR3 Fab to monocytes [13]. Fcγ receptors co-cap with CR3 and fluorescence resonance energy transfer (FRET) can be shown between labelled CR3 and FcγRIIIB in these caps in PMN [19]. The co-capping and FRET can be interrupted by the saccharide N-actyl-D-glucosamine and by some antibodies to CR3. These data have led to the hypothesis that the lateral interaction between FcγRIIIB and CR3 occurs through interaction of the carbohydrates on the Fc receptor with a cryptic lectin binding domain in CR3 [20]. When FcγRIIIB and CR3 were transfected together into fibroblasts, there was FRET between the two molecules without prior capping, and further evidence for direct interaction was obtained from studies of FcγRIIIB diffusion in the transfected cells [21]. These data raise the possibility that spontaneous interaction between this FcγR and CR3 is actively inhibited in normal PMN until capping is induced. The receptor for urokinase, called uPAR, is another glycan phosphatidylinositol-linked membrane protein besides FcγRIIIB that has been shown to interact with CR3, by similar techniques [22]. uPAR also apparently associates with another integrin, αvβ3, in monocytes [23]. This raises the possibility that there may be more general interactions between integrins and gpi-linked proteins. Petty has proposed that interaction with integrins may be a general mechanism by which gpi-linked proteins, which do not span the membrane bilayer, can transmit signals to cells [17,24]. This is an issue of great controversy and intense scrutiny; additional information undoubtedly will be forthcoming in the near future. Moreover, it should be remembered that there is evidence for an interaction between the transmembrane FcγRII and CR3 in PMN [16,25,26] and that monocytes, in which the physical association of FcγR and CR3 was first shown, do not express a gpi-linked FcγR. Thus, there are probably interactions between CR3 and Fc receptors in addition to a possible generic interaction between integrins and gpi-linked proteins.

In both PMN and monocytes, only a subset of CR3 is associated with FcγRs [13,27]. This subset has been defined by its immobility in the plane of the plasma membrane [27]. When mobile CR3 are trapped on the adherent surface of the cell, CR3-dependent, IgG-mediated phagocytosis occurs perfectly normally on the apical surface. This IgG-mediated phagocytosis is still inhibited by anti-CR3 antibodies, even though 60–90% of the CR3 has been removed from the apical surface of the cell. Importantly, when CR3 trapping on the adherent surface occurs in the presence of cytochalasin D, the remaining CR3 are removed from the apical surface and IgG-mediated phagocytosis cannot occur. A subset of CR3 has been identified by monoclonal antibodies which is important in PMN adhesion [27], and it is likely that these are the same CR3 molecules important in IgG-mediated phagocysosis, but this has not yet been proven. Thus, an actin-filament associated CR3 subpopulation may be the molecules most important in cooperation with FcγR. Ligation of CR3 has been shown to lead to the association of a subset of FcγRIIA with cytoskeleton in PMN, and it is tempting to speculate that this association is indirect, with the cytoskeleton-associated CR3 as an intermediate. However, this has not yet been proven.

Understanding the role and mechanism of CR3 association with the cytoskeleton in its cooperation with FcγR is rudimentary. All integrins, including CR3, associate with the actin cytoskeleton through the linking proteins α-actinin, vinculin, and talin which bind to the integrin β chain cytoplasmic tails [28,29]. However, this association of many integrins with actin filament may depend on ligand binding to the integrin, which in leukocytes is regulated by cell activation. While the data cited above suggest that this interaction is essential for function, Griffin *et al.* published a series of papers demonstrating that complement receptor mediated phagocytosis (without IgG) actually requires the release of constitutively actin-associated receptors from the cytoskeleton so they are free to diffuse in the membrane [10,30,31]. A potential resolution of these apparently conflicting data has been suggested by recent experiments examining CD18 integrin diffusion by the technique of nanovid microscopy [32]. These studies showed that the lymphocyte integrin LFA-1, which is closely related to Mac-1, is restricted from free diffusion by association with microfilaments when the cells are in their resting, nonadherent state. This is true only for cells that have activatable integrins [33]. Cell activation by PMA freed the LFA-1 from its association with actin microfilaments, leading to enhanced diffusion, as predicted by the Griffin model. Since LFA-1 dependent adhesion is inhibited by high concentrations of cytochalasin, these experiments predict that the integrin reassociates with cytoskeleton upon ligand binding. Thus, activated leukocyte integrins are freely diffusing until they encounter ligand, when they reassociate with the actin cytoskeleton. A fundamental aspect of leukocyte integrin function is that maintainance of both the low avidity, nonadherent state and the high avidity adherent state requires integrin association with cytoskeleton.

PMN ADHESION TO IMMUNE COMPLEXES (IC)

Examination of PMN adhesion to IC-coated surfaces has brought much insights into cooperation between CR3 and FcγR. When PMN bind to immune complexes, actin rearrangements occur, and the content of F-actin in the cell increases. This effect on cytoskeleton is likely due to activation of the tyrosine kinase syk [34,35] and the increased F-actin is primarily at the lateral borders of the cell in contact with the IC-coated surface. This early adhesion is CR3-independent. However, after initial adhesion, the cell polymerizes more actin and begins to polarize; in cells deficient in CR3 or in the presence of anti-CR3 antibodies, this later phase does not occur. Instead, the PMN originally adherent to the IC-coated surface rapidly assumes a round, unspread morphology and eventually lose adhesion to the IC-coated surface.

It is possible to synthesize these data with the understanding of the regulation of integrin function into an hypothesis for the role of CR3 in FcγR function. The proposed model is that when PMN bind to IC, cell activation leads to release of CR3 from cytoskeleton constraints, leading to its diffusion to sites of cell contact with IC. These CR3 then diffuse to the sites of cell–substrate interaction, where they reassociate with cytoskeleton. Why CR3 should be trapped at adhesion sites is completely unknown, since there is no exogenous ligand for CR3. Perhaps CR3 interacts with the immobilized FcγR, or perhaps a ligand is secreted by the PMN after interaction with IC. The exocytic compartment of PMN is rich in plasma proteins [36], some of which, like fibrinogen, are known to be CR3 ligands [37]. As CR3-mediated adhesion progresses, the PMN polarizes, and FcγR-mediated adhesion decreases. Few data suggest a mechanism for the loss of FcγR-mediated adhesion. It is known that tyrosine phosphorylation of FcγR during IC ligation is transient; perhaps loss of adhesion occurs because of loss of association with syk, which could be required for sustained FcγR-mediated actin polymerization during IC adhesion. In the absence of CR3, this would lead to biphasic actin polymerization, with early adhesion and late loss of adhesion, which is what is observed.

This model has multiple testable features which are subject to further experimentation for understanding the validity of the model. For example, the role of FcγR tyrosine phosphorylation and syk activity in initiation of adhesion to IC can be tested. Other tests of the model depend on understanding the molecular basis of the crosstalk between FcγR ligation and CR3 activation. Data from several laboratories have shown a role for PI3 kinase in integrin activation [38]; our own preliminary data suggest an important role for this enzyme in CR3 activation by FcγR ligation [Jones and Brown, submitted]. This is particularly intriguing in light of the recent discovery that a molecule which binds to the β_2 integrin cytoplasmic tail also binds D3 phospholipids [39,40]. Further dissection of the interplay of these membranes and cytosolic molecules in adhesion may be critical to understanding FcγR cooperation with CR3.

Acknowledgements

The work in the author's laboratory is supported by grants from the United States National Institutes of Health, the American Arthritis Foundation, and the Washington University and Monsanto Cooperative Agreement.

References

1. Ehlenberger AG, Nussenzweig V. The role of membrane receptors for C3b and C3d in phagocytosis. J Exp Med. 1977;145:357–71.
2. Brown EJ, Joiner KA, Frank MM, Paul WE, eds, Fundamental Immunology. New York: Raven Press; 1984:645–68.
3. Fearon DT, Ahearn JM. Complement receptor type 1 (C3b/C4b receptor; CD35) and complement receptor type 2 (C3d/Epstein-Barr virus receptor; CD21). Curr Top Microbiol Immunol. 1990;153:83–98.
4. Ahearn JM, Fearon DT. Structure and function of the complement receptors, CR1 (CD35) and CR2 (CD21). Adv Immunol. 1989;46:183–219.
5. Ross GD. Complement and complement receptors. Curr Opin Immunol. 1989;2:50–62.
6. Brown EJ. Complement receptors and phagocytosis. Curr Opin Immunol. 1991;3:76–82.
7. Myones BL, Dalzell JG, Hogg N, Ross GD. Neutrophil and monocyte cell surface p150,95 has iC3b receptor (CR4) activity. J Clin Invest. 1988;82:640–51.

8. Brown EJ, Lindberg FP, Horton MA, eds, Blood Cell Biochemistry. V. Macrophages and Related Cells. Matrix Receptors of Myeloid Cells. New York: Plenum Press; 1993;11:279–306.
9. Brown EJ, Lindberg FP. Leucocyte adhesion molecule in host defence against infection. Ann Med. 1996;28:201–8.
10. Griffin FM Jr, Mullinax PJ. Augmentation of macrophage complement receptor function in vitro. III. C3b receptors that promote phagocytosis migrate within the plane of the macrophage plasma membrane. J Exp Med. 1981;154:291.
11. Bianco C, Griffin FM Jr, Silverstein SC. Studies of the macrophage complement receptor. Alteration of receptor function upon macrophage activation. J Exp Med. 1975;141:1278–9.
12. Waxman FJ, Hebert LA, Cornacoff JB et al. Complement depletion accelerates the clearance of immune complexes from the circulation of primates. J Clin Invest. 1984;74:1329–40.
13. Brown EJ, Bohnsack JF, Gresham HD. Mechanism of inhibition of immunoglobulin G-mediated phagocytosis by monoclonal antibodies that recognize the Mac-1 antigen. J Clin Invest. 1988;81:365–75.
14. Arnaout MA, Todd RF III, Dana N, Melamed J, Schlossman SF, Colten HR. Inhibition of phagocytosis of complement C3- or immunoglobulin G-coated particles and of C3bi binding by monoclonal antibodies to a monocyte granulocyte membrane glycoprotein (Mo1). J Clin Invest. 1983;72:171–9.
15. Krauss JC, Poo H, Xue W, Mayo-Bond L, Todd RF, Petty HR. Reconstitution of antibody-dependent phagocytosis in fibroblasts expressing Fc gamma receptor IIIB and the complement receptor type 3. J Immunol. 1994;153:1769–77.
16. Graham IL, Lefkowith JB, Anderson DC, Brown EJ. Immune complex-stimulated neutrophil LTB4 production is dependent on beta2 integrins. J Cell Biol. 1993;120:1509–17.
17. Petty HR, Todd RF III. Receptor–receptor interactions of complement receptor type 3 in neutrophil membranes (Review). J Leukoc Biol. 1993;54:492–4.
18. Gresham HD, Graham IL, Anderson DC, Brown EJ. Leukocyte adhesion deficient (LAD) neutrophils fail to amplify phagocytic function in response to stimulation: evidence for CD11b/CD18-dependent and -independent mechanisms of phagocytosis. J Clin Invest. 1991;88:588–97.
19. Zhou M, Todd RF III, Van de Winkel JGJ, Petty HR. Cocapping of the leukoadhesin molecules complement receptor type 3 and lymphocyte function-associated antigen-1 with Fc_{gamma} receptor III on human neutrophils: Possible role of lectin-like interactions. J Immunol. 1993;150:3030–41.
20. Sehgal G, Zhang K, Todd RF, Boxer LA, Petty HR. Lectin-like inhibition of immune-complex receptor-mediated stimulation of neutrophils – effects of cytosolic calcium release and superoxide production. J Immunol. 1993;150:4571–80.
21. Poo H, Krauss JC, Mayobond L, Todd RF, Petty HR. Interaction of Fc-gamma receptor type IIIB with complement receptor type 3 in fibroblast transfectants – evidence from lateral diffusion and resonance energy transfer studies. J Mol Biol. 1995;247:597–603.
22. Sitrin RG, Todd RF, Petty HR et al. The urokinase receptor (CD87) facilitates CD11b/CD18-mediated adhesion of human monocytes. J Clin Invest. 1996;97:1942–51.
23. Wei Y, Lukashev M, Simon DI et al. Regulation of integrin function by the urokinase receptor. Science. 1996;273:1551–5.
24. Petty HR, Todd RF. Integrins as promiscuous signal transduction devices. Immunol Today. 1996;17:209–12.
25. Zhou M-J, Brown EJ. CR3 (Mac-1, $\alpha_M\beta_2$, CD11b/CD18) and FcγRIII cooperate in generation of a neutrophil respiratory burst: requirement for FcγRII and tyrosine phosphorylation. J Cell Biol. 1994;125:1407–16.
26. Graham IL, Anderson DC, Holers VM, Brown EJ. Complement receptor 3 (CR3, Mac-1, integrin alpha-M, beta-2, CD11b/CD18) is required for tyrosine phosphorylation of paxillin in adherent and nonadherent neutrophils. J Cell Biol. 1994;127:1139–47.
27. Graham IL, Gresham HD, Brown EJ. An immobile subset of plasma membrane CD11b/CD18 (Mac-1) is involved in phagocytosis of targets recognized by multiple receptors. J Immunol. 1989;142:2352–8.
28. Hynes RO. Integrins: versatility, modulation, and signaling in cell adhesion. Cell. 1992;69:11–25.
29. Pavalko FM, Laroche SM. Activation of human neutrophils induces an interaction between the integrin β_2-subunit (CD18) and the actin binding protein α-actinin. J Immunol. 1993;151:3795–807.

30. Griffin JA, Griffin FM Jr. Augmentation of macrophage complement receptor function in vitro. I. Characterization of the cellular interactions required for the generation of a T-lymphocyte product that enhances macrophage complement receptor function. J Exp Med. 1979;150:653.
31. Griffin FM Jr, Griffin JA. Augmentation of macrophage complement receptor function in vitro. II. Characterization of the effects of a unique lymphokine upon the phagocytic capabilities of macrophages. J Immunol. 1980;125:884.
32. Kucik DF, Dustin ML, Miller JM, Brown EJ. Adhesion activating phorbol ester increases the mobility of leukocyte integrin LFA-1 in cultured lymphocytes. J Clin Invest. 1996;97: 2139–44.
33. Schmidt CE, Horwitz AF, Lauffenburger DA, Sheetz MP. Integrin-cytoskeletal interactions in migrating fibroblasts are dynamic, asymmetric, and regulated. J Cell Biol. 1993;123:977–91.
34. Greenberg S, Chang P, Wang DC, Xavier R, Seed B. Clustered syk tyrosine kinase domains trigger phagocytosis. Proc Natl Acad Sci USA. 1996;93:1103–7.
35. Indik ZK, Park JG, Hunter S, Schreiber AD. The molecular dissection of Fc gamma receptor mediated phagocytosis. Blood. 1995;86:4389–99.
36. Borregaard N, Kjeldsen L, Rygaard K et al. Stimulus-dependent secretion of plasma proteins from human neutrophils. J Clin Invest. 1992;90:86–96.
37. Altieri DC, Bader R, Mannucci PM, Edgington TS. Oligospecificity of the cellular adhesion receptor. Mac-1 encompasses an inducible recognition specificity for fibrinogen. J Cell Biol. 1988;107:1893–900.
38. Shimizu Y, Mobley JL, Finkelstein LD, Chan ASH. Role for phosphatidylinositol 3-kinase in the regulation of β1 integrin activity by the CD2 antigen. J Cell Biol. 1995;131:1867–80.
39. Kolanus W, Nagel W, Schiller B et al. Alpha L beta 2 integrin/LFA-1 binding to ICAM-1 induced by cytohesin-1, a cytoplasmic regulatory molecule. Cell. 1996;86:233–42.
40. Klarlund JK, Guilherme A, Holik JJ, Virbasius JV, Chawla A, Czech MP. Signaling by phosphoinositide-3,4,5-trisphosphate through proteins containing pleckstrin and sec7 homology domains. Science. 1997;275:1927–30.

14
Commentary

H. METZGER

The six chapters in this section deal with the mechanisms and consequences of a similar phenomenon: stimulation of cellular responses upon interaction of antigen with an antibody. Unlike the clonotypic antibody-like molecules synthesized by B- and T-lymphocytes which are themselves transmembrane proteins, in the cases reviewed in this section the antibodies are soluble, are likely to be polyclonal, are synthesized by cells other than the one they will stimulate, and act through an intermediary Fc receptor. Without apparent exception, these Fc receptors use a common device. They initiate a response when two or more of them are approximated after they simultaneously interact with multiple antibodies in a polyvalent antigen–antibody complex. When the immunoglobulin isotype has only a relatively weak affinity for its receptor, the reaction begins when the Fc region on one of the antibodies is an antigen–antibody aggregate binds to the corresponding Fc receptor; when the isotype has a high affinity, the monomeric immunoglobulin may be pre-bound so that the initiating bond will be between a multivarient antigen and the cell-bound antibodies. In either case, the subsequent crosslinking during which multiple Fc receptors become clustered, proceeds much more rapidly [1]. In those instances studied most intensively, the consequence of this approximation is the phosphorylation of receptor tyrosines through the action of a src-family kinase. Considerable evidence implicates the tyrosines in the ITAM motifs [2,3] as the sites modified, but recent direct studies suggest that the two canonical tyrosines in these homologous sequences are not always equivalently phosphorylated [4].

Two models have been proposed in order to explain how the topological change (approximation) is converted into a covalent modification (phosphorylation). Holowka and Baird and their colleagues at Cornell University have presented evidence that aggregated receptors redistribute to specialized micro-

J.G.J. van de Winkel and P.M. Hogarth (eds.), The Immunoglobulin Receptors and their Physiological and Pathological Roles in Immunity. 151–154.

domains in the plasma membrane [5,6]. These domains, which are relatively resistant to solubilization by detergents such as the Triton X-100, are enriched in sphingolipids, cholesterol, and extracellular proteins anchored into the outer leaflet of the plasma membrane by a glycosyl inositolphosphate moiety [7]. Most importantly, they are also enriched in intracellular proteins such as the src-family kinases that are anchored into the inner leaflet of the plasma membrane by one or more fatty acid moieties [8]. The Cornell group proposes that the receptors may contain sites that bind the special lipids in these domains with a low intrinsic affinity; when made effectively multivalent because of clustering the avidity is enhanced and this drives the cluster into the kinase-enriched domains thereby tipping the balance between the action of kinases and phosphatases (see below).

My own group has proposed an alternative model. We have suggested that at any instant in time, a small fraction of unaggregated receptors is associated with kinase. Upon aggregation, the constitutively associated kinase 'transphosphorylates' the newly adjacent receptor(s) principally because they are now proximal and not because of any marked activation of the kinase [9,10]. Thereafter, the phosphorylated receptors are thought to serve as a nidus to which additional molecules of kinase are recruited and this promotes additional rounds of phosphorylation. Our data appear to rule out the possibility that the aggregation-induced phosphorylation occurs either because of a marked increase in the specific activity of the src-family kinases [10] or indirectly, because of a sharp decrease in the susceptibility of the phosphotyrosines to the action of phosphatases on aggregated receptors [11].

The two models are not incompatible: the mechanism we have proposed could occur principally or exclusively in the special milieu on which Holowka and Baird and their colleagues have focused. Both groups have performed their experiments on the rodent Fc receptor with high affinity for IgE (FcεRI) but there is every reason to believe that other Fc receptors are likely to initiate the biochemical cascade by the same mechanism.

The initial aggregate of phosphorylated receptors is a dynamic structure; all other factors remaining equal, the relative times individual receptors remain clustered is determined by the intrinsic rate constants for the antigen–antibody and antibody–Fc receptor interactions. Receptors that are released are rapidly dephosphorylated [12,13] because they leave the proximity of the src-family kinase and are subject to the constitutive action of (probably membrane-bound) phosphatases [11]; if they rebind, such receptors are presumably susceptible to rephosphorylation although that has not been explicitly verified. Experimentally, receptors can be stably clustered with chemically crosslinked oligomers of the immunoglobulin. Such receptors undergo repeated round of phosphorylation and dephosphorylation and at least when the receptor aggregates are small, the dynamically phosphorylated clustered receptors persist for long periods [14]. It is likely that one factor contributing to the dephosphorylation of clustered receptors is the need for the aggregates to compete with one another for limited amounts of kinase [15].

The implied simplicity of the alternative models is deceptive. Consider the matter of structure: we know nothing about the three-dimensional form of any of the cytoplasmic domains on which the initial covalent changes take place. The small size of these extensions suggests that by themselves and in solution they may lack a stable conformation, but the presence of the charged inner leaflet of the plasma membrane and the effect of clustering may induce a more defined structure. Detailed information is developing about the src-family kinases [16], but it is disappointing that that part of the enzymes that appears to be critical for at least their initial interaction with receptors and the plasma membrane – the 'unique' domains [17,18] – was not part of the constructs examined in the initial studies.

From a functional point of view also, the initial events are more complicated than might at first be thought. It is a straightforward calculation that even for a simple active cluster such as a dimer of Fc receptors that use only γ chains for signalling, there will be four ITAMS and therefore eight canonical tyrosines that may or may not be phosphorylated [cf. Reference 19]. How many of these 2^8 (=256) discrete phosphorylation states of the clustered receptors will have distinctive functional attributes is still unknown.

The phosphorylated receptors are thought to serve as a locus for the recruitment of additional components such as Syk kinase and may in addition develop interactions with other cellular components such as other kinases, cytoskeletal proteins and the like. It is these later steps that serve as the principal focus of the reviews grouped in this section and it is exciting to read of the rapid accumulation of knowledge about these events. Nevertheless, as my accounting of the initial steps has attempted to highlight, the complexity of the events in real space and time exceed by orders of magnitude the flow diagrams (e.g. Figure 8.1) we are constrained to use by the current state of our knowledge. The transient and dynamic nature of the myriad interactions need to be quantitatively explored in order to identify those points that are most decisive for regulating the cellular response(s). Dealing with these complex assemblies of interactants at the structural and functional level represents a major challenge in the years ahead.

References

1. Wofsy C, Kent UM, Mao S-Y, Metzger H, Goldstein B. Kinetics of tyrosine phosphorylation when IgE dimers bind to Fc receptors on rat basophilic leukemia cells. J Biol Chem. 1995;270:20264–72.
2. Reth M. Antigen receptor tail clue [letter]. Nature. 1989;338:383–4.
3. Cambier JC. Antigen and Fc receptor signaling: the awesome power of the immunoreceptor tyrosine-based activation motif (ITAM). J Immunol. 1995;155:3281–5.
4. Pribluda VS, Pribluda C, Metzger H. Biochemical evidence for the phosphorylated tyrosines, serines, and threonines on the aggregated high affinity receptor for IgE are in the immunoreceptor tyrosine-based activation motifs. J Biol Chem. 1997;272:11185–92.
5. Field KA, Holowka D, Baird B. FcεRI-mediated recruitment of p53/56^{lyn} to detergent-resistant membrane domains accompanies cellular signaling. Proc Natl Acad Sci USA. 1995;92:9201–5.

6. Field KA, Holowka D, Baird B. Compartmentalized activation of the high affinity immunoglobulin E receptor within membrane domains. J Biol Chem. 1997;272:4276–80.
7. Melkonian KA, Chu T, Tortorella LB, Brown DA. Characterization of proteins in detergent-resistant membrane complexes from Madin-Darby canine kidney epithelial cells. Biochemistry. 1995;34:16161–70.
8. Resh MD. Myristylation and palmitylation of Src family members: the fats of the matter. Cell. 1994;76:411–13.
9. Pribluda VS, Pribluda C, Metzger H. Transphosphorylation as the mechanism by which the high-affinity receptor for IgE is phosphorylated upon aggregation. Proc Natl Acad Sci USA. 1994;91:11246–50.
10. Yamashita T, Mao S-Y, Metzger H. Aggregation of the high-affinity IgE receptor and enhanced activity of $p53/56^{lyn}$ protein-tyrosine kinase. Proc Natl Acad Sci USA. 1994;91: 11251–5.
11. Mao S-Y. Metzger H. Characterization of protein-tyrosine phosphatases that dephosphorylate the high affinity IgE receptor. J Biol Chem. 1997;272:14067–73.
12. Paolini R, Jouvin M-H, Kinet J-P. Phosphorylation and dephosphorylation of the high-affinity receptor for immunoglobulin E immediately after receptor engagement and disengagement. Nature. 1991;353:855–8.
13. Pribluda VS, Metzger H. Transmembrane signaling by the high affinity IgE receptor on membrane preparations. Proc Natl Acad Sci USA. 1992;89:11446–50.
14. Kent UM, Mao S-Y, Wofsy C, Goldstein B, Ross S, Metzger H. Dynamics of signal transduction after aggregation of cell-surface receptors: Studies on the type I receptor for IgE. Proc Natl Acad Sci USA. 1994;91:3087–91.
15. Torigoe C, Goldstein B, Wofsy C, Metzger H. Shuttling of initiating kinase between discrete aggregates of the high affinity receptor for IgE regulates the cellular response. Proc Natl Acad Sci USA. 1997;94:1372–7.
16. Sicheri F, Moarefi I, Kuriyan J. Crystal structure of the Src family tyrosine kinase Hck. Nature. 1997;385:602–9.
17. Timson Gauen LK, Kong AN, Samelson LE, Shaw AS. p59fyn tyrosine kinase associates with multiple T-cell receptor subunits through its unique amino-terminal domain. Mol Cell Biol. 1992;12:5438–46.
18. Vonakis BM, Chen H, Haleem-Smith H, Metzger H. The unique domain as the site on lyn kinase for its constitutive association with the high affinity receptor for IgE. J Biol Chem. 1997;272:24072–80.
19. Wofsy C, Torigoe C, Kent UM, Metzger H, Goldstein B. Exploiting the difference between extrinsic and intrinsic kinases: Implications for regulation of signaling by immunoreceptors. J Immunol. 1997;159:5985–92.

15
FcγR and IgG-mediated negative regulation of immune responses

M. DAËRON and B. HEYMAN

INTRODUCTION

In 1892, Behring reported that protective immunity could be induced in guinea pigs by injecting mixtures of diphtheria toxin and antitoxin, whereas toxin alone elicited severe local lesions and insignificant protection [1]. It was subsequently shown that induction of immunity was prevented if the mixtures contained an excess of antitoxin [2]. This was the first indication that preformed antibodies (Ab) can modulate immune responses and seeded the concept of a feedback regulation of Ab production.

In a seminal paper in 1968, Henry and Jerne demonstrated that anti-sheep erythrocytes (SRBC) immune serum induced a profound suppression of specific Ab responses and that molecules responsible for the suppression, in hyper-immune sera, resided in the 7S fraction (IgG), whereas the 19S fraction (IgM), instead, had enhancing properties [3]. IgG is the only Ab class that reproducibly inhibits Ab responses. IgG Ab to immunoglobulins (Ig) were subsequently found to be responsible for other immunoregulatory phenomena such as allotype suppression [4] and idiotype suppression [5]. Likewise, the injection of IgG anti-µ antibodies into newborn mice induced a profound B cell deficiency that abrogated all subsequent Ab responses [6].

Since these early observations, the immunoregulatory properties of IgG Ab have been intensively investigated. *In vivo* studies demonstrated that mouse IgG Ab against epitopes present at high density on an antigen (Ag), suppress Ab responses of all classes to all epitopes borne by that Ag, provided they have an intact Fc portion. *In vitro* studies conferred a prominent role to a subclass of low-affinity receptors for the Fc portion of IgG (FcγRIIB). These were found to

J.G.J. van de Winkel and P.M. Hogarth (eds.), The Immunoglobulin Receptors and their Physiological and Pathological Roles in Immunity. 155–167.

negatively regulate not only B cell receptor (BCR)-mediated B cell activation, but also T cell receptor (TCR)-mediated T cell activation and FcR-mediated mast cell activation. FcγRIIB therefore appeared as general negative coreceptors for all receptors whose cell-activating properties depend on immunoreceptor tyrosine-based activation motifs (ITAMs). An immunoreceptor tyrosine-based inhibition motif (ITIM) was identified in the intracytoplasmic domain of FcγRIIB, and similar ITIMs were found in other molecules with immunoregulatory properties. Following tyrosine phosphorylation, ITIMs recruit cytoplasmic phosphatases that can interfere with signal transduction at different steps.

While they permitted the identification of ITIMs and the elucidation of their mode of action, *in vitro* models progressively drifted away from *in vivo* phenomena. The analysis of suppression of Ab response was successively substituted for that of inhibition of B cell proliferation, for that of inhibition of B cell activation measured by cytokine production or increase in intracellular Ca^{2+} concentration, and recently, for the recruitment of cytoplasmic phosphatases. Other *in vitro* models revealed inhibitory effects of FcγRIIB that still need to be validated *in vivo*. The present review aims at discussing how accurately ITIM-dependent FcγRIIB-mediated negative regulation accounts for the *in vivo* immunosuppressive effects of IgG.

IgG-MEDIATED SUPPRESSION OF ANTIBODY RESPONSES

In vivo analysis

Most *in vivo* studies aiming at analysing the suppressive effects of passive Ab on Ab responses have been performed in mice injected i.v. with Ag and with IgG specific for this Ag, in saline. The response was assessed by enumerating specific Ab-producing cells in spleens or by measuring the concentration of specific Ab in serum.

Using such experimental conditions, a single injection of μg amounts of Ab is sufficient to suppress 95–99% of a primary Ab response [7,8]. IgG inhibit IgM [3,7,8], IgG [9,10] and IgE [11] responses (IgA responses have not been tested). In spite of such a profound suppression of the Ab response, priming of T helper cells [12], induction of DTH reactions [13] and priming for immunological memory (although less efficient) [10,14] were not abolished.

Suppression affects only the response to antigens to which IgG Ab bind [8]. However, IgG binding to one epitope on SRBC suppresses the Ab response to all epitopes borne by SRBC [7–9,15]. Suppression is therefore Ag-specific, but not epitope-specific. Not all IgG Ab of the same specificity, however, are equally suppressive. Monoclonal IgG anti-hapten with a low affinity [7] or to rare epitopes on SRBC [8], were unable to suppress. Finally, monoclonal IgG anti-trinitrophenyl (TNP) could suppress the *in vivo* response to SRBC with high, but not low, conjugation ratios of TNP [15]. Suppression therefore is observed when high-affinity Ab bind to epitopes present at a high density on Ag.

IgG may have opposite effects, depending on the nature of Ag. Monoclonal IgG anti-TNP Ab suppressed the response to TNP-SRBC, but enhanced the response to keyhole limpet haemocyanine conjugated to the same hapten (TNP-KLH) administered i.v. in saline [9,15]. As a general rule, IgG suppress the response to particulate Ag such as erythrocytes [3,8] and to proteins administered in complete Freund's adjuvant (CFA) [11], whereas they enhance the response to proteins administered in soluble form [16]. In contrast to suppression, enhancement of Ab responses can be induced not only by IgG [9,15,16], but also by IgM [3] and IgE [17].

One critical requirement for suppression is that IgG have an intact Fc portion. $F(ab')_2$ fragments of polyclonal IgG anti-SRBC were found to be 100–1000-fold less suppressive than intact IgG [14]. Likewise, allotype suppression and idiotype suppression, induced by IgG anti-allotype and IgG anti-idiotype, respectively, were not induced by $F(ab')_2$ fragments of the same antibodies [4,18]. Two major effector functions depend on the Fc portion of IgG: binding to FcγR and activation of complement. A role of complement was made unlikely by the finding that two monoclonal IgG_1 anti-TNP Ab suppressed the Ab response to SRBC-TNP equally well in spite of the fact that only one of them was able to activate complement [19]. No direct evidence for a role of FcγR in passive IgG-mediated suppression of Ab responses was however obtained *in vivo*.

In vitro analysis

In the most commonly used *in vitro* assay, spleen cells from mice primed *in vivo* with SRBC are cultured with SRBC or SRBC-TNP, in the presence of IgG specific for SRBC or TNP, and the resulting direct PFC response is measured 4–6 days later. Such studies confirmed most of the *in vivo* findings described above. Thus, suppression of Ab responses *in vitro* is Ag specific but not epitope-specific [20,21], requires that IgG bind to an epitope present at a high density on Ag [15] and that IgG have an intact Fc portion [22]. In addition, *in vitro* experiments suggested that FcγR are involved in suppression. Deglycosylated monoclonal IgG, unable both to activate complement and to bind to FcγR, were not suppressive [20], whereas mAbs deficient only in classical complement activation were [21]. Suppression could not be induced in spleen cell cultures depleted of FcγR-positive B cells [23]. A direct demonstration of the role of FcγR was obtained by showing that suppression was decreased by the rat mAb 2.4G2 [24], which blocks the IgG-binding site of mouse low-affinity FcγR [25].

FcγRIIB-DEPENDENT INHIBITION OF B CELL ACTIVATION

Ag-induced stimulation can be mimicked by Ab that aggregate BCR. $F(ab')_2$ fragments of anti-μ antibodies induced splenic B cells to secrete polyclonal Ig and to proliferate, whereas intact IgG-anti-μ did not [26]. Failure of intact IgG antibodies to stimulate B cells was understood to result from an active inhibition since B cells that had been incubated with IgG anti-μ failed to proliferate in response to a subsequent stimulation by a B cell mitogen [27]. Inhibition lasted for several days. The observation that preincubation of B cells with 2.4G2 permitted intact IgG anti-μ to activate murine B cells as efficiently as F(ab')2 fragments [28] confirmed the model proposed earlier by Sinclair, according to which coaggregation of BCR (*via* the Fab portions of anti-μ antibodies) and B cell FcγR (*via* the the Fc portion) prevented B cell activation [29].

Murine B cells express two isoforms of low-affinity receptors for IgG: FcγRIIB1 [30] and FcγRIIB1' [31]. A molecular analysis of FcγRIIB-mediated inhibition of B cell activation was made in IIA1.6, a FcγRIIB-negative variant of the murine lymphoma B cell line A20/2J [32]. These cells express membrane IgG_{2a} whose aggregation, either by polyclonal rabbit IgG anti-mouse Ig or by $F(ab')_2$ fragments of the same antibodies, triggers the secretion of IL-2. When stably transfected with cDNA encoding any of the three murine FcγRIIB isoforms (FcγRIIB1, FcγRIIB1' or FcγRIIB2), IIA1.6 cells secreted IL-2 upon challenge with anti-Ig $F(ab')_2$ but not upon challenge with intact IgG of the same specificity [31,33]. IgG anti-Ig, however, triggered IL-2 secretion by IIA1.6 cells expressing FcγRIIB whose intracytoplasmic (IC) domain had been deleted. When assessed by the secretion of a cytokine, B cell activation could therefore be inhibited by coaggregating BCR with FcγRIIB provided these had an intact IC domain. Mutational analysis of FcγRIIB IC domain led to the identification of the AENTITYSLLKHP sequence, present in all three FcγRIIB isoforms, as being necessary [33] and sufficient [34] for inhibition. This sequence is highly conserved in the IC domain of human FcγRIIB1 and FcγRIIB2, which both inhibited BCR-mediated B cell activation when expressed in IIA1.6 cells [35].

FcγRIIB-DEPENDENT INHIBITION OF CELL ACTIVATION BY ALL ITAM-BEARING RECEPTORS

IgG-induced inhibition of cell activation was for long understood as a B cell-specific negative feedback process by which antibodies negatively regulate their own production. FcR that trigger cell activation, BCR and TCR were however soon recognized as members of the same family of immunoreceptors. These receptors, involved in the (direct or Ab-mediated) recognition of Ag, possess signal transduction subunits the IC domain(s) of which contain(s) one or several ITAMs [36] consisting of a twice repeated YxxL sequence flanking 7 variable residues [37]. The aggregation of all ITAM-bearing receptors triggers cell activation *via* similar pathways, involving the sequential activation of the *src* and *syk* family protein tyrosine kinases (PTK) [38]. It thus became an attractive possibility that FcγRIIB, which are widely expressed by many cells of hematopoietic origin that also express ITAM-bearing receptors, might negatively regulate activation of these cell types. This proved to be the case.

Anti-CD3-induced IL-2 secretion by FcγRIIB1-expressing lymphoma and hybridoma T cells was inhibited when the CD3 complex was co-aggregated with FcγRIIB [39]. Inhibition was confirmed and analysed in BW5147 thymoma cells with a functional TCR, in which wild-type and mutated FcγRIIB were expressed [39]. Likewise, the coaggregation of FcεRI with FcγRIIB1, constitutively expressed by mouse bone marrow-derived mast cells (BMMC), inhibited IgE-induced mediator release [40]. Inhibition was also seen in the rat mast cell line RBL-2H3 transfected with cDNAs encoding murine FcγRIIB of the three isoforms, provided these had an intact IC domain [40]. Human FcγRIIB1 and FcγRIIB2 exerted the same inhibitory effects on IgE-induced reactions as murine FcγRIIB, when expressed in RBL cells, and the coaggregation of FcεRI with FcγRIIB, constitutively expressed by human blood basophils, inhibited IgE-induced histamine release [39]. Human FcγRIIB also inhibited mediator release induced by the aggregation of FcγRIIA, a single-chain low-affinity IgG receptor unique to humans and whose IC domain contains an ITAM [39].

Finally, FcγRIIB could inhibit cell activation triggered by chimeric molecules made of the extracellular and transmembrane domains of IgM and the IC domain of Igα [34], or by chimeras made of the extracellular and transmembrane domains of the a chain of the human IL-2 receptor (CD25) and the IC domain of either TCRζ or FcRγ [39]. This led to the prediction that all immunoreceptors associated with any of these ITAM-bearing subunits should be susceptible to inhibition by FcγRIIB. It was indeed confirmed for FcγRIIIA-mediated mouse mast cell activation using FcγRIIB-deficient mice [41]. It follows that IgG antibodies may be capable of regulating the activation (via native Ag, peptides on Ag-presenting cells or Ag complexed to IgG, IgE and IgA antibodies) of the many cell types which coexpress FcγRIIB and any ITAM-bearing receptor. These include B and T lymphocytes, mast cells and macrophages, Langerhans cells and other antigen-presenting cells, polymorphonuclear cells of the three types.

FcγRIIB AS MEMBERS OF A GROWING FAMILY OF ITIM-BEARING NEGATIVE CO-RECEPTORS

Mutational analysis of murine and human FcγRIIB sequences showed that the same highly conserved 13-amino acid sequence that was required for inhibition of BCR-mediated cell activation was required for inhibition of FcεRI- and TCR-mediated cell activation [39]. The mutation of the tyrosine residue contained in that sequence abrogated inhibition of cell activation via the three ITAM-bearing receptors [34,39]. The third residue following this tyrosine being a leucine, the inhibitory sequence appeared to contain a motif which was reminiscent of the two YxxL motifs constitutive of ITAMs. For this reason, it was referred to as ITIM, and ITIMs were searched for in other molecules.

Killer cell inhibitory receptors (KIRs) bind to MHC class I molecules. They are expressed on natural killer (NK) cells and on a subset of T cells [42]. When they bind to their ligands on target cells, they inhibit cell-mediated cytotoxicity of the three types: NK cytotoxicity, ADCC and CTL-mediated cytotoxicity, triggered by NK cell receptors, FcγRIIIA and TCR, respectively [43,44]. KIRs comprise immunoglobulin superfamily (IgSF) molecules and C-type lectin molecules [reviewed in Reference 45]. Their IC domain contains one or two YxxL motifs. Murine NK cells also express gp49B1 [46], an IgSF molecule with two IC YxxL motifs previously described on murine mast cells [47]. When transfected in NK cells, chimeric molecules having a KIR extracellular domain and a gp49B1 IC domain inhibited cytotoxicity [48].

Other ITIM-bearing negative receptors were found on B and T lymphocytes, mast cells and Ag-presenting cells (APC). They comprise IgSF members, such as CD22 expressed by B cells, CTLA-4 expressed by T cells, gp49B1 expressed by murine mast cells, and ILT3 expressed by dendritic cells, monocytes and macrophages, as well as a lectin-like molecule named mast cell associated function antigen (MAFA), also expressed by mast cells. CD22 binds to sialic acid-rich molecules expressed by a variety of cells, CTLA-4 to B7 molecules (CD80 and CD84) on APC; gp49B1, ILT3 and MAFA ligands are not known. All these molecules were shown to negatively regulate cell activation by ITAM-bearing receptors [49–53]. Their IC domain contains one or several ITIM-like motifs [47,49,52,54,55].

The comparison of inhibitory sequences in these negative coreceptors makes it possible to propose I/VxYxxL/V as a consensus sequence for ITIMs [56]. Two negative co-receptors, however, have ITIMs which do not fit with the consensus sequence. These are GxYxxM, in CTLA-4, and SxYxxL, in MAFA.

MECHANISMS OF INHIBITION OF CELL ACTIVATION BY ITIM-BEARING NEGATIVE CORECEPTORS

One common property of ITIMs is to become tyrosine phopshorylated and to recruit SH2 domain-bearing phosphatases in the vicinity of ITAM-bearing receptors. FcγRIIB must be co-aggregated with ITAM-bearing receptors for their ITIM to be phosphorylated [57] by kinases associated with cell-activating receptors (Malbec *et al.* submitted). Phosphorylated FcγRIIB ITIMs bound *in vitro* to two SH2 domain-bearing protein tyrosine phosphatases, SHP-1 and SHP-2 [58,59], and to the inositol polyphosphate 5-phosphatase SHIP [57,60–62]. Interestingly, the three phosphatases were differentially recruited *in vivo.* Phosphorylated FcγRIIB recruited SHIP in both B cells [61,63] and mast cells [57,62], but SHP-1 [60] and SHP-2 [61] in B cells only. FcγRIIB-dependent negative signalling was lost in B cells [60] but not in mast cells [57,62] derived from SHP-1-deficient motheaten mice. Once recruited, SHIP may act on not yet identified 5-phopshorylated second messenger precursors such as phosphatidyl inositol phosphates or second messengers such as inositol phosphates, which contribute to increase the intracellular Ca^{2+} concentration.

By contrast, human IgSF KIRs are associated with the src family PTK lck in NK cells [64] and, when aggregated by extracellular ligands, their ITIMs become tyrosine phosphorylated [65]. Phosphorylated KIR ITIM peptides bound *in vitro*, and recruited *in vivo* SHP-1 and SHP-2, but not SHIP [58,59,65]. The ability of phosphorylated ITIMs to bind SHP-1 and SHP-2 *in vitro* was recently shown to depend on the I or V residue, at position Y-2 in phosphorylated FcγRIIB and KIR ITIMs (F. Vély *et al.* submitted). Binding to phosphorylated ITIMs increases the catalytic activity of SHP-1 [60] which dephosphorylates ITAMs and ZAP-70 in NK cells [65], thereby inhibiting early steps of signal transduction.

ITIMs may therefore use two mechanisms to inhibit cell activation. When expressed in the same (mast) cell, and when coaggregated with the same activating receptors (FcεRI), both FcγRIIB and KIRs inhibited serotonin release. KIRs, however, inhibited both the early transient rise in intracellular Ca^{2+} concentration and the delayed sustained increase in Ca^{2+} concentration, whereas FcγRIIB inhibited only the late Ca^{2+} response [66,67]. So far, all ITIM-bearing negative receptors other than FcγRIIB seem to function like KIRs.

DOES ITIM-DEPENDENT NEGATIVE REGULATION ACCOUNT FOR IgG-MEDIATED SUPPRESSION OF Ab RESPONSES *IN VIVO*?

As knowledge on mechanisms by which the many ITIM-bearing negative co-receptors regulate cell activation of a variety of cell types increases, one may wonder whether ITIM-dependent negative regulation accounts for the effects of IgG antibodies observed *in vivo*. The role of FcγRIIB ITIM in inhibition of B cell, T cell and mast cell activation was demonstrated only with *in vitro* tumour models that made possible the mutational analysis of FcγRIIB and biochemical studies. The requirement of FcγRIIB for inhibition of B cell activation was demonstrated using 2.4G2 on spleen B cells *in vitro*. IgG-mediated *in vivo* suppression was shown to depend on the Fc portion of Ab, but it was not demonstrated to involve FcγRIIB. The most convincing evidence for a role of FcγRIIB in the *in vivo* regulation of Ab responses was the observation that, after immunization with high doses of SRBC or TNP-KLH, approximately 4-fold higher titres of specific Ab were induced in FcγRIIB-deficient mice compared with littermate controls [41]. Whether passively injected IgG can suppress Ab responses in these knock-out mice needs to be investigated.

Different parameters, such as antibody titres in serum, PFC responses, secretion of polyclonal Ig, thymidine incorporation, secretion of IL-2 by A20/2J cells or their variants, and increase in intracellular Ca^{2+} concentration, were used to assess inhibition (Figure 15.1). These responses are not equivalent and conclusions drawn from the use of one experimental system may not apply to another. Finally, anti-Ig $F(ab')_2$ fragments were assumed to aggregate BCR in a similar way as Ag, and intact anti-Ig IgG were thought to co-aggregate BCR and FcγRIIB like Ag-IgG complexes. It follows that, even though ITIM-dependent inhibition of B cell activation is likely to take place during IgG-mediated suppression of Ab responses, it may be reductionist to assume that it is the only mechanism involved *in vivo*.

The possibility that other negative regulatory mechanisms may exist was suggested by the finding that mAbs of all four IgG subclasses, i.e. IgG_1, IgG_{2a}, IgG_{2b} and IgG_3, could suppress an Ab response [7,8,15]. Murine natural [68] and recombinant [69] FcγRIIB bind mouse IgG_1, IgG_{2a} and IgG_{2b}, but not

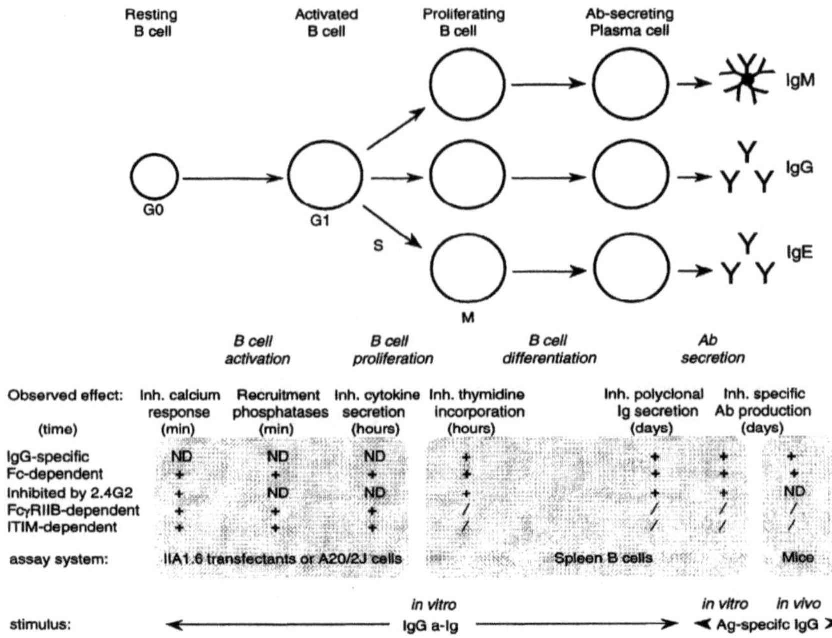

Figure 15.1 *In vivo* and *in vitro* analysis of negative regulation of B cell activation by IgG Ab. The figure summarizes the effects observed with the different assay systems used to study the mechanism of the suppressive effects of IgG Ab, in relation to the differentiation of B cells into Ab-producing cells. ND: not done; /: not testable with the experimental model used

IgG_3, and IgG_3 receptors bind IgG_3, but not IgG_1, IgG_{2a} or IgG_{2b}. Suppression by IgG_3 Ab therefore cannot be mediated by FcγRIIB. As FcR for IgG_3 were not cloned, one cannot exclude, however, that they also have an ITIM in their IC domain.

An alternative mechanism might be the masking of epitopes by Ab. Masking of epitopes may account for suppression induced by high concentrations of specific IgM [7], for some reports that $F(ab')_2$ fragments were also suppressive [70,71] and that suppression was sometimes epitope-specific [72]. That suppression is induced primarily by one Ab class, i.e. IgG, is 100–1000 times more efficient when IgG have an intact Fc portion, and although Ag-specific, is not epitope-specific, are however strong arguments against masking of epitopes being a major explanation for IgG-induced suppression.

Another possibility is that, when complexed with IgG, particulate Ag are targeted to phagocytes whose FcγRI, FcγRIIB2 and FcγIIIA mediate both phagocytosis and endocytosis. Increased degradation of particulate Ag via phagocytosis, indeed, decreases Ag presentation (C. Bonnerot, personnal communication), whereas an increased endocytosis of soluble Ag enhances Ag presentation [33]. Since FcγRIIB1 expressed by B cells mediate neither endocytosis nor phagocytosis [31,33], the complexation with IgG does not increase Ag presentation by B cells [33]. Endocytosis of soluble Ag complexed with IgG *via* FcγR expressed by macrophages and dendritic cells, on the contrary, would be enhanced, as well as their subsequent presentation to T cells. This may explain that anti-TNP IgG suppressed Ab responses to SRBC or to TNP-KLH in CFA but enhanced Ab responses to TNP-KLH in saline.

Finally, no firm conclusion as to the mechanism behind IgG-mediated negative feedback regulation of Ab responses *in vivo* can be drawn at present, and it is reasonable to assume that several mutually nonexclusive mechanisms concur to suppression.

IgG-MEDIATED INHIBITION OF CELL ACTIVATION IN PHYSIOLOGY AND PATHOLOGY

Whatever the mechanism of suppression seen in experimental models, IgG Ab are likely to possess general immunoregulatory properties, in both physiology and pathology. A physiological negative feedback of Ab production is suggested by the following observation. When rabbits are immunized with bacteriophages (ϕX 174) and, a few weeks later, depleted of endogenous ϕX 174-specific Ab by exchange transfusion, an increase in serum levels of anti-ϕX 174 Ab is detected [73]. The interpretation is that serum Ab could inhibit immunization by persisting endogenous Ag. An interesting therapeutic application of Ab-mediated suppression stemmed from the observation that feto-maternal immunization of Rh-negative mothers with Rh-positive fetal blood is less common if mother and child are ABO-incompatible [74]. An interpretation was that the presence of natural maternal Ab against A or B blood group Ag suppressed the anti-Rh Ab response in an epitope-nonspecific fashion. Subsequently, IgG purified from Rh-negative donors, immunized with Rh-positive blood, proved to be an efficient preventive treatment of haemolytic disease of the newborn [75] and became widely used in clinics.

Evidence for IgG-mediated negative regulation of T cell activation *in vivo* is scarce. Activated T cells express FcγRIIB [76], which can conceivably be co-aggregated to TCR when T cells recognize MHC-associated peptides on APC or target cells in the presence of IgG antibodies directed to epitopes expressed on the same cell. This might happen when allogeneic cells are protected from rejection by passive antibodies [77] with intact Fc portions [78]. A similar situation might be viral infections. The as yet unexplained beneficial effects of intravenous injections of polyclonal immunoglobulins from a large number of normal individuals in autoimmune patients [79] might be partly accounted for by a similar mechanism. It is indeed possible that Ig preparations contain protective natural autoantibodies to target cells of autoreactive T cell clones or to epitopes of the TCR complex that might be lacking in patients.

Finally, IgG-mediated inhibition of mast cell activation might explain the intriguing fact that 80% of individuals are not allergic although they possess everything that is necessary for immediate hypersensitivity reactions. FcγRIIB expressed by mast cells and basophils are indeed likely to be coaggregated with FcεRI when an IgG-complexed allergen binds to receptor-bound IgE and, thus, to maintain physiological mast cell degranulation below the level where pathological manifestations develop. This regulatory mechanism may already be used empirically when allergic patients are 'desensitized' by repeated injections of increasing doses of allergens. The therapeutical effects of such an immunotherapy are usually correlated with the appearance of high titres of IgG Ab in the serum [80]. It might be deliberately exploited with appropriate bispecific ligands as a new therapeutic approach of allergic diseases.

References

1. von Behring E, Wernick K. Über Immunisierung und Heilung von Versuchstieren bei der Diphterie. Z Hyg Infektionskrankheit. 1892;12:10–44.
2. Smith T. Active immunity produced by so-called balanced or neutral mixtures of diphtheria toxin and antitoxin. J Exp Med. 1909;11:241–60.
3. Henry C, Jerne NK. Competition of 19S and 7S antigen receptors in the regulation of the primary immune response. J Exp Med. 1968;128:133–52.
4. Shek PN, Dubiski S. Allotypic suppression in rabbits:competition for target cell receptors between isologous and heterologous antibody and between native antibody and antibody fragments. J Immunol. 1975;114:621–8.
5. Kohler H, Richardson BC, Smyk S. Immune response to phosphorylcholine. IV. Comparison of homologous and isologous anti-idiotypic antibody. J Immunol. 1978;120:233–8.
6. Manning DD, Jutila JW. Immunosuppression of mice injected with heterologous anti-immunoglobulin heavy chain antisera. J Exp Med. 1972;135:1316–23.
7. Brüggemann M, Rajewsky K. Regulation of the antibody response against hapten-coupled erythrocytes by monoclonal anti-hapten antibodies of various isotypes. Cell Immunol. 1982; 71:365–73.
8. Heyman B, Wigzell H. Immunoregulation by monoclonal sheep erythrocyte specific IgG antibodies. Suppression is correlated to level of antigen binding and not to isotype. J Immunol. 1984;132:1136–43.
9. Enriques-Rincon F, Klaus GGB. Differing effects of monoclonal anti-hapten antibodies on humoral responses to soluble or particulate antigens. Immunology. 1984;52:129–36.
10. Heyman B, Wigzell H. IgM enhances and IgG suppresses immunological memory in mice. Scand J Immunol. 1985;21:255–66.
11. Strannegård Ö, Belin L. Suppression of reagin synthesis in rabbits by passively administered antibody. Immunology. 1970;20:775–85.
12. Kappler JW, Hoffman M, Dutton RW. Regulation of the immune response. I. Differential effect of passively administered antibody on the thymus-derived and bone marrow-derived lymphocytes. J Exp Med. 1971;134:577–87.
13. Liew FY, Parish CR. Regulation of the immune response by antibody. I. Suppression of antibody formation and concomitant enhancement of cell-mediated immunity by passive antibody. Cell Immunol. 1972;4:66–85.
14. Sinclair NRSC. Regulation of the immune response. I. Reduction in ability of specific antibody to inhibit longlasting IgG immunological priming after removal of the Fc fragment. J Exp Med. 1969;129:1183–201.
15. Wiersma EJ, Coulie PG, Heyman B. Dual immunoregulatory effects of monoclonal IgG-antibodies:suppression and enhancement of the antibody response. Scand J Immunol. 1989; 29:439–48.
16. Terres G, Stoner RD. Specificity of enhanced immunological sensitization of mice following injections of antigens and specific antisera. Proc Soc Exp Biol Med. 1962;109:88–91.

17. Heyman B, Liu T, Gustavsson S. In vivo enhancement of the specific antibody response via the low affinity receptor for IgE. Eur J Immunol. 1993;23:1739–42.
18. Kohler H, Richardson B, Rowley DA, Smyk S. Immune response to phosphorylcholine. III. Requirement of the Fc portion and equal effectiveness of IgG subclasses in anti-receptor antibody-induced suppression. J Immunol. 1977;119:1979–86.
19. Wiersma EJ, Nose M, Heyman B. Evidence of IgG-mediated enhancement of the antibody response in mice without classical pathway complement activation. Eur J Immunol. 1990;20: 2585–9.
20. Heyman B, Nose M, Weigle WO. Carbohydrate chains on IgG_{2b}. A requirement for efficient feedback immunosuppression. J Immunol. 1985;134:4018–23.
21. Heyman B, Wiersma E, Nose M. Complement activation is not required for IgG-mediated suppression of the antibody response. Eur J Immunol. 1988;18:1739–43.
22. Heyman B. Fc-dependent IgG-mediated suppression of the antibody response: fact or artefact? Scand J Immunol. 1990;31:601–7.
23. Stockinger B, Lemmel EM. Fc Receptor dependency of antibody-mediated feedback regulation: on the mechanism of inhibition. Cell Immunol. 1978;40:395–403.
24. Heyman B. Inhibition of IgG-mediated immunosuppression by a monoclonal anti-Fc receptor antibody. Scand J Immunol. 1989;29:121–6.
25. Unkeless JC. Characterization of monoclonal antibody directed against mouse macrophage and lymphocyte Fc receptors. J Exp Med. 1979;150:580–96.
26. Sidman CL, Unanue ER. Requirements for mitogenic stimulation of murine B cells by soluble anti-IgM antibodies. J Immunol. 1979;122:406–13.
27. Sidman CL, Unanue ER. Control of B-lymphocyte function I. Inactivation of mitogenesis by interactions with surface immunoglobulin and Fc receptor molecules. J Exp Med. 1976;144: 882–96.
28. Phillips NE, Parker DC. Cross-linking of B lymphocyte Fcγ receptors and membrane immunoglobulin inhibits anti-immunoglobulin-induced blastogenesis. J Immunol. 1984; 132:627–32.
29. Sinclair NRSC, Chan PL. Regulation of the immune responses. IV. The role of the Fc fragment in feedback inhibition by antibody. Adv Exp Med Biol. 1971;12:609–15.
30. Amigorena S, Bonnerot C, Choquet D, Fridman WH, Teillaud JL. FcγRII expression in resting and activated B lymphocytes. Eur J Immunol. 1989;19:1379–85.
31. Latour S, Fridman WH, Daëron M. Identification, molecular cloning, biological properties and tissue distribution of a novel isoform of murine low-affinity IgG receptor homologous to human FcγRIIB1. J Immunol. 1996;157:189–97.
32. Jones B, Tite JP, Janeway Jr CA. Different phenotypic variants of the mouse B cell tumour A20/2J are selected by antigen- and mitogen-triggered cytotoxicity of L3T4-positive, I-A-restricted T cell clones. J Immunol. 1986;136:348–56.
33. Amigorena S, Bonnerot C, Drake J et al. Cytoplasmic domain heterogeneity and functions of IgG Fc receptors in B-lymphocytes. Science. 1992;256:1808–12.
34. Muta T, Kurosaki T, Misulovin Z, Sanchez M, Nussenzweig MC, Ravetch JV. A 13-amino-acid motif in the cytoplasmic domain of FcγRIIB modulates B-cell receptor signalling. Nature. 1994;368:70–73.
35. Van den Herik-Oudijk IE, Westerdaal NAC, Henriquez NV, Capel PJA, Van de Winkel JGJ. Functional analysis of human FcγRII (CD32) isoforms expressed in B lymphocytes. J Immunol. 1994;152:574–85.
36. Cambier JC. New nomenclature for the Reth motif (or ARH1/TAM/ARAM/YXXL). Immunol Today. 1994;16:110.
37. Reth MG. Antigen receptor tail clue. Nature. 1989;338:383–4.
38. Cambier JC. Antigen and Fc receptor signaling: the awsome power of the immunoreceptor tyrosine based activation motif (ITAM). J Immunol. 1995;155:3281–5.
39. Daëron M, Latour S, Malbec O et al. The same tyrosine-based inhibition motif, in the intracytoplasmic domain of FcγRIIB, regulates negatively BCR-, TCR-, and FcR-dependent cell activation. Immunity. 1995;3:635–46.
40. Daëron M, Malbec O, Latour S, Arock M, Fridman WH. Regulation of high-affinity IgE receptor-mediated mast cell activation by murine low-affinity IgG receptors. J Clin Invest. 1995;95:577–85.
41. Takai T, Ono M, Hikida M, Ohmori H, Ravetch JV. Augmented humoral and anaphylactic responses in FcγRII-deficient mice. Nature. 1996;379:346–9.

42. Nakajima H, Tomiyama H, Takiguchi M. Inhibition of γδ T cell recognition by receptors for MHC class I molecules. J Immunol. 1995;155:4139–42.
43. Mingari MC, Vitale C, Cambiaggi A et al. Cytolytic T lymphocytes displaying natural killer (NK)-like activity. Expression of NK-related functional receptors for HLA class I molecules (p58 and CD94) and inhibitory effect on the TCR-mediated target cell lysis or lymphokine production. Int J Immunol. 1995;7:697–703
44. Moretta A, Bottino C, Vitale M et al. Receptors for HLA class I molecules in human natural killer cells. Annu Rev Immunol. 1996;14:619–48.
45. Chambers WH, Brissette-Storkus CS. Hanging in the balance:natural killer cell recognition of target cells. Chem Biol. 1995;2:429–35.
46. Wang LL, Mehta IK, LeBlanc PA, Yokoyama WM. Mouse natural killer cells express gp49B1, a structural homologue of human killer cell inhibitory receptors. J Immunol. 1997; 158:13–17.
47. Castells MC, Wu X, Arm JP, Austen KF, Katz HR. Cloning of the gp49B gene of the immunoglobulin superfamily and demonstration that one of its two products is an early-expressed mast cell surface protein originally described as gp49. J Biol Chem. 1994;269: 8393–101.
48. Rojo S, Burshtyn N, Long E, Wagtmann N. Type I transmembrane receptor with inhibitory function in mouse mast cells and NK cells. J Immunol. 1997;158:9–12.
49. Sato S, Miller AS, Inaoki M et al. CD22 is both a positive and negative regulator of B lymphocyte antigen receptor signal transduction: altered signaling in CD22-deficient mice. Immunity. 1996;5:551–62.
50. Waterhouse P, Marengère LEM, Mittrücker H-W, Mak TW. CTLA-A, a negative regulator of T-lymphocyte activation. Immunol Rev. 1996;153:183–207.
51. Katz HR, Vivier E, Castells MC, McCormick MJ, Chambers JM, Austen KF. Mouse mast cell gp49B1 contains two immunoreceptor tyrosine-based inhibition motifs and suppresses mast cell activation upon coligation with FcεRI. Proc Natl Acad Sci USA. 1996;93:10809–14.
52. Colonna M. Immunoglobulin-superfamily inhibitory receptors: from natural killer cells to antigen-presenting cells. Immunol Res. 1997;148:169–71.
53. Guthmann MD, Tal M, Pecht I. A new member of the C-type lectin family is a modulator of mast cell secretory response. Int Arch Allergy Clin Immunol. 1995;107:82–6.
54. Dariavach P, Mattei MG, Golstein P, Lefranc M-P. Human Ig superfamily CTLA-4 gene:chromosomal localization and identity of protein sequence between murine and human CTLA-4 cytoplasmic domain. Eur J Immunol. 1988;18:1901–5.
55. Guthmann MD, Tal M, Pecht I. A secretion inhibitory signal transduction molecule on mast cells is another C-type lectin. Proc Natl Acad Sci USA. 1995;92:9397–401.
56. Vivier E, Daëron M. Immunoreceptor tyrosine-based inhibition motifs. Immunol Today. 1997;18:286–91.
57. Fong DC, Malbec O, Arock M, Cambier JC, Fridman WH, Daëron M. Selective in vivo recruitment of the phosphatidylinositol phosphatase SHIP by phosphorylated FcγRIIB during negative regulation of IgE-dependent mouse mast cell activation. Immunol Lett. 1996;54: 83–91.
58. Burshtyn DN, Scharenberg AM, Wagtmann N et al. Recruitment of tyrosine phosphatase HCP by the killer cell inhibitory receptor. Immunity. 1996;4:77–85.
59. Olcese L, Lang P, Vély F et al. Human and mouse killer-cell inhibitory receptors recruit PTP1C and PTP1D protein tyrosine phosphatases. J Immunol. 1996;156:4531–4.
60. D'Ambrosio D, Hippen KH, Minskoff SA et al. Recruitment and activation of PTP1C in negative regulation of antigen receptor signaling by FcγRIIB1. Science. 1995;268:293–6.
61. D'Ambrosio D, Fong DC, Cambier JC. The SHIP phosphatase becomes associated with FcγRIIB1 and is tyrosine phosphorylated during 'negative' signaling. Immunol Lett. 1996; 54:73–82.
62. Ono M, Bolland S, Tempst P, Ravetch JV. Role of the inositol phosphatase SHIP in negative regulation of the immune system by the receptor FcγRIIB. Nature. 1996;383:263–6.
63. Chacko GW, Tridandapani S, Damen JE, Liu L, Krystal G, Coggeshall KM. Negative signaling in B lymphocytes induces tyrosine phosphorylation of the 145-kDa inositol polyphosphate 5-phosphatase, SHIP. J Immunol. 1996;157:2234–8.
64. Bottino C, Vitale M, Olcese L et al. The human natural killer cell receptor for major histocompatibility complex class I molecules:surface modulation of p58 molecules and their

linkage to CD3 ζ chain, FcεRI γ chain and the p56 lck kinase. Eur J Immunol. 1994;24:2527–34.

65. Binstadt BA, Brumbaugh KM, Dick CJ et al. Sequential involvement of lck and SHP-1 with MHC-recognizing receptors on NK cells inhibits FcR-initiated tyrosine kinase activation. Immunity. 1996;5:629–38.
66. Choquet D, Partiseti M, Amigorena S, Bonnerot C, Fridman WH, Korn H. Cross-linking of IgG receptors inhibits membrane immunoglobulin-stimulated calcium influx in B lymphocytes. J Cell Biol. 1993;121:355–63.
67. Bléry M, Delon J, Trautmann A et al. Reconstituted killer-cell inhibitory receptors for MHC class I molecules control mast cell activation induced via immunoreceptor tyrosine-based activation motifs. J Biol Chem. 1997;272:8989–96.
68. Daëron M, Yodoi J, Néauport-Sautès C, Moncuit J, Fridman WH. Receptors for immunoglobulin isotypes (FcR) on murine T cells. I. Multiple expression on T lymphocytes and hybridoma T cell clones. Eur J Immunol. 1985;15:662–7.
69. Daëron M, Sautès C, Bonnerot C et al. Murine type II Fcγ receptors and IgG-Binding Factors. Chem Immunol. 1989;47:21–78.
70. Cerottini J-C, McConahey PJ, Dixon FJ. The immunosuppressive effect of passively administered antibody IgG fragments. J Immunol. 1969;102:1008–15.
71. Tao TW, Uhr JW. Capacity of pepsin-digested antibody to inhibit antibody formation. Nature. 1966;212:208–9.
72. Möller G. Antibody-mediated suppression of the immune response is determinant specific. Eur J Immunol. 1985;15:409–12.
73. Bystryn J-C, Graf MW, Uhr JW. Regulation of antibody formation by serum antibody. II. Removal of specific antibody by means of exchange transfusion. J Exp Med. 1970;132:1279–87.
74. Levine P. The influence of the ABO system on Rh hemolytic disease. Hum Biol. 1958;30:14–28.
75. Clarke CA, Donohoe WTA, Woodrow JC et al. Further experimental studies on the prevention of Rh haemolytic disease. Br Med J. 1963;1:979–84.
76. Leclerc JC, Plater C, Fridman WH. The role of Fc receptors (FcR) on thymus-derived lymphocytes. I. Presence of FcR on cytotoxic lymphocytes and absence of direct role in cytotoxicity. Eur J Immunol. 1977;7:543–8.
77. Voisin GA. Immunological facilitation, a broadening of the concept of the enhancement phenomenon. Progr Allergy. 1971;15:328–75.
78. Capel PJA, Tamboer WPM, De Waal RMW, Jansen JLJ, Koene RAP. Passive enhancement of skin grafts by alloantibodies is Fc dependent. J Immunol. 1979;122:421–9.
79. Kazatchkine MD, Dietrich G, Hurez V et al. V region-mediated selection of autoreactive repertoires by intravenous immunoglobulin (i.v. Ig). Immunol Rev. 1994;139:79–107.
80. Gleich GJ, Zimmermann EM, Henderson LL, Yunginger JW. Effect of immunotherapy on immunoglobulin E and immunoglobulin G antibodies to ragweed antigens: a six-year prospective study. J Allergy Clin Immunol. 1978;62:261.

16
FcγR on T cells

M. SANDOR and R. G. LYNCH

INTRODUCTION

T cells regulate most aspects of immunity. As an example, specificity, class selection and the quantity of produced antibodies are all affected by T cells. Theoretically Fc receptors on T cells offer an interface between humoral and cellular immunity, an idea that is simple and appealing from several points of view. Immunoglobulin binding to Fc receptors displayed on the same T cells that regulate antibody production could bring about an end product feedback regulatory circuit. In most immune responses there is a shift that brings forth the apparent dominance of either the cellular or humoral response [1]. Fc receptors on the T cells regulating cellular responses offer a possible mechanism for that phenomenon. In the 1970s, when the mechanism of isotype switching and selection was not understood, the most seductive part of this hypothesis was that the presence of isotype-specific receptors on T cells offered a logical explanation for immunoglobulin isotype regulation. At that time a series of observations were made, that fitted well with this hypothesis:

(1) Fc receptors were detected on some T cells [2,3];

(2) T cell activation up-regulated the expression of Fc receptors [4,5];

(3) This expression was further enhanced in an isotype-specific manner by the presence of high concentration of the immunoglobulin of respective class [6,7].

For a decade a large effort was focused on testing the hypothesis that FcR on T cells were involved in isotype-specific antibody regulation [8]. In the mid-1980s it became clear that T cell-produced lymphokines can regulate antibody

J.G.J. van de Winkel and P.M. Hogarth (eds.), The Immunoglobulin Receptors and their Physiological and Pathological Roles in Immunity. 169–183.

production in an isotype-specific manner and that the lymphokine-mediated crosstalk of various T cell subpopulations can regulate the balance of humoral and cellular immune response [9]. These findings did not exclude an immunoregulatory role for the T cell-expressed Fc receptors, but having a lymphokine solution for the mechanism of Ig isotype regulation lessened the interest in studies of T cell receptors.

In the 1990s a renewed interest followed the finding of close similarities in the structure and signalling of TCR and some Fc receptors. The shared motifs regulating the function of both receptors [10] are discussed in detail in chapters 9, 10, 14 and 19 of this book. Another intruiging new finding is that Fc receptor expression is tightly regulated during T cell development. This issue is the major focus of the present chapter.

FcγR EXPRESSION ON T CELLS IS DEVELOPMENTALLY REGULATED

T cell development in the thymus

The major pathway of T cell differentiation occurs in the thymus. In the adult thymus the proportion of T cell populations corresponding to a given stage of differentiation is low (except the CD4, CD8 double-positive stage). On the contrary, T cells present in the fetal thymus represent large populations of various stages of T cell development, depending on fetal age. The thymus of the 12-day fetus contains CD44+, ckit+, Thy-, FcγR–, CD3–, CD4–, CD8– cells that include progenitors for T, B and myeloid cells [11]. These cells and probably some already pre-thymically committed Thy low, ckit+ cells [12] present in the fetal blood, can differentiate into FcγR+ pro-T cells. The phenotype of these cells is summarized in Table 16.1. The triple-negative (CD3,CD4,CD8), CD44+ thymic population can be divided into IL-2Rα (CD25) positive and negative fractions. The TCR genes are not rearranged yet in the less differentiated CD25–, FcγR+ fraction [13] and these cells contain precursors for both T cells and NK cells [14]. These cells are in cell cycle [11] and for expansion they require signalling initiated by either the c-kit (SCF) receptor or the cytokine receptor (IL-2, IL-4, IL-7, IL-9 and IL-15) common gamma subunit (γ_c). The FcγR+ pro-T cells express GATA-3 and TCF-1 transcription factors, intracellular

Table 16.1 The phenotype of 2.4G2+ pro-T cell

FcγRIII and FcγRII+
No pluripotent potential: precursor only for T and NK cells
Triple negative (CD3,CD4,CD8), intracellular CD3ε+
CD44+, CD2–
IL-2R (CD25)–, IL-7R+, ckit+
TCR rearrangement negative
TCF-1+, GATA-3+
lck and pTα mRNA+
Most in cell cycle

CD3ε and lck and pTα message [11]. When these pro-T cells mature into pre-T cells the level of FcγR, CD44 and c-kit is down-regulated and the pre-T cells start to express IL-2α, CD2, some surface CD3, most associated with the pre-TCR, and they intensively begin to rearrange the TCRβ genes [13]. In fact, because of their restricted expression, FcγR becomes an important marker to identify murine pro-T cells in T cell development (Figure 16.1). Unlike the day 12–15 fetal thymus, the proportion of pro-T cells in the adult thymus is very low. There are two lines of evidence that pro-T cells also express FcγR in the adult thymus. First, after gating out the CD8, CD4 and CD3 positive cells, a CD44+, Thy low, FcγR+ population of adult thymocytes can be demonstrated by cytofluorometry. Second, in adult SCID or RAG mice in which T cell differentiation is blocked at the CD44–CD25+ stage by the absence of TCR rearrangement (see Figure 16.1), up to 15% of thymocytes display a Thy+ FcγR phenotype.

The surface expression of FcγR on pro-T cells was assessed by the binding of 2.4G2 monoclonal antibody, that recognizes both FcγRII (CD32) and FcγRIII (CD16). This raised the question as to the type of FcγR displayed by pro-T cells. As CD16 recruits and activates protein kinases but CD32 interacts with phosphatases it was important to determine the type of FcγR expressed on the developing thymocytes. PCR studies [13,15–17] and 2.4G2 antibody staining of fetal thymocytes from CD16 [18], γ chain [17] or CD32 KO mice clearly established that murine pro-T cells express both CD16 and CD32. These studies also demonstrated that the developmental windows of CD16 and CD32 expression during T cells development totally overlapped.

A population of T cells express FcγR after activation through TCR. Ontogenetically the first T cells that appear in mice represent a small population of γ/δ T cells that express monoclonal Vγ3 Vδ1 TCR and which migrate to the skin from the fetal thymus. There is some indication that the thymic development of these cells involves TCR-mediated activation and selection. It has been shown that the intrathymic maturation of these cells induces the expression of activation markers and CD16 [19]. In the day 17–19 fetal thymus a large population of α/β T cells undergoes positive and negative selection that requires TCR signalling. It is evident that in the case of this major lineage of T cells selection associated activation signals do not induce FcγR expression [14,15] (Figure 16.1).

In humans, the expression pattern of FcγR in T cell development is less clear. Ontogenic studies of human fetal thymus revealed the presence of FcγR on thymocytes at the tenth week of gestation, with a continuously decreasing level of expression during fetal development and beyond [20]. In the adult thymus surface CD4 and CD3 negative, but intracellular CD3 and CD16 positive cells were detected [21] that can give rise to NK cells. Human FcγR negative, triple-negative cells that differentiate into T cells have been demonstrated [22], but it is evident that the FcγR expression on human pro-T cells has not been carefully studied as in the murine system.

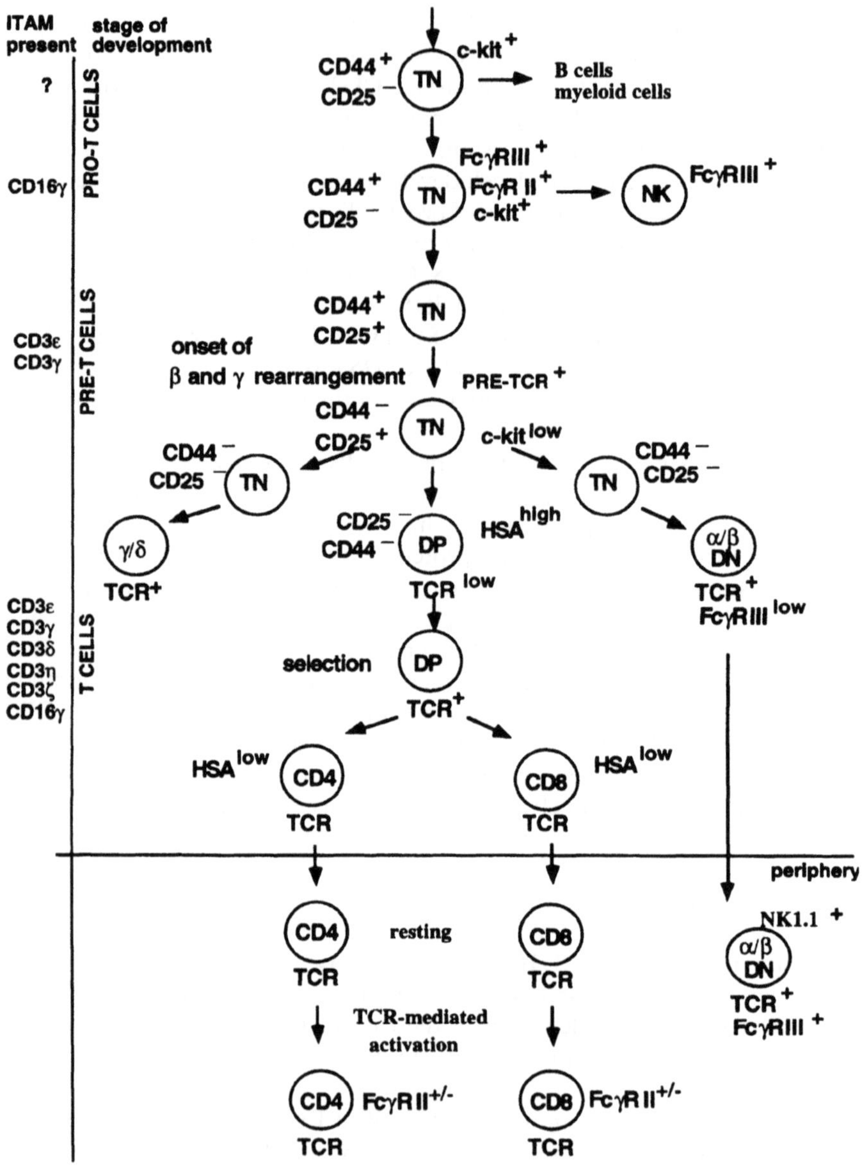

Figure 16.1 Expression of Fcγ receptors and ITAMs during T cell development

T cell development in the fetal liver and yolk sac

The fetal liver also hosts lymphoid precursors that are identified by the expression of the AA4.1 marker. A portion of these cells express CD16 [23]. However these cells in differentiation assays yield B cells and macrophages but not T cells [23]. The precursor cells that can differentiate into T cells are AA4.1+, FcγR–. That result may indicate that the expression of FcγR in T cell development is site specific, however it is possible that in the fetal thymus an earlier T cell precursor is present, that in the pathway of T cell differentiation will transiently express FcγR. AA4.1+ CD16+ cells are also present in the day 10 yolk sac (L. Lu, R. Auerbach, M Sandor manuscript in preparation) but their potential to differentiate into lymphocytes is not yet clarified.

Extrathymic T cell development in the gut

She *et al.* presented cell transfer studies which suggested that FcγR are expressed during the early T cell development in the intestinal epithelium. In the RAG or the fyn/lck double knock out mice T cell differentiation is blocked and an early T cell precursor with the CD3–, CD8α/α+, CD2–, CD25–, FcγR+ phenotype accumulates, representing 95% of the intraepithelial lymphocytes [24].

CD16 PROVIDES THE FIRST ITAM MOTIF IN T CELL DEVELOPMENT

The tyrosine kinase activating ITAM motifs play a crucial role in T cell activation and development. Mice deficient in TCR, pre-TCR [25] or in the associated signalling CD3ε [26] or CD3ζ [27] molecules have no, or few, T cells. TCR appears first on CD4+, CD8+ T cells, while pre-TCR is expressed first on CD4–, CD8–, CD25+ cells (Figure 16.1). It is likely that the composition of the CD3 complex and the ITAM-induced responses are different in mature and pre-T cells. The function of the pre-T cell receptor is thought to be the expansion of pre-T cells, while the TCR is required for both the expansion and the differentiation of T cells. Pro-T cells also express ITAM motifs that are associated with the γ chain of CD16. It is interesting that the appearance of the ITAM-associated pre-TCR CD16 is down-regulated similarly as in the pre-TCR with the expression of TCR. During different phases of T cell development different ITAM-associated molecules provide activation sites for tyrosine kinase and it seems that T-lymphocytes tend to express only one type of ITAM-associated molecular complex at a time. Fc receptors express either ITAM or ITAM-regulating ITIM motifs and that might be the reason, that unlike any other lymphoid-myeloid lineage, the majority of resting T cells do not express any Fc receptors. The presence of CD16 on mature T cell potentially could initiate responses and by-pass the requirement of antigen to initiate T cell responses. That point is highlighted by the existence of a T cell population that uses the CD16γ chain in its TCR-associated CD3 complex instead of CD3ζ. It is interesting that, similar to T cell development, the ckit+, CD19+ pro-B cells also express both CD16 and CD32 and that CD16 (but not CD32) is down-regulated at the time when pre-BCR (also an ITAM-associated molecule) appears on pre-B cells [28].

CD16 message can be detected in day 12 thymus, but it is present even earlier in murine development at sites of embryonic haematopoiesis (Sandor *et al.* manuscript in preparation). It is possible that CD16 and ITAM motifs play a role in embryogenesis prior to lymphopoiesis.

A POSSIBLE ROLE FOR FcγR IN T CELL DEVELOPMENT

The expression of both CD16 and CD32 on pro-T cells and their down-regulation on pre-T cells suggest that these receptors might have a role in early T cell development. In CD16 [18] or CD32 [30] gene-deleted mice the T cell development and function seems to be unaltered. T cells are also not changed in CD16γ KO mice, except for a lower TCR level on a population of intestinal epithelial T cells and the absence of vaginal epithelial γ/δ T cells [17,29,31]. These aberrant features are not FcγR but TCR related. CD16 preferentially activates the protein kinase syk. In RAG mice that were repopulated by syk-deficient haematopoeitic cells, B cell and epithelial γ/δ T cell development was blocked but most of the T cells (and the lineage that expresses CD16 at the pro-T cell stage) were similar to the wild type both in number and function [32,33]. These data indicate that FcγR has either no role in T cell development or if it has a role, it is redundant and the loss is compensated by other pathways.

The presence of the anti-FcγR antibody 2.4G2 influenced T cell differentiation in fetal thymic organ culture. The effect required very early thymocytes as it was detected in cultures initiated from day 13 [15] but not from day 14.5 [17] fetal thymus lobes. The effect was an enhancement and acceleration of α/β T cell differentiation, although the γ/δ lineage was unaffected. The percentage of single positive α/β T cells was higher and that phenotype appeared faster in the cultures. It was sufficient to have 2.4G2 present only during the first 3 days of the 12-day culture, another indication that an early T cell precursor is involved. A similar effect was detected in the presence of soluble recombinant Fcγ receptor [15]. At the same time when FcγR+ pro-T cells are dominant in the fetal thymus, a Thy–CD44– cell type was detected that bound the recombinant soluble FcγRII (that is almost identical with the extracellular portion of FcγRIII). The model that is suggested on the basis of these experiments, proposes that the FcγR of pro-T cells interacts with a non-immunoglobulin ligand on thymic stromal cells and that interaction slows down T cell development [34]. Selective lineage-specific breakpoints in differentiation have been shown in later phases of T cell development. Such pauses could provide time to expose the maturing cells to regulatory factors. Inhibition of this interaction or the deletion of the FcγR gene would result in a faster rate of development. As the interaction does not change the end phenotype of the differentiating T cells, no altered phenotype in adult FcγR KO animals, and an earlier maturation of T cells in fetal thymus is predicted. Conversely, overexpression of FcγR on differentiating T cells should slow down T cell development. A mouse has been recently generated [16] expressing a CD16γ transgene with a CD2 promoter. The phenotype of these transgenic mice show an overexpression of CD16 on T cells that is not down-regulated at the pro-T cell and pre-T cell transition and a slower development of T cells and NK cells. The older animals are similar to the wild type, but the CD4,CD8 staining pattern of the thymus of the young animals is similar to the one that is seen in the late fetal thymus. It is not proven yet that the CD16 overexpression and the delayed development are in a causative relationship and it is possible that this effect is related to the incorporation of CD16γ chain into the TCR complex. However in this case one expects similar phenotypes in CD16γ transgenic CD3ζ KO mice [35] or in mice expressing the single ITAM containing ζ transgene with the same promoter [16], and that is not the case.

It is clear that more experiments are required to prove and understand the role of FcγR in T cell development. A similar scheme, in which blocking FcγR can speed up differentiation without changing the phenotype of the matured cells is also observed during murine B cell development (de Andres and Lynch manuscript in preparation) and with *in vitro* development of human bone marrow monocytes [35]. There are clinical conditions when acceleration of lymphoid and myeloid cells repopulation could be of value. Further studies might clarify whether manipulating FcγR in such conditions can change the speed of reappearance of these cells.

FcγR ON MATURE T CELLS

At the time when T cell purification protocols were less developed and anti-FcγR antibodies were not yet available the reported frequency of T cells in murine spleen and human blood was as high as 15 and 30% respectively [36]. In the pre-flow cytometry era EA-rosetting, a measure of FcγR expression, was used as a marker of suppressor T cells in a large number of studies analysing various clinical conditions [37]. Using antibodies to T cell markers and FcγRs this number is under 5%, however it can be much higher in the blood of some human donors. Most resting T cells are FcγR negative and that is different from all other lymphoid and myeloid lineages. As we discussed before, the absence of FcγR receptor expression on T cells might be a result of down-regulation to avoid interference with TCR signalling [38]. However some T cell lineages do express FcγR constitutively. A subpopulation of NK cells, the so-called TNK lineage expressing NK1.1, CD3+, CD2+, and α/β TCR+ are also CD16+ and can kill antibody-coated target cells [39]. A subpopulation of these TNK cells, that are present both in humans and mice, displays CD4 and has a restricted TCR repertoire. Recently a lot of attention has been focused on these CD1 restricted cells as they might be the first source of regulatory IL-4 in the early immune response.

Most γ/δ TCR+ cells in human (but not in murine) blood also display CD16 that is functional in antibody-dependent cell killing [40]. Murine intestinal intraepithelial CD8α/α+ T cells (both the α/β and the γ/δ lineage) express low levels of CD16, especially in young animals [24]. Similarly a low constitutive expression of CD16 is described on murine epithelial dendritic T cells located in the skin [41,42]. The above mentioned T cells are interesting exceptions as they express two different ITAM-containing receptor complexes at the same time. The signalling of the two ITAM-associated receptors (CD16 and TCR) has been compared on transfectants [43], but studies on these rare cells, where the downstream elements of signalling pathways are differentiated to use different ITAM motifs, may yield a better understanding of differences in signalling by CD16 and TCR-associated ITAM motifs. It is also interesting to note that most of the mature T cells expressing both TCR and CD16 display TCRs that are either restricted in V gene usage or have a non-variant TCR. These T cells can mediate antibody-mediated effector functions and may be more similar to NK cells and macrophages than to classical T cells. CD32 can also be present on adult T cells after activation (reviewed later).

A SUBPOPULATION OF T CELLS USE CD16γ CHAIN IN THEIR TCR

A close relationship between TCR and FcγR is further demonstrated by the finding that in a population of T cells CD16γ replaces CD3ζ in the CD3 complex. In CD16γ KO animals there are no intraepithelial γ/δ T cells in the vagina [31], indicating that this lineage of T cells uses γ chains exclusively in their CD3 complex. Analysis of the composition of the CD3 complex of wild type and CD3 ζ chain KO animals by immunoprecipitation clearly shows that a population of intestinal epithelial T cells uses both CD16γ and CD3ζ in their TCR [44,45]. Splenic transgenic T cells in G8 γ/δ TCR transgenic mice contain CD16γ in their TCR [46]. It is reported that the same T cell CD3ζ can be replaced by CD16γ in TCR tumour infiltrating CD4 and CD8 T cells [47].

There is some controversy about the functional consequences of the presence of CD16γ in the CD3 complex. Studies with cells transfected with chimeric molecules containing the cytoplasmic region of γ or ζ chain showed a similarity of signalling by the two chains [48] and Jurkat T cells transfected with CD16 also exhibited a similar signalling by the two receptors [43]. The response of murine intestinal intraepithelial T cells to TCR-induced signals is limited [49] and it was suggested that this is because of the presence of CD16γ in the CD3 complex. When G8 γ/δ TCR transgenic mice were bred to a CD3ζ background all G8 cells in these mice use CD16γ in their TCR. These cells were not reactive with their antigen or Concanavalin A, but gave a proliferative response to anti-CD3ε or anti-γ/δ TCR antibodies [50]. There are also data indicating that the CD16γ TCR expressing tumour infiltrating lymphocytes are hyporesponsive [47]. On the other side of the coin, human intraepithelial T lymphocytes are also hyporesponsive, even though they do not have CD16γ in their TCR [44], suggesting that other factors might be involved in the altered T cell response of these cells. Furthermore, recently a CD16 γ chain transgene was introduced into CD3 ζ chain KO animals, and both normal T cell development and TCR-induced T cell responses were reported in these animals [35]. At this time a simple conclusion on the role of CD16γ in the TCR complex is not available, but it seems that quantitative factors and T cell differentiation stage can alter the consequence of replacement of CD16γ with CD3 ζ in the CD3 complex. The study of the signalling properties of CD16+ T cells might help to solve this problem.

FcγR ON T CELLS IS UP-REGULATED BY ACTIVATION THROUGH TCR RECEPTORS

TCR and FcγR are share structural and functional similarities. Another level of relation between the family of these two receptors is that TCR-induced signals can induce the expression of FcγR on a subpopulation of T cells [42]. As early as 30 years ago it was shown that alloactivation or Concanavalin A activation brings about the appearance of FcγRs [51–54] that, in most cases, are CD32. In fact murine FcγR (a form of CD32) was cloned first from a T cell lymphoma S49 [55], and the human CD32 was cloned first from a PHA-induced T cell library [56]. There are some interesting features of the up-regulation of FcγR. Only a subpopulation of T cells is induced to express CD32. Even in the case of TCR transgenic animals that express monoclonal TCR only a fraction of T cells will be FcγR+ upon *in vitro* and *in vivo* activation [54] (Sandor, unpublished), indicating that factors other than TCR activation are required for the up-regulation. The expression is down-regulated in resting T cells and can be repeatedly up-regulated by restimulation of the T cells. TCR-mediated signals during thymic development do not induce FcγR expression [13,15]. The T cell activation up-regulation of FcγR can be further increased by the presence of IgG. The ligand-induced increase can be detected both *in vitro* [57,58] and *in vivo* [59]. There is a similar pattern of FcγR on T and B cells. Both pro-T cells and pro-B cells express CD16 that is down-regulated with the appearance of cognate antigen receptor. The mature cells express CD32, but expression is constitutive on all B cells, while CD32 appears only transiently on a subpopulation of T cells. The rare exceptions are some CD16+ T cells: activation of dendritic epithelial T cells increases the level of CD16 on these cells while activation of human peripheral γ/δ cell brings about the loss of CD16 on these cells. The expression of FcγR on T cells after TCR activation may be important as this is the time when T cells achieve functions such as helping, killing and regulation. However it is also possible that the increased FcγR expression is a reflection of less need to not express FcγRs that can interfere with the TCR signal process after the TCR signal has been delivered.

T CELL FcγR AND DISEASE

Twenty years ago CD4+ helper T cells were detected by their binding of IgM-sensitized red blood cells (Tμ cells). FcγR+ T cells (Tγ-cells) identified by rosetting both with sheep red blood cells and IgG-coated ox red blood cells, were thought to be CD8+ suppressor cells. For 10 years a lot of work focused on the frequency and characteristics of Tγ-cells in various clinical conditions [37,60]. Since then, detection methods both for T cells and FcγR have changed, and the concept of suppression has evolved. While some of the principles of the early investigations are not accepted today, it is still true that FcγR expression is more frequent on post-activated CD8+ T cells than on CD4+ T cells and that there are conditions in which the frequency of FcγR+ T cells significantly increases. In myeloma patients and in plasmacytoma-bearing mice a high concentration of monoclonal immunoglobulin appears in the serum: this is accompanied by an increase of the proportion of T cells expressing Fc receptors specific for the isotype of the restrictive monoclonal protein. These T cells were shown to down-regulate the immunoglobulin production of myeloma or plasmacytoma cells [61]. One of the mechanisms proposed that the down-regulation is mediated by T cell-derived TGF-β [62]. Activated T cells produce soluble FcγR, and the purified FcγR or immunoglobulin binding factor (IBF) [63] has been reported to inhibit the production of IgG by both myeloma cells [64] and normal B cells [65]. Recently, it has been reported that this suppressor activity may come from TGF-β that co-purifies with the IBF on IgG columns [63]. In a form of cyclic neutropenia that is sometimes called Tγ disease, the cyclic increase in the proportion of CD8+, FcγR+ cells correlates with the cyclic decrease in neutrophils [66]. The significance of FcγR in the disease is not understood. An evaluation of FcγR+ T cells in the blood can reflect a transient increase of TNK cells which is known to accompany a variety of clinical conditions. Some T cell lymphomas can also express FcγR, resulting in a high frequency of FcγR+ T cells in the blood.

FUNCTIONS OF FcγR ON T CELLS

As mentioned above, some intraepithelial T cells and a subset of TNK cells constitutively express CD16. It is well documented that CD16 can induce the effector functions of these cells, such as antibody-mediated killing and lymphokine production. In the presence of antibodies these T cells can be important in the early responses at epithelial sites. Early refers to the FcγR-mediated T cell response that precedes the cognate T cell response, and it is also early in ontogeny. In mice dendritic epithelial T cells are the first T cells to appear in the embryo and they are present in the skin of the newborn. The FcγR expressed might use maternal antibodies in the protection of newborns.

The role of CD32 on a subpopulation of activated T cells is less understood. In principle these receptors give a potential interface for immune complexes to interfere with activated T cells. On B cells, crosslinking BCR and CD32 brings about a down-regulation of the B cell activation by recruiting a phosphatase to the proximity of an activating tyrosine-phosphates motif that is important for BCR-induced activation. CD32 crosslinked with the TCR by antibodies results in a similar inhibition [66]. However, when activated FcγR+ T cells interact with class I- or class II-expressing cells coated with antibodies, a similar crosslinking of TCR and FcγR would require a close proximity of the antibodies to the antigen-presenting histocompatibility antigens. KIR molecules express an ITAM motif similar to CD32 and might be better suited for inhibitory effects because they react with the antigen-presenting class I molecule directly. Nevertheless, antibodies reacting directly with TCR (like anti-TCR idiotype) might down-regulate T cell activation by interaction with FcγR. A careful study of this scenario is merited by the ongoing clinical trials where anti-TCR antibodies are used to promote or block T cell responses.

Activated FcγR+ T cells [68,69] have been shown to be a source of soluble FcγR. Activated T cells tend to accumulate at a site of an ongoing immune response, so they can dispense soluble FcγR at strategically important sites. That might be of significance as soluble FcγR has been shown to block the formation of large immune complexes [70], inhibit local inflammation [71] and influence antigen presentation by interacting with MAC1 antigen on dendritic cells [72].

SUMMARY

Cells that belong to the lymphoid-myeloid lineages constitutively express one or more type of FcγR. Mature T cells are the only exception to this principle and a large majority of resting T cells do not express any kind of FcγR. The absence of or an improper transition of resting T cells to the activated T cell stage can have life threatening consequences for the host and accordingly this transition is tightly regulated. FcγRs display various cytoplasmic motifs that have been shown to interfere with T cell activation and it is possible that this accounts for why most resting T cells do not express FcγR.

FcγRs are T cell differentiation antigens. CD16 and CD32 appear on late pro-T cells and are down-regulated on pre-T cells. ITAM motifs are crucial for the differentiation and activation of T cells and CD16 on pro-T cells provides the first ITAM motif in T cell development. There are some data indicating that CD16 is not only present but is functional on pro-T cells, and ongoing investigations may provide a more decisive answer to the question as to whether FcγRs have a role in early T cell development.

FcγR is also an activation antigen on a subpopulation of T cells. Activated T cell are the ones that help, kill, suppress or regulate and the presence of FcγR at this stage may represent an interface by which immune complexes modulate T cell function. Additionally, activated T cells tend to accumulate at immunologically important sites and can secrete soluble FcγR at these locations. It is an interesting new direction of FcγR research that soluble FcγR might serve as a regulator of inflammation.

A small population of T cells, most of which belong to the TNK or the intraepithelial T cell lineages, constitutively expresses CD16, and that expression is further up-regulated by various activating signals. Antibodies binding to these cells can induce T cell functions such as killing and lymphokine secretion. By this mechanism these cells, unlike most T cells, can participate in the very early innate response to various challenages.

We started this review with the statement that T cells are the master regulators of the immune response. This review has communicated that while knowledge about the expression pattern of FcγRs on T cells is substantial, the function of these receptors on T cells is still not well understood. Further studies that clarify the function of FcγR on T cells might unmask some new principles of immunoregulation.

References

1. Mosmann TR, Coffman RL. TH1 and TH2 cells: different patterns of lymphokine secretion lead to different functional properties. Ann Rev Immunol. 1989;7:145–73.
2. Paraskevas F, Lee ST, Israels LG. Cell surface associated globulins in lymphocytes. I. Reverse immune cytoadherence: a technique for their detection in mouse and human lymphocytes. J Immunol. 1971;106:160–70.
3. Basten A, Miller JF, Warner NL, Abraham R, Chia E, Gamble J. A subpopulation of T cells bearing Fc receptors. J Immunol. 1975;115:1159–65.
4. Hudson L, Sprent J, Miller JF, Playfair JH. B cell-derived immunoglobulin on activated mouse T lymphocytes. Nature. 1974;251:60–2.
5. Klein M, Neauport-Sautes C, Ellerson JR, Fridman WH. Binding site of human IgG subclasses and their domains for Fc receptors of activated murine T cells. J Immunol. 1977; 119:1077–83.
6. Fridman WH, Neauport-Sautes C, Daëron M et al. Induction of Fc receptors and immunoglobulin-binding factors in T-cell clones. Mol Immunol. 1984;21:1243–51.
7. Lynch RG, Mathur A, Williams KR, Mueller A. Isotype-specific interactions between regulatory T cells and secreted and membrane-bound monoclonal immunoglobulins. Curr Top Microbiol Immunol. 1985;122:200–4.
8. Fridman WH, Daëron M, Amigorena S, Rabourdin-Combe C, Neauport-Sautes C. Bases for an isotypic network. Mol Immunol. 1986;23:1141–8.
9. Finkelman FD, Holmes J, Katona IM et al. Lymphokine control of in vivo immunoglobulin isotype selection. Annu Rev Immunol. 1990;8:303–33.
10. Cambier JC. Antigen and Fc receptor signaling. The awesome power of the immunoreceptor tyrosine-based activation motif (ITAM). J Immunol. 1995;155:3281–5.
11. Hattori N, Kawamoto H, Katsura Y. Isolation of the most immature population of murine fetal thymocytes that includes progenitors capable of generating T, B, and myeloid cells. J Exp Med. 1996;184:1901–8.
12. Rodewald HR, Kretzschmar K, Takeda S, Hohl C, Dessing M. Identification of pro-thymocytes in murine fetal blood: T lineage commitment can precede thymus colonization. EMBO J. 1994;13:4229–40.
13. Rodewald HR, Awad K, Moingeon P et al. Fc gamma RII/III and CD2 expression mark distinct subpopulations of immature CD4-CD9- murine thymocytes: in vivo developmental kinetics and T cell receptor beta chain rearrangement status. J Exp Med. 1993;177:1079–82.
14. Rodewald HR, Moingeon P, Lucich JL, Dosiou C, Lopez P, Reinherz EL. A population of early fetal thymocytes expressing Fc gamma RII/III contains precursors of T lymphocytes and natural killer cells. Cell. 1992;69:139–50.
15. Sandor M, Galon J, Takacs L et al. An alternative Fc gamma-receptor ligand: potential role in T-cell development. Proc Natl Acad Sci USA. 1994;91:12857–61.
16. Flamand V, Shores EW, Tran T et al. Delayed maturation of CD4(–)CD8(–) Fc gamma-RII/III+ T and natural killer cell precursors in Fc-epsilon-RI-gamma transgenic mice. J Exp Med. 1996;184:1725–35.
17. Heiken H, Schulz RJ, Ravetch JV, Reinherz EL, Koyasu S. T lymphocyte development in the absence of Fc epsilon receptor I gamma subunit: analysis of thymic-dependent and independent alpha beta and gamma delta pathways. Eur J Immunol. 1996;26:1935–43.
18. Hazenbos WLW, Gessner JE, Hofhuis FMA et al. Impaired IgG-dependent anaphylaxis and arthus reaction in Fc-gamma-RIII (CD16) deficient mice. Immunity. 1996;5:181–8.

19. Leclercq G, Plum J. Developmentally ordered appearance of a functional Fc gamma receptor on TCR V gamma 3 thymocytes. Evidence for stimulation-induced expression. J Immunol. 1994;153:2429–35.
20. Gilhus NE, Matre R. Development and distribution of HLA-DR antigens and Fcγ receptors in the human thymus. Acta Pathol Microbiol Immunol Scand. 1983;91:35–42.
21. Mingari MC, Poggi A, Biassoni R et al. In vitro proliferation and cloning of CD3-CD16+ cells from human thymocyte precursors. J Exp Med. 1991;174:21–6.
22. Spits H, Lanier LL, Phillips JH. Development of human T and natural killer cells. Blood. 1995;85:2654–70.
23. Carlsson L, Candeias S, Staerz U, Keller G. Expression of Fc gamma RIII defines distinct subpopulations of fetal liver B cell and myeloid precursors. Eur J Immunol. 1995;25:2308–17.
24. Page ST, Van Oers NSC, Perlmutter RM, Weiss A, Pullen AM. Differential contribution of Lck and Fyn protein tyrosine kinases to intraepithelial lymphocyte development. Eur J Immunol. 1997;27:554–62.
25. Malissen B, Malissen M. Functions of TCR and pre-TCR subunits – lessons from gene ablation. Curr Opin Immunol. 1996;8:383–93.
26. Malissen M, Gillet A,j Ardouin L et al. Altered T cell development in mice with a targeted mutation of the CD3-epsilon gene. EMBO J. 1995;14:4641–53.
27. Love PE, Shores EW, Johnson MD et al. T cell development in mice that lack the zeta chain of the T cell antigen receptor complex. Science. 1993;261:918–21.
28. Sandor M, Michael H, Andres BD, Lynch RD. Developmentally regulated Fcγ receptor expression in lymphopoiesis FcγR III (CD16) provides an ITAM motif for pro-T and pro-B cells. Immunol Lett. 1996;54:123–7.
29. Takai T, Li M, Sylvestre D, Clynes R, Ravetch JV. FcR gamma chain deletion results in pleiotrophic effector cell defects. Cell. 1994;76:519–29.
30. Takai T, Ono M, Hikida M, Ohmori H, Ravetch JV. Augmented humoral and anaphylactic responses in Fc gamma RII-deficient mice. Nature. 1996;379:346–9.
31. Park SY, Arase H, Wakizaka K et al. Differential contribution of the FcR gamma chain to the surface expression of the T cell receptor among T cells localized in epithelia: analysis of FcR gamma-deficient mice. Eur J Immunol. 1995;25:2107–10.
32. Cheng AM, Rowley B, Pao W, Hayday A, Bolen JB, Pawson T. Syk tyrosine kinase required for mouse viability and B-cell development. Nature. 1995;378:303–6.
33. Mallick-Wood CA, Pao W, Cheng AM et al. Disruption of epithelial gamma delta T cell repertoires by mutation of the Syk tyrosine kinase. Proc Natl Acad Sci USA. 1996;93:9704–9.
34. Lynch RG, Hagen M, Mueller A, Sandor M. Potential role of FcγR in early development of murine lymphoid cells: evidence for functional interaction between FcγR on pre-thymocytes and an alternative, non-Ig ligand on thymic stromal cells. Immunol Lett. 1995;44:105–9.
35. Liu CP, Lin WJ, Huang M, Kappler JW, Marrack P. Development and function of T cells in T cell antigen receptor/CD3 ζ knockout mice reconstituted with FcεRIγ. Proc Natl Acad Sci USA. 1997;94:616–21.
36. Lynch RG, Sandor M. Fc receptors on T and B lymphocytes. In: Metzger H, ed., Fc Receptors and the Action of Antibodies. Washington: American Society of Microbiology; 1990:305–34.
37. Gupta S, Good RA. Human T cell subpopulations as defined by Fc receptors. Thymus. 1979; 1:135–49.
38. Sandor M, Lynch RG. Lymphocyte Fc receptors: the special case of T cells. Immunol Today. 1993;14:227–31.
39. Koyasu S, D'Adamio L, Arulanandam AR, Abraham S, Clayton LK, Reinherz EL. T cell receptor complexes containing Fc epsilon RI gamma homodimers in lieu of CD3 ζ and CD3 eta components: a novel isoform expressed on large granular lymphocytes. J Exp Med. 1992; 175:203–9.
40. Braakman E, van de Winkel JG, van Krimpen BA, Jansze M, Bolhuis RL. CD16 on human gamma delta T lymphocytes: expression, function, and specificity for mouse IgG isotypes. Cell Immunol. 1992;143:97–107.
41. Kuziel WA, Lewis J, Nixon-Fulton JETR, Trucker PW. Murine epidermal γ/δ T cells express FcγR alpha gene. Eur J Immunol. 1991;21:1563.
42. Sandor M, Cook GA, Sacco RE et al. TCR induced expression of Fc receptors on murine T cell subsets in vitro and in vivo. Immunology. 1992;185:268–80.
43. Wirthmueller U, Kurosaki T, Murakami MS, Ravetch JV. Signal transduction by Fc gamma RIII (CD16) is mediated through the gamma chain. J Exp Med. 1992;175:1381–90.

44. Balk SP, Polischuk JE, Probert C et al. Composition of TCR-CD3 complex in human intestinal intraepithelial lymphocytes: lack of Fc epsilon RI gamma chain. Int Immunol. 1995;7:1237–41.
45. Liu CP, Ueda R, She J et al. Abnormal T cell development in CD3ζ mutant mice and identification of a novel T cell population in the intestine. EMBO J. 1993;12:4863.
46. Qian D, Sperling AI, Lancki DW et al. The γ chain of the high-affinity receptor for IgE is a major functional subunit of T-cell antigen receptor complex in γδ T lymphocytes. Proc Natl Acad Sci USA. 1993;90:11875.
47. Mizoguchi H, O'Shea JJ, Longo DL, Loeffler CM, McVicar DW, Ochoa AC. Alterations in signal transduction molecules in T lymphocytes from tumor-bearing mice. Science. 1992;258: 1795–8.
48. Romeo C, Seed B. Cellular immunity to HIV activated by CD4 fused to T cell or Fc receptor polypeptides. Cell. 1991;64:1037–46.
49. Malissen M, Gillet A, Rocha B, Trucy J, Vivier E, Boyer C. T cell development in mice lacking the CD3-ζ/η gene. EMBO J. 1993;12:4347–55.
50. Khattri R, Sperling AI, Qian D et al. TCR-gamma delta cells in CD3 zeta-deficient mice contain Fc epsilon RI gamma in the receptor complex but are specifically unresponsive to antigen. J Immunol. 1996;157:2320–7.
51. Fridman WH, Nelson RA Jr, Liabeuf A. Production of an immunoglobulin-binding factor (IBF) by antigen-stimulated lymph node lymphocytes. J Immunol. 1974;113:1008–16.
52. Sprent J, Hudson L. Surface immunoglobulin on H-2-activated T lymphocytes. Transplant Proc. 1973;5:1731–3.
53. Krammer PH, Hudson L, Sprent J. Fc-receptors, Ia-antigens, and immunoglobulin on normal and activated mouse T lymphocytes. J Exp Med. 1975;142:1403–15.
54. Sandor M, Houlden B, Bluestone J, Hedrick SM, Weinstodk J, Lynch RG. *In vitro* and *in vivo* activation of murine γ/δ T cells induces the expression of IgA, IgM, and IgG Fc receptors. J Immunol. 1992;148:2363–9.
55. Ravetch JV, Luster AD, Weinshank R et al. Structural heterogeneity and functional domains of murine immunoglobulins G Fc receptors. Science. 1986;234:718–25.
56. Hibbs ML, Bonadonna L, Scott BM, McKenzie IF, Hogarth PM. Molecular cloning of human immunoglobulin G Fc receptors. Proc Natl Acad Sci USA. 1988;85:2240–4.
57. Daëron M, Neauport-Sautes C, Yodoi J, Moncuit J, Fridman WH. Receptors for immunoglobulin isotypes (FcR) on murine T cells. II. Multiple FcR induction on hybridoma T cell clones. Eur J Immunol. 1985;15:668–74.
58. Daëron M, Bonnerot C, Sandor M et al. Molecular mechanisms regulating the expression of murine T-cell Fc gamma receptor II. Mol Immunol. 1988;25:1143–50.
59. Mathur A, Lynch RG. Increased T gamma and T mu cells in BALB/c mice with IgG and IgM plasmacytomas and hybridomas. J Immunol. 1986;136:521–5.
60. Moretta L, Mingari MC, Moretta A. Human T cell subpopulations in normal and pathologic conditions. Immunol Rev. 1979;45:163–93.
61. Lynch RG. Immunoglobulin-specific suppressor T cells. Adv Immunol. 1987;40:135–51.
62. Berg DJ, Lynch RG. Immune dysfunction in mice with plasmacytomas. I. Evidence that transforming growth factor-beta contributes to the altered expression of activation receptors on host B lymphocytes. J Immunol. 1991;146:2865–72.
63. Bouchard C, Galinha A, Tartour E, Fridman WH, Sautes C. A transforming growth factor beta-like immunosuppressive factor in immunoglobulin G-binding factor. J Exp Med. 1995; 182:1717–26.
64. Hoover RG, Lary C, Page R et al. Autoregulatory circuits in myeloma. Tumor cell cytotoxicity mediated by soluble CD16. J Clin Invest. 1995;95:241–7.
65. Gisler RH, Fridman WH. Suppression of in vitro antibody synthesis by immunoglobulin-binding factor. J Exp Med. 1975;142:507–17.
66. Bom-van Noorloos AA, Pegels HG, van Oers RH et al. Proliferation of T gamma cells with killer-cell activity in two patients with neutropenia and recurrent infections. N Engl J Med. 1980;302:933–7.
67. Daëron M, Latour S, Malbec O et al. The same tyrosine-based inhibition motif, in the intracytoplasmic domain of Fc gamma RIIB, regulates negatively BCR-, TCR-, and FcR-dependent cell activation. Immunity. 1995;3:635–46.
68. Fridman WH, Rabourdin-Combe C, Neauport-Sautes C, Gisler RH. Characterization and function of T cell Fc gamma receptor. Immunol Rev. 1981;56:51–88.

69. Daëron M, Neauport-Sautes C, Blank U, Fridman WH. 2.4G2, a monoclonal antibody to macrophage Fc gamma receptors, reacts with murine T cell Fc gamma receptors and IgG-binding factors. Eur J Immunol. 1986;16:1545–50.
70. Gavin AL, Wines BD, Powell MS, Hogarth PM. Recombinant soluble Fc gamma RII inhibits immune complex precipitation. Clin Exp Immunol. 1995;102:620–5.
71. Ierino FL, Powell MS, Mckenzie IF, Hogarth PM. Recombinant soluble human Fc gamma RII: production, characterization, and inhibition of the arthus reaction. J Exp Med. 1993; 178:1617–28.
72. Galon J, Gauchat JF, Mazieres N et al. Soluble Fc gamma receptor type III (FcgammaRIII, CD16) triggers cell activation through interaction with complement receptor. J Immunol. 1996;157:1184–92.

17
Role of IgG and IgE FcR in antigen presentation

P. M. GUYRE and J. K. O'SHEA

INTRODUCTION

The first critical step in the chain of events leading to activation of T and B lymphocytes by an exogenous antigen, whether a vaccine molecule or a pathogenic organism, is efficient uptake of the antigen by professional antigen presenting cells (APCs). Antigens are concentrated by APCs both by high-rate fluid phase pinocytosis referred to as macropinocytosis, and by receptor-mediated uptake via, for example, mannose receptors on dendritic cells or surface immunoglobulin on B cells [reviewed in Reference 1]. The mechanism of uptake determines the intracellular compartment in which the antigen is processed, and this in turn influences the repertoire of peptides produced and the types of T cells stimulated. Pioneering studies by Cohen *et al.* in 1972 suggested that Fc receptors can play a critical role in the uptake of antigen by APCs. These authors demonstrated that the dose of antigen required for T cell activation could be reduced over 100-fold in the presence of antigen-specific IgG antibodies [2]. As detailed elsewhere in this volume, Fc receptors comprise an extensive and well characterized family of molecules that specifically bind IgG, IgA or IgE. Since individual Fc receptors are selectively expressed on some APCs but not on others, and exhibit differential binding of immunoglobulins (reviewed in Chapter 2), the presence, amount, class and subclass of antigen-specific antibodies can play a critical role in determining which APCs participate most efficiently in lymphocyte activation. This chapter provides an overview of the capacity of particular IgG and IgE Fc receptors to mediate antibody-enhanced antigen presentation, and lays the foundation for the possible use of Fc receptor-targeted antigens to enhance the potency of vaccines.

J.G.J. van de Winkel and P.M. Hogarth (eds.), The Immunoglobulin Receptors and their Physiological and Pathological Roles in Immunity. 185–194.

INDUCTION OF AN IMMUNE RESPONSE

The development of high titre and long-lasting protective immunity to an infectious agent requires the activation of T lymphocytes specific for the organism. T cells directly kill infected cells, activate myeloid cells to ingest and destroy pathogens, help B cells to produce antibodies, and control inflammation. The activation of T cells to proliferate and carry out the foregoing functions requires the T cell to interact via its T cell receptor (TCR) with antigen-derived peptides in association with major histocompatibility (MHC) molecules on the surface of an APC. The most effective APCs appear to be dendritic cells, macrophages and B lymphocytes, all of which express high numbers of both class I and class II MHC molecules, and can efficiently endocytose antigens and degrade them into their component peptides. Although our understanding is incomplete, it is becoming clear that different APCs can stimulate qualitatively different responses, probably as a consequence of the particular cytokines that they release [3]. In some cases cellular immunity predominates, while in others antibody production is most prevalent. Moreover, the dominant isotype of antibody produced is strongly influenced by cytokines. Thus, targeting of antigen to a particular APC has the potential to significantly influence the type of response to that antigen.

Targeting of an antigen may also determine whether T cell activation or T cell anergy is elicited. For example, an optimal antibody response to a thymus-dependent antigen normally requires that the B cell obtain help from a CD4+ helper T cell. However, in order to provide help, the CD4+ T cell must first be activated by an APC. The B cell should be a particularly effective APC because it contains antigen-specific antibody on its surface which allows it to bind, internalize and process antigen for very efficient presentation to the helper T cell. However, antigen presentation by resting B cells to resting T cells may lead not to T cell activation, but rather to T cell tolerance [reviewed in Reference 4], apparently because the resting B cell lacks key co-stimulatory molecules required for activation of the resting T cell. This implies that in the naive individual, the resting T cell must first interact with antigen presented by a macrophage or dendritic cell before interacting with the resting B cell, and suggests that primary responses could be enhanced by targeting antigen to specifically interact with non-B cell APCs. Indeed, the examples cited in this chapter show the potential for antigen targeting to selectively modulate the immune response. However, many studies have been done only in cell culture systems, and none have yet compared the influence of targeting to different cell types in a systematic way. Such studies are needed before rational design of targeted vaccines can become a reality.

TARGETING TO FcR ON SPECIFIC ANTIGEN-PRESENTING CELLS

A variety of leukocyte surface molecules have been examined as targets for antigen delivery, including MHC Class II, surface immunoglobulin, and receptors for GM-CSF, complement, α2 macroglobulin, mannose, IgG, and IgE [4]. To be effective, the targeted molecule should be selectively expressed on professional APCs and should mediate efficient antigen uptake into intracellular compartments that lead to peptide presentation by MHC class I and/or class II molecules. Following is a brief overview of studies in which antigen was targeted to IgG or IgE Fc receptors (summarized in Table 17.1), resulting in efficient T cell activation.

Mononuclear phagocytes

The enhanced T cell activation reported by Cohen *et al.* [2] was traced to antigen-specific cytophilic IgG, which, by binding to IgG Fc receptors (FcγR) facilitated the uptake of antigen for processing and presentation to T cells. Later studies by Chang and co-workers [5] and by Manca *et al.* [6] demonstrated that T cell activation usually required several hundred-fold less antigen when antigen-specific antibody was present. Heyman *et al.* showed that *in vivo*, antigen-specific IgG could either enhance or suppress the antibody response to an immunogen [7].

As noted earlier in this volume, monocytes and macrophages express three major classes of FcγR (FcγRI, FcγRII and FcγRIII). In studies to evaluate the ability of the individual FcγR to induce antigen-specific T cell activation, Gosselin *et al.* showed that targeting tetanus toxoid (TT) to both FcγRI and

Table 17.1 Antigen targeting to Fc receptors

Construct	*Ag targeted*	*Cell targeted*	*Fc receptor targeted*	*In vitro*	*In vivo*	*Ref.*
mAb	TT	Monocytes	FcγRI, FcγRII, FcγRIII	√		8
	TT peptide	Monocytes	FcγRI	√		29
	TT	B cells	FcγRII	√		16
	OVA	B cells	FcεRII	√	√	20
	TT peptide	Dendritic cells	FcγRI	√		23
	Human IgG	Monocytes/DCs	FcγRI		√	30
Anti-hapten	F(ab)2 IgG	B cells	FcεRII	√		18
Ab +	TT	B cells	FcεRII	√		19
Hapten-Ag	BSA	B cells	FcεRII		√	32
	rBet vI	ME-PBMC	FcεRI	√		12
	rBet vI	Dendritic cells	FcεRI	√		26
Antigenized antibody	Peptide from *falciparum*, HIV, influenza	Various APCs	FcγR (at least partial)	√	√	33–37

Abbreviations: Ag, antigen; TT, tetanus toxoid; OVA, ovalbumin; BSA, bovine serum albumin; rBet vI, recombinant birch pollen allergen; ME-PBMC, monocyte-enriched peripheral blood mononuclear cells

FcγRII on human monocytes resulted in enhanced activation of cultured T cells [8]. Whereas non-targeted antigen required 3 μg/ml to achieve half-maximal T cell proliferation, targeting to either the type I or type II FcγR reduced the half-maximal antigen dose to less than 0.01 μg/ml. Thus, targeting to specific FcγR on monocytes can dramatically decrease the amount of antigen required to achieve a significant T cell response.

In addition to FcγR, studies have shown that peripheral blood monocytes of atopic individuals express functional FcεRI, the high affinity receptor for IgE [9]. More recent studies have also shown expression of FcεRI in monocytes from non-atopic individuals, and suggest that the previously observed difference in expression of monocyte FcεRI between atopic and non-atopic individuals may have been due to the lower levels of serum IgE in non-atopic individuals [10,11].

Maurer *et al.* have shown that presentation of allergen to T cells by monocytes from atopic donors is enhanced 100–1000-fold by targeting to FcεRI via anti-allergen IgE, suggesting that FcεRI-mediated allergen presentation by monocytes may have a role in induction and/or maintenance of allergy in atopic individuals [12]. In a recent review, however, Mudde *et al.* noted that freshly isolated monocytes from non-atopic donors are not capable of IgE-mediated antigen presentation. They theorized that monocyte IgE-mediated antigen uptake may lead to processing and presentation identical to that associated with IgG-enhanced antigen presentation, in addition to removing IgE from circulation [11]. Other studies have shown that IgE-bound antigen may be directed to B cell FcεRII, and away from monocyte FcεRI in non-atopic individuals [13]. These data, coupled with studies that show B cells tend to foster a Th2 response while monocytes direct a Th1 response [14], suggest that the role of monocyte FcεRI in allergy may be minimal, and that further studies are required to completely elucidate its role.

B cells

Studies by Snider and Segal indicated that targeting antigen to FcγRII on murine B cells did not result in enhanced antigen presentation, whereas targeting to macrophage FcγRII did [15]. However, Liu *et al.* showed that when tetanus toxoid (TT) was targeted to human B cell FcγRII, activation of TT-specific Th cells was enhanced by up to 1000-fold [16]. These differences are probably explained by the differential expression of FcγRII isoforms on macrophages, human B cells and murine B cells. For example, FcγRIIb1, which is present on both murine and human B cells is able to bind immune complexes but does not trigger endocytosis. FcγRIIb2 which is expressed on human but not mouse B cells, and FcγRIIa which is expressed on monocytes and macrophages, both mediate endocytosis [17] and are likely to lead to enhanced antigen presentation. A definitive analysis of the association between FcγRII isoforms and the capacity for antigen presentation is needed.

The low affinity receptor for IgE (FcεRII or CD23) is another surface molecule which has been studied for mediating enhanced antigen presentation. This receptor is primarily expressed on B cells and follicular dendritic cells in mice and the corresponding human FcεRIIa is found primarily on human B cells. The finding that FcεRII on B cells can mediate the endocytosis of IgE-bound antigen promoted studies which demonstrated that antigen presentation by murine B cells to an antigen-specific autologous T cell line could be enhanced by targeting antigen to FcεRII on B cells [18]. Blocking the binding of IgE to FcεRII on B cells with a monoclonal antibody to FcεRII completely eliminated this enhanced presentation. A similar study demonstrated human FcεRII-enhanced antigen presentation [19].

Squire *et al.* showed that FcεRII targeting is effective *in vivo*. BALB/c mice, when immunized with ovalbumin (OVA)-anti-FcεRII conjugates, made a significant OVA-specific IgG_1 response and a detectable IgE response, whereas no response occurred to OVA alone. Targeting to FcεRII in this study appeared to be 100-fold more effective than targeting to FcγRII, and to favor a Th2-type response [20]. IgE-mediated enhancement of IgE responses may thus provide a positive feedback mechanism that plays a role in allergic responses.

Dendritic cells

Dendritic cells (DCs) play a critical role in antigen presentation *in vivo* and have been shown to be more potent APCs than macrophages. This has stimulated an interest in targeting antigens to DCs as a potentially very effective way to increase the immunogenicity of weak antigens. Sallusto and Lanzavecchia used granulocyte/macrophage colony-stimulating factor (GM-CSF) and interleukin 4 to establish monocyte-derived DCs that develop the capacity to capture, process and present antigen very efficiently [21]. These cells had typical dendritic morphology and expressed high levels of MHC class I and class II molecules, CD1, CD80 (B7-1) and CD86 (B7-2), as well as FcγRII (CD32). These DCs were as efficient as antigen-specific B cells in presenting tetanus toxoid (TT) to specific T cell clones. Moreover, their efficiency of antigen presentation was further enhanced by specific antibodies, apparently due to FcγRII-mediated antigen uptake.

While most early studies indicated that DC express FcγRII, there have been conflicting conclusions regarding expression of FcγRI. Fanger *et al.* used countercurrent elutriation and magnetic bead selection to isolate human peripheral blood DC and showed that they express both FcγRI and FcγRII, but not FcγRIII [22,23]. These DCs were able to perform both FcγRI- and FcγRII-mediated phagocytosis. Antigen targeting studies showed that targeting of antigen to DC FcγRI enhanced antigen presentation to T cells by 50- to 500-fold over antigen alone, showing an important role for FcγRI in DC antigen presentation. In addition, DCs stimulated T cell proliferation up to 10-fold more efficiently than similarly treated monocytes (Figure 17.1).

Another subset of DCs, follicular dendritic cells (FDCs), have been shown to trap IgG and IgE immune complexes via FcγRII and FcεRII respectively, which

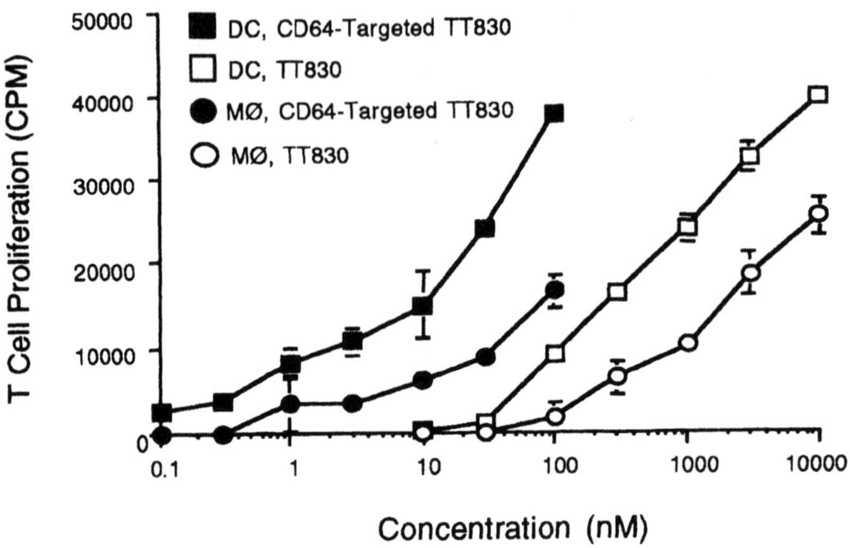

Figure 17.1 Targeting a tetanus toxoid peptide (TT830) to CD64 enhances T cell stimulation. A fusion protein incorporating TT830, in place of the Fc domain of the anti-CD64 antibody H22, enhanced T cell stimulation by both human monocytes (MØ) and blood dendritic cells (DC). Adapted from Reference 23

can lead to B cell proliferation and enhancement of B cell memory [reviewed in Reference 24]. In a review by Maurer and Stingl, the role of Fc receptors for IgE on Langerhans cells (LCs) has also been discussed [25]. They reported the presence of FcεRI on LCs from atopic individuals, and showed that LCs can concentrate allergens via FcεRI, suggesting a possible role for allergen presentation by LCs in atopy. A further study has shown the presence of FcεRI on peripheral blood DCs from atopic individuals [26]. These DCs were able to stimulate allergen-specific T cell proliferation up to 10-fold more efficiently than monocytes when cultured with haptenated allergen and anti-hapten IgE; this presentation could be blocked by addition of anti-FcεRIα mAb. Thus, various subsets of DCs appear capable of IgE-mediated enhancement of allergen presentation in atopic individuals. It remains to be determined whether the effect of these DC is to maintain or to diminish atopy.

Fc RECEPTOR-TARGETED FUSION PROTEINS

Antigenized antibodies

Vaccines based on synthetic peptides representing antigenic epitopes of disease-causing molecules have several theoretical advantages over traditional vaccines consisting of either killed or attenuated pathogens. By selection of only the peptides which confer protective immunity, peptide-based vaccines can exclude the peptides that elicit a deleterious immune response, such as the epitopes that cause an autoimmune response through molecular mimicry. Furthermore, the chemically defined nature of peptide-based vaccines avoids the potentially infectious materials contained in traditional vaccines. For viral and parasitic diseases in which no vaccines are yet available, most notably AIDS, a peptide-based vaccine may be the only safe choice. Peptide-based vaccines against tumour antigens and antigens involved in autoimmunity also have special significance in the development of immunotherapy, where the goal is either to establish or break tolerance toward the antigen. However, the development of peptide vaccines has been confounded by the usually poor immunogenicity of peptides, and various approaches have been taken to increase the potency of peptide vaccines.

One method to enhance peptide immunogenicity is to incorporate the peptide into so-called antigenized antibodies (AgAb). Peptide epitopes derived from diverse antigens can be genetically engineered into the complementarity-determining regions (CDR) of IgG, resulting in longer half life and the potential for interaction with FcγR. The AgAb can thereby be more effectively presented to the immune system [27]. AgAb have been shown to be 100- to 1000-times more efficient than free synthetic peptides for priming antigen-specific T cells *in vivo*, and the enhanced presentation was, at least in part, mediated by Fc receptors [28]. However, due to binding of the IgG Fc domain to multiple FcR, the AgAb approach does not result in selective targeting of antigen to a specific APC type, and has the potential to target to non-APCs that express FcR, including platelets and endothelial cells.

FcγRI-targeted fusion proteins

Targeting of antigens to surface molecules using monoclonal antibody-based fusion proteins lacking an Fc domain is an approach which can be used to direct antigen selectively to a subset of APC. This has been done for targeting to FcγRI (CD64), which, as noted above, is expressed on monocytes, macrophages and DCs but not on B cells [22]. Liu *et al.* replaced the Fc domain of the humanized anti-CD64 monoclonal antibody H22 with a defined Th-specific peptide of tetanus toxoid (TT) and showed this construct to markedly enhance activation of peptide-specific T cell lines [29]. Figure 17.1 shows the relative efficiency of T cell activation for peptide and targeted peptide presented by either human DC or monocytes [23]. Additional studies showed a similar construct incorporating an antagonist peptide to be more effective than the non-targeted antagonist peptide for inhibition of T cell activation.

In addition to selective APC targeting, H22-derived fusion proteins differ from AgAb in two other ways (Table 17.1). First, since the targeting molecule binds outside the ligand binding domain of FcγRI, its binding is not inhibited by serum IgG and would be expected to efficiently target to monocytes and DCs *in vivo*. Indeed, IgG did not block the effectiveness of these targeted TT peptide constructs, and the targeted H22 molecule was found to be a much more potent immunogen than non-targeted IgG in transgenic mice expressing human CD64 [30]. Second, by incorporating the antigenic moiety in place of the Fc domain, relatively large intact antigens can be targeted as well as peptides. These results indicate that selective targeting of antigenic fusion proteins to the non-ligand binding domain of FcγRI represents a novel and potentially useful strategy for stimulation of immunity *in vivo.*

SUMMARY

The studies reviewed in this chapter demonstrate that targeting of antigens to specific FcR molecules on APCs can markedly reduce the concentration of antigen required for a significant immunological response. In nearly all cases, antigen targeted to APC FcR was 10- to 1000-fold more potent than the equivalent non-targeted antigen. Thus, Fc receptor targeting may be particularly useful for immunization with antigens that are toxic or difficult to produce.

Until now, antigen targeting has been used primarily in the context of humoral responses and activation of CD4+, MHC class II-restricted T cells. Recent studies have shown that exogenous antigens, when taken up by macrophages via phagocytosis, can also elicit a CTL response mediated by CD8+ T cells [31]. This type of phagocytosis-mediated presentation of extra-cellular antigen to CD8+ CTL, although incompletely understood, might be further enhanced by targeting the vector to specific Fc receptors on monocytes, macrophages and DC. As we have only begun to explore the capacity of specific antigen targeting to shape the immune response, the coming years should bring substantial new insights into the most relevant receptors for targeted vaccine approaches.

References

1. Lanzavecchia A. Mechanisms of antigen uptake for presentation. Curr Opin Immunol. 1996; 8:348 54.
2. Cohen BE, Rosenthal AS, Paul WE. Antigen-macrophage interaction II. Relative roles of cytophilic antibody and other membrane sites. J Immunol. 1973;111:820–8.
3. Lanzavecchia A. Identifying strategies for immune intervention. Science. 1993;260:937–44.
4. Liu C and Guyre PM. Antigen targeting. In: Fanger, MW, ed., Bispecific Antibodies. Austin: R.G. Landes; 1995:133–45.
5. Chang TW. Regulation of immune response by antibodies: the importance of antibody and monocyte Fc receptor interaction in T cell activation. Immunol Today. 1985;6:245.
6. Manca F, Fenoglio D, Kunkl A et al. Differential activation of T cell clones stimulated by macrophages exposed to antigen complexed with monoclonal antibodies. J Immunol. 1988; 140:2893–8.
7. Wiersma EJ, Coulie PG, Heyman B. Dual immunoregulatory effects of monoclonal IgG-antibodies: suppression and enhancement of the antibody response. Scand J Immunol. 1989; 29:439–48.
8. Gosselin EJ, Wardwell K, Gosselin DR et al. Enhanced antigen presentation using human Fcγ receptor (monocyte/macrophage)-specific immunogens. J Immunol. 1992;149:3477–81.
9. Maurer D, Fiebiger E, Reiniger B et al. Expression of functional high affinity immunoglobulin E receptors (FcεRI) on monocytes of atopic individuals. J Exp Med. 1994;179:745–50.

10. Reischl IG, Corvaia N, Effenberger F et al. Function and regulation of FcεRI expression on monocytes from non-atopic donors. Clin Exp Allergy. 1996;26:630–41.
11. Mudde GC, Reischl IG, Corvaia N et al. Antigen presentation in allergic sensitization. Immunol Cell Biol. 1996;74:167–73.
12. Maurer D, Ebner C, Reiniger B et al. The high affinity IgE receptor (FcεRI) mediates IgE-dependent allergen presentation. J Immunol. 1995;154:6285–90.
13. Bheekra Escura R, Wasserbauer E, Hammerschmid F et al. Regulation and targeting of T-cell immune responses by IgE and IgG antibodies. Immunology. 1995;86:343-50.
14. Gajewski TF, Pinnas M, Wong T et al. Murine Th1 and Th2 clones proliferate optimally in response to distinct antigen-presenting cell populations. J Immunol. 1991;146:1750–8.
15. Snider DP, Segal DM. Efficiency of antigen presentation after antigen targeting to surface IgD, IgM, MHC, FcγRII, and B220 molecules on murine splenic B cells. J Immunol. 1989; 143:59–65.
16. Liu C, Gosselin EJ, Guyre PM. FcγRII on human B cells can mediate enhanced antigen presentation. Cell Immunol. 1996;167:188–194.
17. Van Den Herik-Oudijk IE, Westerdaal NA, Henriquez NV et al. Functional analysis of human FcγRII (CD32) isoforms expressed in B lymphocytes. J Immunol. 1994;152:574–85.
18. Kehry MR, Yamashita LC. Low-affinity IgE receptor (CD23) function on mouse B cells: role in IgE-dependent antigen focusing. Proc Natl Acad Sci USA. 1989;86:7556–60.
19. Pirron U, Schlunck T, Prinz JC et al. IgE-dependent antigen focusing by human B lymphocytes is mediated by the low-affinity receptor for IgE. Eur J Immunol. 1990;20: 1547–51.
20. Squire CM, Studer EJ, Lees A et al. Antigen presentation is enhanced by targeting antigen to the FcεRII by antigen-anti-FcεRII conjugates. J Immunol. 1994;152:4388–96.
21. Sallusto F, Lanzavecchia A. Efficient presentation of soluble antigen by cultured human dendritic cells is maintained by granulocyte/macrophage colony-stimulating factor plus interleukin 4 and downregulated by tumor necrosis factor-α. J Exp Med. 1994;179:1109–18.
22. Fanger NA, Wardwell K, Shen L et al. Type I (CD64) and type II (CD32) Fcγ receptor-mediated phagocytosis by human blood dendritic cells. J Immunol. 1996;157:541–8.
23. Fanger NA, Vogitlaender D, Liu C et al. Characterization of expression, cytokine regulation, and effector function of the high affinity IgG receptor FcγRI (CD64) expressed on human blood dendritic cells. J Immunol. 1997;158:3090–8.
24. Van den Berg TK, Yoshida K, Dijkstra CD. Mechanism of immune complex trapping by follicular dendritic cells. Curr Topics Microbiol Immunol. 1995;201:49–67.
25. Maurer D, Stingl G. Immunoglobulin E-binding structures on antigen-presenting cells present in skin and blood. J Invest Dermatol. 1995;104:707–10.
26. Maurer D, Fiebiger E, Ebner C et al. Peripheral blood dendritic cells express FcεRI as a complex composed of FcεRIα- and FcεRIγ-chains and can use this receptor for IgE-mediated allergen presentation. J Immunol. 1996;157:607–16.
27. Zanetti M. Antigenized antibodies. Nature. 1992;355:476–7.
28. Brumeanu TD, Swiggard WJ, Steinman RM et al. Efficient loading of identical viral peptide onto class II molecules by antigenized immunoglobulin and influenza virus. J Exp Med. 1993; 178:1795–9.
29. Liu C, Goldstein J, Graziano RF et al. FcγRI-targeted fusion proteins result in efficient presentation by human monocytes of antigenic and antagonist T cell epitopes. J Clin Invest. 1996;98:2001–7.
30. Heijnen IA, Van Vugt MJ, Fanger NA et al. Antigen targeting to myeloid-specific human FcγRI/CD64 triggers enhanced antibody responses in transgenic mice. J Clin Invest. 1996; 97:331–8.
31. Kovacsovics-Bankowski M, Clark K, Benacerraf B et al. Efficient major histocompatibility complex class I presentation of exogenous antigen upon phagocytosis by macrophages. Proc Natl Acad Sci USA. 1993;90:4942–6.
32. Gustavsson S, Hjulstrom S, Liu T et al. CD23/IgE-mediated regulation of the specific antibody response in vivo. J Immunol. 1994;152:4793–800.
33. Billetta R, Hollingdale MR, Zanetti M. Immunogenicity of an engineered internal image antibody. Proc Natl Acad Sci USA. 1991;88:4713–7.
34. Lanza P, Billetta R, Antonenko S et al. Active immunity against the CD4 receptor by using an antibody antigenized with residues 41-55 of the first extracellular domain. Proc Natl Acad Sci USA. 1993;90:11683–7.

35. Zaghouani H, Steinman R, Nonacs R et al. Presentation of a viral T cell epitope expressed in the CDR3 region of a self immunoglobulin molecule. Science. 1993;259:224–7.
36. Brumeanu TD, Swiggard WJ, Steinman RM et al. Efficient loading of identical viral peptide onto class II molecules by antigenized immunoglobulin and influenza virus. J Exp Med. 1993; 178:1795–9.
37. Kuzu H, Kuzu Y, Zaghouani H et al. In vivo priming effect during various stages of ontogeny of an influenza A virus nucleoprotein peptide. Eur J Immunol. 1993;23:1397–400.

18
Structure and function of CD23

D. H. CONRAD

The low affinity receptor for IgE (FcεRII) which is more commonly designated as CD23 is the most unusual of the Fc receptors. Unlike the other receptors discussed in this book, CD23 has no Ig-like domains and is a member of the calcium-dependent animal lectin family. The object of this chapter is to briefly discuss some of the characteristics of CD23, both with regard to structure and function. More extensive reviews have been published elsewhere and the reader is referred to these for additional information [1,2].

STRUCTURE

The low affinity receptor for IgE (FcεRII) was initially discovered in 1975 by Lawrence *et al.* [3] using aggregated IgE on human peripheral blood mononuclear cells. A 45 kDa B cell activation antigen was designated CD23 at the 2nd International workshop on human leukocyte differentiation antigen in 1984. CD23 was identified by using several monoclonal antibodies prepared by immunization of EBV-transformed human B cells. CD23 was not detected on resting B cells isolated from peripheral blood, lymphoid tissues, or mantle zone B cells, but on activated B cells and germinal centre (GC) B cells [4]. Subsequently it was shown that CD23 and FcεRII are the same protein [5], and the term CD23 will be used from this point. Cloning studies demonstrated that CD23 is a type II integral membrane glycoprotein [6,7]. Human CD23 consists of a 23 amino acid (a.a.) N-terminal intracytoplasmic domain (IC), 21 a.a. transmembrane domain (TM), and 277 a.a. C-terminal extracellular domain (EC), resulting in a 321 a.a. protein [5]. The gene for huCD23 is found on chromosome 19 [8], and is present as a single copy consisting of 11 exons [9]. The 5′ flanking region has two TATA boxes and one complete CCAAT box, and these boxes exhibit typical promoter activity [10].

J.G.J. van de Winkel and P.M. Hogarth (eds.), The Immunoglobulin Receptors and their Physiological and Pathological Roles in Immunity. 195–206.

In humans, this gene encodes two transcripts, FcεRIIa (CD23a) and FcεRIIb (CD23b), which differ in their 5′ untranslated region and in the first 6 a.a. which are found in the intracytoplasmic region [11]. The cellular distribution of each transcript is also different. CD23a is present only on B cells and possibly follicular dendritic cells (FDC), but CD23b is expressed on a much wider number of hematopoietic cell types, such as $CD5^+$ B cells, T cells, eosinophils, bone marrow derived mast cells, platelets, monocytes, and Langerhans cells [reviewed in References 1,2].

A model for the structure of the human and mouse CD23 is shown in Figure 18.1. As mentioned earlier, CD23 is a glycoprotein with two potential sites for N-linked glycosylation, and may also have several O-linked carbohydrates [12]. It belongs to the calcium type (C-type) animal lectin family which is due to a requirement for calcium to bind the respective ligand [14]. The conserved cysteines are noted in Figure 18.1; the prototypic member of this family is the asialoglycoprotein receptor [15]. The superfamily contains both membrane and soluble proteins in which this lectin homology region is maintained. Notably, a number of family members belong to the selectin adhesion family. The inverted RGD sequence present in the C terminal domain of human CD23 has been proposed to be important in adhesion functions [16]. In comparison with human CD23, the murine homologue (mCD23) has very similar characteristics, but there are some differences. Murine CD23 is 49 kDa type II membrane glycoprotein [7]. Analysis of the cDNA indicated that it consists of 331 a.a. and the core mw is 37 kDa [7,17]. Murine CD23 also belongs to C-type animal lectin family [7]. Unlike human CD23, mCD23 does not contain the inverted RGD sequence, but does have one more additional repeat region in the stalk area. The

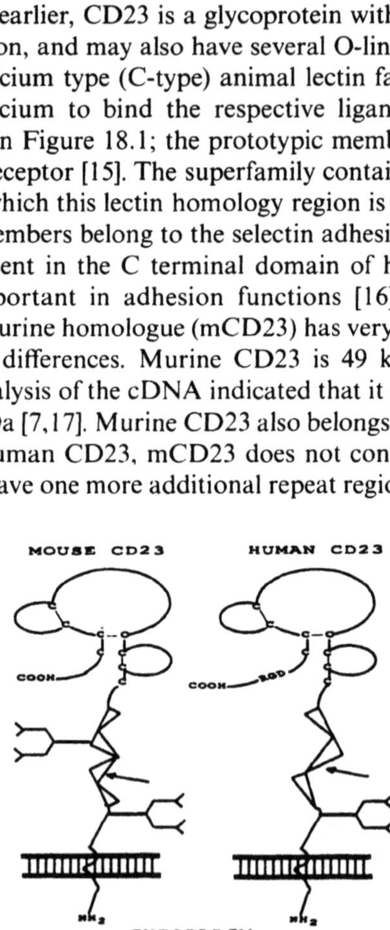

Figure 18.1 Models for the structure of human and mouse CD23 (from Conrad [13], with permission). The lectin homology region is the carboxyl terminal 'head' with the conserved cysteines shown. The 'stalk' region contains several regions, each encoded by a separate exon, which have homology with each other. These regions are represented by triangles (three in man, four in mouse). The arrow represents the site of initial cleavage in the generation of soluble CD23

gene for mCD23 is found on chromosome 8 [18] and consists of 12 exons which span 12.9 kb. The 5′ flanking site has TATA and CCAAT boxes, and contains sequences homologous with MHC class II genes i.e. X box, Y box and TRE element [19]. Similar motifs are also found on the 5′ flanking region of the huCD23 gene. This region of the gene has been implicated in promoter activity [20,21]. Although mCD23b mRNA (detected by RT-PCR) analogous to human CD23b is expressed at very low level after activation with LPS and IL-4 [22], only one form of mCD23, analogous to human CD23a, is expressed at detectable protein levels [18]. Analogous to the huCD23a, the single mCD23 isoform is relatively B cell specific with the exception of FDCs [23]. Ontogenic studies have indicated that expression correlates with IgD expression on mature B cells [24]. Cell surface CD23 has been shown to associate with MHC class II antigen [25] and with B cell sIg [26] in mouse and human systems, respectively.

CD23 has a somewhat unusual metabolic pathway; upon arrival at the cell surface [27] and possibly also internally [28], CD23 is cleaved with the majority of the COOH-terminus being released. The initial fragment, termed sCD23 or IgE binding factor is 37 kDa (human) and is not N-glycosylated [12]. This fragment is then further broken down to the size limit of 12–16 kDa, consisting of the lectin homology domain only [12,29]. Although some evidence suggested that the initial proteolysis is autoproteolytic [30], current data suggest that an as yet unidentified metalloprotease is responsible for at least the initial cleavage (L.Marshall and D.Conrad, unpublished observations). Except for the 12 kDa fragment, the human fragments have been shown to bind IgE [29]. Mouse CD23 is initially cleaved into 38 kDa, and is then cleaved into smaller fragments, termed mouse soluble CD23, (msCD23) [31]. The cleavage site of mCD23 for msCD23 is in the third and fourth repeat domains of the α-helical coiled coil (stalk region) [32]. Recently a second pathway for the product of sCD23 has been reported in both mouse and human where alternative transcripts lacking the membrane domains were described. In view of the protocols used for their detection (RT-PCR), whether this 'alternative pathway' is biologically significant remains uncertain [33,34].

LIGAND BINDING

As mentioned earlier, the binding site of IgE for CD23 is on the Cε3 domain [35,36]. The binding is Ca^{2+} dependent, but does not require carbohydrate as deglycosylation of IgE did not significantly influence the binding [14,37]. Mutations in the lectin domain change IgE binding, indicating that the interaction site of CD23 with IgE is in the lectin cassette [38,39]. Thus, the lectin module has evidently evolved to interact with a protein rather than a carbohydrate moiety in this situation. Recent evidence indicates that the stalk region of the molecule has homology with tropomyosin and other similar molecules that exhibit α-helical coiled-coil motifs [40,41]. Such molecules are anticipated to form trimers and even higher oligomers and such situations have been reported for both the asialoglycoprotein receptor and mannose binding protein, two members of the C-lectin family [15,42]. Dierks *et al.* [43], using chemical cross-linkers, demonstrated that the intact form of mCD23 oligomerized to at least a trimeric form and also showed that the stalk repeat region mediated this oligomerization. In addition, the fact that the multimeric form was exclusively isolated by IgE affinity chromatography, supported the hypothesis that the oligomerization resulted in a higher affinity for IgE, presumably via multipoint binding. This finding has led to a new model for CD23 (Figure 18.2) and helps explain the dual affinity seen between IgE and CD23 [43]. Unlike huCD23, mouse sCD23 does not exhibit this oligomerization capacity and only interacts with IgE with a single low affinity ($10^6 M^{-1}$) [32]. In a study with huCD23, Beavil *et al.* [44] found that sCD23 could be crosslinked to trimers and higher oligomers suggesting that the multimeric structure was more stable in human CD23. This may explain the increased biological activity reported for human sCD23 (see below). All of the above data implicate oligomerization in mediating a high affinity interaction with IgE. Recent evidence has suggested that the interaction may not be with symmetrical sites on IgE (H. Gould, personal communication); however, with this change, the current model for the structure of the membrane form of CD23 continues to be the structure shown in Figure 18.2.

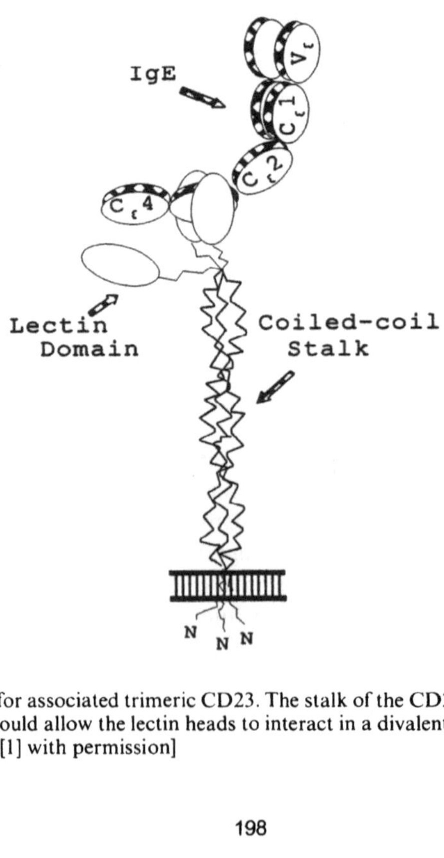

Figure 18.2 Model for associated trimeric CD23. The stalk of the CD23 molecule forms a coiled-coil structure. This would allow the lectin heads to interact in a divalent (or trivalent) manner with IgE [from Reference [1] with permission]

A second ligand was identified for CD23, namely CD21 [45], which is also a receptor for C3d, C3dg and EBV [reviewed in Reference 46]. The affinity between CD23 and CD21 is low since a highly multivalent CD23/liposome complex is required for detection. Interestingly, the interaction with CD21 may be via a more classical lectin modality, since fucose-1-phosphate will block this interaction [47]. SCR(short consensus repeat) deletional studies implicated SCR 5 and 6 as the regions that interact with CD23 via the N-linked carbohydrate on these SCRs [48]. In addition, SCR 1 and 2, which have been implicated in C3 binding [49] appear to increase the affinity of interaction of CD23/CD21, presumably via a protein–protein interaction. Additional ligands for human CD23 have been found. CD23 can bind CD11b/CD18, and CD11c/CD18 on monocytes [50,51]. Interaction with these integrins presumably explains the CD23 induction of TNF and IL-6 release from monocytes [52].

REGULATION OF CD23

Several cytokines regulate the expression of both huCD23 and mCD23. IL-4 increases both huCD23 and mCD23 expression on B cells [53,54], as well as increasing soluble CD23. This increase in expression is accompanied by an increase in transcriptional activity [10]. With B cell activation agents, such as lipopolysaccharide (LPS) or CD40L, IL-4 induces a very high expression of mCD23 (termed CD23 superinduction [55]). IL-13 has a similar effect as IL-4, and increases CD23 as well as IgE production [56], at least in humans. Based on gene deletion studies, the IL-4 (and IL-13) action is mediated by the transcription factor STAT6 [57,58]. The STAT6 binding site in human and mouse CD23 has been identified [59,60]. Recent data using linker scanning mutagenesis identified other regions of the CD23a promoter that are necessary to give IL-4 mediated CD23 up-regulation [61]. In humans, EBV infection induces constitutive CD23 expression due to an EBV responsive element in the 5′ region of CD23 (–229 to +305) and also an EBV responsive enhancer (+248 to +284 in intron 1) [62]. Regulation is complicated somewhat by the presence of the two isoforms, CD23a and CD23b. Interferon (both α and γ) has an inhibitory effect on CD23 expression on B cells [63]. On haematopoietic cells other than B cells, the CD23b isoform is up-regulated by these same interferons. In contrast, IL-4 upregulates both the a and b isoforms and thus, induces upregulation on all cell types capable of expressing CD23. There is essentially no constitutive expression of CD23b, therefore, its total expression requires IL-4 (or IL-13) [53,56]. There is some evidence that regulation can occur as a result of influencing the breakdown of membrane CD23 into sCD23. Thus, TNF-α decreases membrane CD23 and increases sCD23, apparently by increasing CD23 cleavage [64]. IgE bound to CD23 especially in the murine system, inhibits cleavage, resulting in higher membrane and lower soluble CD23, respectively. Finally, some parasite infections alter CD23 expression levels, again presumably by influencing breakdown. An excretory component from *Dirofilaria immitis*, a type of filarial parasite increases CD23 expression on human splenic B cells and T cells [65].

The modulation of CD23 expression by the parasite, *Leishmania chagasi* is also observed in infected human and mouse B cells. This infestation induces loss of membrane CD23 expression, and increases soluble CD23. However, CD23 mRNA levels are not affected by these parasite infections [66].

FUNCTION OF CD23 – MEMBRANE CD23

CD23 has several potential functions which are related to B cell proliferation and differentiation, regulation of IgE production, antigen presentation and cell adhesion [reviewed in Reference 13]. Since CD23 was initially found as the receptor for IgE, many functional studies were undertaken to determine its role in IgE regulation. Addition of anti-CD23 mAbs to mononuclear cells from either tonsils or peripheral blood cells results in the inhibition of IL-4 induced IgE secretion; the ability of anti-CD23 mAbs to inhibit IgE secretion was restricted to those which are identical, or are close to the IgE binding site on CD23 indicating that anti-CD23 mAbs have their effect either by steric hindrance or by binding to the same site [67]. The addition of anti-IgE mAb or IgE–anti-IgE immune complexes to a human IgE plasmacytoma cell line, or to B cells of highly atopic individuals also inhibits IgE production, while the production of other isotypes, such as IgA or IgG are not altered [68]. In *in vivo* studies, injection of a polyclonal antibody specific to the lectin homology region of CD23, inhibits IgE synthesis in a rat model [69]. Recent studies have demonstrated that transgenic mice which overexpress CD23 have a drastic decrease in IgE production to both antigen and parasite infections [70] and this inhibition occurs subsequent to isotype switching, since germline-ε transcripts were not effected [71]. Culture of B cells *in vitro* with cells overexpressing CD23 also resulted in an inhibition of IgE production [71]; these findings have renewed interest in the use of CD23 as a mediator for type I allergy [72]. Data with CD23 knockout animals has been more controversial. In one study, an enhancement of IgE production was reported [73], but this was not supported by other studies [74,75]. There is some support for the interaction of CD23 and CD21 playing a role in B cell activation and IgE production. Co-ligation of sIg and CD21 using anti-sIg and FcγRII and CD23-transfected fibroblasts resulted in enhanced B cell activation [76], a result predicted by the earlier studies of Carter and Fearon [77]. In addition, activation of B cells by ligation of CD21 by CD23 liposome complex enhances IgE production in the presence of IL-4 [45]. The CD21–CD23 interaction in mouse is not as clear as in the human system [78].

When IgE is bound to antigen and CD23, CD23 mediates endocytosis of the Ag–IgE complex, and causes an increase in antigen specific IgE production [79]; this phenomenon is not seen in CD23 knockout animals [75]. This results in greatly enhanced antigen processing/presentation in both the human [80] and mouse [81] systems. This IgE and CD23-dependent antigen presentation is considered to increase allergen presentation to specific T cells and exacerbate allergy in atopic patients. In addition to endocytosis, CD23 on the monocyte/macrophage can mediate phagocytosis and IgE-dependent cytotoxicity. This function was observed by the killing of IgE-coated schistosomes by eosinophils, macrophages and platelets. IgE depletion inhibited this killing function, while addition of IgE anti-schistosome antibody increased the killing function [82]. The phagocytic aspect of CD23 is more evident with the CD23b isoform [83]. Additional investigation of the role of CD23 in monocyte function revealed that ligation of CD23 with IgE–anti-IgE complexes caused monocytes to produce nitric oxide (NO), and also enabled monocytes to kill tumour cells as well as cells infected with *Leishmania* [84,85]. CD23 has also been suggested to play a role in homotypic cell adhesion, although the data is somewhat contradictory [78]. CD23 cDNA-transfected cells were shown to have an increase in cell aggregation and cell adhesion [86]. However, CD23 knockout animals have no decrease in homotypic adhesion [78], lending question to this role.

The relationship of CD23 to immune functions has been examined in several diseases. $CD23^+$ B cells are increased in allergic children [87] and sCD23 and CD23 are increased in atopic patients [88]. In rheumatoid arthritis (RA) patients, B cells expressing CD23 are increased and sCD23 is increased [89,90]. Arguing for a role for CD23 in RA is the finding that anti-CD23 antibodies ameliorate collagen-induced arthritis in a mouse model system [91].

FUNCTION OF CD23 – SOLUBLE CD23

Soluble CD23, the proteolytic product of membrane CD23, has been implicated with several functions, especially in the human system. With regard to IgE synthesis, several studies in human have indicated a small but significant increase in IgE, in the presence of sCD23 [92,93]. Deletion of the 33 a.a. on the COOH terminus caused an inhibition of IgE synthesis [94]. Soluble CD23 can act as an autocrine growth factor. Human sCD23 (25 kDa) is reportedly mitogenic for normal B cells activated with PMA, and promotes the survival of germinal center B cells [95] . Soluble CD23 in synergy with IL-1 promotes growth of haematopoietic progenitor cells [96] and thymocyte maturation [97]. IL-1 with sCD23 also potentiates the secretion of IL-6 and IL-1 receptor antagonists from human monocytes [98]. In addition, in conjunction with IL-1, human sCD23 acts as a growth factor for pre-T cells and myeloid cells. It has not been determined whether this autocrine effect is mediated by signal transduction through CD21 or via an alternative ligand for sCD23. In the murine system, no effects of sCD23 on IgE synthesis and autocrine activity were found [99]. The rationale for the species difference is unclear, although a potential explanation relates to the increased stability of the multimeric structure in human sCD23 discussed above. Soluble CD23 directly activates monocytes and induces the production of monokines, which can contribute to the antigen independent stimulation of resting T cells. This activity is presumably mediated via interaction of sCD23 with CD11/CD18 [50].

References

1. Conrad DH. FcεRI, ε-BP and FcεRII: Structure and involvement in allergic disease. In: Townley RG, Agrawal DK, eds, Immunopharmacology of Allergic Diseases. New York: Marcel Dekker; 1996:79–98.
2. Delespesse G, Sarfati M, Wu CY, Fournier S, Letellier M. The low-affinity receptor for IgE. Immunol Rev. 1992;125:77–97.
3. Lawrence DA, Weigle WO, Spiegelberg HL. Immunoglobulins cytophilic for human lymphocytes, monocytes, and neutrophils. J Clin Invest. 1975;55:268–75.
4. Nahm MH, Takes PA, Bowen MB, Macke KA. Subpopulations of B lymphocytes in germinal centers, II. A germinal center B cell subpopulation expresses sIgD and CD23. Immunol Lett. 1989;21:201–8.
5. Yukawa K, Kikutani H, Owaki H et al. A B cell-specific differentiation antigen CD23 is a receptor for IgE (FcεR) on lymphocytes. J Immunol. 1987;138:2576–80.
6. Kikutani H, Inui S, Sato R et al. Molecular structure of human lymhocyte receptor for immunoglobulin E. Cell. 1986;47:657–65.
7. Bettler B, Hofstetter H, Rao M, Yokoyama WM, Kilchherr F, Conrad DH. Molecular structure and expression of the murine lymphocyte low affinity receptor for IgE (FcεRII). Proc Natl Acad Sci USA. 1989;86:7566–70.
8. Wendel-Hansen V, Rivière M, Uno M et al. The gene encoding CD23 leukocyte antigen (FCE2) is located on human chromosome 19. Somat Cell Mol Genet. 1990;16:283–5.
9. Suter U, Bastos R, Hofstetter H. Molecular structure of the gene and the 5′-flanking region of the human lymphocyte immunoglobulin E receptor. Nucleic Acids Res. 1987;15:7295–308.
10. Suter U, Texido G, Hofstetter H. Expression of human lymphocyte IgE receptor (FcεRII/CD23) identification of FcεRIIa promoter and its functional analysis in B lymphocytes. J Immunol. 1989;143:3087–92.
11. Yokota A, Kikutani H, Tanaka T et al. Two species of human Fc receptor II (FcεRII/CD23): tissue-specific and IL-4-specific regulation of gene expression. Cell. 1988;55:611–18.
12. Letellier M, Nakajima T, Delespesse G. IgE receptor on human lymphocytes. IV. Further analysis of its structure and of the role of N-linked carbohydrates. J Immunol. 1988;141: 2374–81.
13. Conrad DH. FcεRII/CD23: the low affinity receptor for IgE. Annu Rev Immunol. 1990;8: 623–45.
14. Richards ML, Katz DH. The binding of IgE to murine FcεRII is calcium-dependent but not inhibited by carbohydrate. J Immunol. 1990;144:2638–46.
15. Halberg DF, Wager RE, Farrell DC et al. Major and minor forms of the rat liver asialoglycoprotein receptor are independent galactose-binding proteins. Primary structure and glycosylation heterogeneity of minor receptor forms. J Biol Chem. 1987;262:9828–38.
16. Moulder K. The role of RGD in CD23-mediated cell adhesion. Immunol Today. 1996;17: 198–9.
17. Conrad DH, Peterson LH. The murine lymphocyte receptor for IgE. I. Isolation and characterization of the murine B cell Fc epsilon receptor and comparison with Fc epsilon receptors from rat and human. J Immunol. 1984;132:796–803.
18. Conrad DH, Kozak CA, Vernachio J, Squire CM, Rao M, Eicher EM. Chromosomal location and isoform analysis of mouse FcεRII/CD23. Mol Immunol. 1993;30:27–33.
19. Richards ML, Katz DH, Liu F-T. Complete genomic sequence of the murine low affinity Fc receptor for IgE: demonstration of alternative transcripts and conserved sequence elements. J Immunol. 1991;147:1067–74.
20. Richards ML, Katz DH. Regulation of the murine FcεRII (CD23) gene: Functional characterization of an IL-4 enhancer element. J Immunol. 1994;152:3453–66.
21. Dierks SE, Campbell KA, Studer EJ, Conrad DH. Molecular mechanisms of murine FcεRII/CD23 regulation. Mol Immunol. 1994;31:1181–9.
22. Kondo H, Ichikawa Y, Nakamura K, Tsuchiya S. Cloning of cDNAs for new subtypes of murine low-affinity Fc receptor for IgE (FcεRII/CD23). Int Arch Allergy Immunol. 1994; 105:38–48.
23. Maeda K, Burton GF, Padgett DA et al. Murine Follicular dendritic cells (FDC) and low affinity Fc-receptors for IgE (FcεRII). J Immunol. 1992;148:2340–7.

24. Waldschmidt TJ, Conrad DH, Lynch RG. The expression of B cell surface receptors. I. The ontogeny and distribution of the murine B cell IgE Fc receptor. J Immunol. 1988;140:2148–54.
25. Bonnefoy J, Guillot O, Spits H, Blanchard D, Ishizaka K, Banchereau J. The low-affinity receptor for IgE (CD23) on B lymphocytes is spatially associated with HLA-DR antigens. J Exp Med. 1988;167:57–72.
26. Lee WT, Conrad DH. The murine lymphocyte receptor for IgE. III. Use of chemical cross-linking reagents to further characterize the B lymphocyte Fc epsilon receptor. J Immunol. 1985;134:518–25.
27. Nakajima T, Delespesse G. Relationship between human IgE-binding factors (IgE–BF) and lymphocyte receptors for IgE. J Immunol. 1987;139:848–54.
28. Lee BW, Simmons CF, Wileman T, Geha RS. Intracellular cleavage of newly synthesized low affinity Fc receptor (FcεRII) provides a second pathway for the generation of the 28-kDa soluble FcεRII fragment. J Immunol. 1989;142:1614–20.
29. Letellier M, Sarfati M, Delespesse G. Mechanisms of formation of IgE-binding factors (soluble CD23) – I. FcεR II bearing B cells generate IgE-binding factors of different molecular weights. Mol Immunol. 1989;26:1105–12.
30. Letellier M, Nakajima T, Pulido-Cejudo G, Hofstetter H, Delespesse G. Mechanism of formation of human IgE-binding factors (Soluble CD23): III. Evidence for a receptor (FcεRII)-associated proteolytic activity. J Exp Med. 1990;172:693–700.
31. Keegan AD, Conrad DH. The murine lymphocyte receptor for IgE V. Biosynthesis, transport, and maturation of the B cell Fcε Receptor. J Immunol. 1987;139:1199–205.
32. Bartlett WC, Kelly AE, Johnson CM, Conrad DH. Analysis of murine soluble FcεRII: sites of cleavage and requirements for dual affinity interaction with IgE. J Immunol. 1995;154: 4240–9.
33. Matsui M, Nunez R, Sachi Y, Lynch RG, Yodoi J. Alternative transcripts of the human CD23/FcεRII: a possible novel mechanism of generating a soluble isoform in the type-II cell surface receptor. FEBS Lett. 1993;335:51–6.
34. Nunez R, Lynch RG. CD23 isoforms in murine T and B lymphocytes. Pathobiology. 1993;61: 128–37.
35. Chretein I, Helm B, Marsh P, Padlan E, Wijdenes J, Banchereau J. A monoclonal anti-IgE anitbody against an epitope (amino acids 367-376) in the CH3 domain inhibits IgE binding to the low affinity IgE receptor (CD23). J Immunol. 1988;141:3128–34.
36. Keegan AD, Fratazzi C, Shopes B, Baird B, Conrad DH. Characterization of new rat anti-mouse IgE monoclonals and their use along with chimeric IgE to further define the site that interacts with $Fc_{\varepsilon}RII$ and $Fc_{\varepsilon}RI$. Mol Immunol. 1991;28:1149–54.
37. Vercelli D, Helm B, Marsh P, Padlan E, Geha RS, Gould H. The B-cell binding site on human immunoglobulin E. Nature. 1989;338:649–51.
38. Bettler B, Maier R, Ruegg D, Hofstetter H. Binding site for IgE of the human low affinity Fcε receptor FcεRII/CD23) is confined to the domain homologous with animal lectins. Proc Natl Acad Sci USA. 1989;86:7118–22.
39. Bettler B, Texido G, Raggini S, Rüegg D, Hofstetter H. Immunoglobulin E-binding site in Fcε receptor (FcεRII/CD23) identified by homolog-scanning mutagenesis. J Biol Chem. 1992;267:185–91.
40. Beavil AJ, Edmeades RL, Gould HJ, Sutton BJ. α-Helical coiled-coil stalks in the low-affinity receptor for IgE (FcεRII/CD23) and related C-type lectins. Proc Natl Acad Sci USA. 1992; 89:753–7.
41. Gould H, Sutton B, Edmeades R, Beavil A. CD23/FcεRII: C-type lectin membrance protein with a split personality? In: Gordon J, ed., Monographs in Allergy. Basel: Karger; 1991:28–49.
42. Drickamer K. Two distinct classes of carbohydrate-recognition domains in animal lectins. J Biol Chem. 1988;263:9557–60.
43. Dierks SE, Bartlett WC, Edmeades RL, Gould HJ, Rao M, Conrad DH. The oligomeric nature of the murine FcεII/CD23: implications for function. J Immunol. 1993;150:2372–82.
44. Beavil RL, Graber P, Aubonney N, Bonnefoy J-Y, Gould HJ. CD23/FcεRII and its soluble fragments can form oligomers on the cell surface and in solution. Immunology. 1995;84:202–6.
45. Aubry J-P, Pochon S, Graber P, Jansen KU, Bonnefoy J-Y. CD21 is a ligand for CD23 and regulates IgE production. Nature. 1992;358:505–7.

46. Fearon DT, Ahearn JM. Complement receptor type 1 (C3b/C4b receptor; CD35) and complement receptor type 2 (C3d/Epstein-Barr virus receptor; CD21). Curr Top Microbiol Immunol. 1990;153:83–98.
47. Pochon S, Graber P, Yeager M et al. Demonstration of a second ligand for the low affinity receptor for immunoglobulin E (CD23) using recombinant CD23 reconstituted into fluorescent liposomes. J Exp Med. 1992;176:389–97.
48. Aubry J-P, Pochon S, Gauchat J-F et al. CD23 interacts with a new functional extracytoplasmic domain involving *N*-linked oligosaccharides on CD21. J Immunol. 1994;152:5806–13.
49. Delcayre AX, Lotz M, Lernhardt W. Inhibition of Epstein-Barr virus-mediated capping of CD21/CR2 by alpha interferon (IFN-alpha): immediate antiviral activity of IFN-alpha during the early phase of infection. J Virol. 1993;67:2918–21.
50. Lecoanet-Henchoz S, Gauchat J, Aubry J et al. CD23 regulates monocyte activation through a novel interaction with the adhesion molecules CD11b-CD18 and CD11c-CD18. Immunity. 1995;3:119–25.
51. Lecoanet-Henchoz S, Plater-Zyberk C, Graber P et al. Mouse CD23 regulates monocyte activation through an interaction with the adhesion molecule CD11b/CD18. Eur J Immunol. 1997;27:2290–4.
52. Armant M, Rubio M, Delespesse G, Sarfati M. Soluble CD23 directly activates monocytes to contribute to the antigen-independent stimulation of resting T cells. J Immunol. 1995;155: 4868–75.
53. Kikutani H, Suemura M, Owaki H et al. Fcε receptor, a specific differentiation marker transiently expressed on mature B cells prior to isotype switching. J Exp Med. 1986;164: 1455–69.
54. Conrad DH, Waldschmidt TJ, Lee WT et al. Effect of B cell stimulatory factor-1 (interleukin 4) on Fcε and Fcτ receptor expression on murine B lymphocytes and B cell lines. J Immunol. 1987;139:2290–6.
55. Keegan AD, Snapper CM, VanDusen R, Paul WE, Conrad DH. Superinduction of the murine B cell FcεRII by T helper cell clones. Role of interleukin-4. J Immunol. 1989;142: 3868–74.
56. Punnonen J, Aversa G, Cocks BG et al. Interleukin 13 induces interleukin 4-independent IgG4 and IgE synthesis and CD23 expression by human B cells. Proc Natl Acad Sci USA. 1993;90:3730–4.
57. Shimoda K, van Deursen J, Sangster MY et al. Lack of IL-4-induced Th2 response and IgE class switching in mice with disrupted *Stat6* gene. Nature. 1996;380:630–3.
58. Takeda K, Tanaka T, Shi W et al. Essential role of Stat6 in IL-4 signalling. Nature. 1996;380: 627–30.
59. Kotanides H, Reich NC. Requirement of tyrosine phosphorylation for rapid activation of a DNA binding factor by IL-4. Science 1993;262:1265–7.
60. Tinnell SB, Jacobs-Helber SM, Sterneck E, Sawyer S, Conrad DH. STAT6, NF-κB and C/EBP$_\beta$ in CD23 expression: STAT6 enhances CD40 induced CD23 expression, but is not required for CD40/IL-4 induced superinduction. 1998 (Submitted for publication).
61. Richards ML, Katz DH. Analysis of the promoter elements necessary for IL-4 and anti-CD40 antibody induction of murine FcεRII: Comparison with the germline ε promoter. J Immunol. 1997;158: 263–72.
62. Lacy J, Roth G, Shieh B. Regulation of the human IgE receptor (FcεRII/CD23) by EBV: localization of an intron EBV-responsive enhancer and characterization of its cognate GC-box binding factors. J Immunol. 1994;153:5537–48.
63. Mayumi M, Kawabe T, Kim KM et al. Regulation of Fc epsilon receptor expression on a human monoblast cell line U937. Clin Exp Immunol. 1988;71:202–6.
64. Hashimoto S, Koh K, Tomita Y et al. TNF-α regulates IL-4-induced FcεRII/CD23 gene expression and soluble FcεRII release by human monocytes. Int Immunol. 1995;7:705–13.
65. Yamaoka KA, Kolb JP, Miyasaka N, Inuo G, Fujita K. Purified excretory-secretory component of filarial parasite enhances Fc epsilon RII/CD23 expression on human splenic B and T cells and IgE synthesis while potentiating Th 2-related cytokine generation from T cells. Immunology. 1994;81:507–12.
66. Noben NN, Wilson ME, Lynch RG. Modulation of the low-affinity IgE Fc receptor (Fc epsilon RII/CD23) by Leishmania chagasi. Int Immunol. 1994;6:935–45.
67. Bonnefoy J, Shields J, Mermod J-J. Inhibition of human interleukin 4-induced IgE synthesis by a subset of anti-CD23/FcεRII monoclonal antibodies. Eur J Immunol. 1990;20:139–44.

68. Sherr E, Macy E, Kimata H, Gilly M, Saxon A. Binding the low affinity FcεR on B cells suppresses ongoing human IgE synthesis. J Immunol. 1989;142:481–9.
69. Flores-Romo L, Shields J, Humbert Y et al. Inhibition of an *in vivo* antigen-specific IgE response by antibodies to CD23. Science. 1993;261:1038–41.
70. Texido G, Eibel H, Le Gros G, Van der Putten H. Transgene CD23 expression on lymphoid cells modulates IgE and IgG_1 responses. J Immunol. 1994;153:3028–42.
71. Cho S, Kilmon MA, Studer EJ, Conrad DH. B cell activation and Ig – especially IgE – production is inhibited by high CD23 levels *in vivo* and *in vitro*. Cell Immunol. 1997;180:36–46.
72. Conrad DH, Kilmon MA, Studer EJ, Cho S-W. The low affinity receptor for IgE as a therapeutic target. Biochem Soc Trans. 1997;25:393–7.
73. Yu P, Kosco-Vilbois M, Richards M, Köhler G, Lamers MC, Kohler G. Negative feedback regulation of IgE synthesis by murine CD23. Nature. 1994;369:753–6.
74. Stief A, Texido G, Sansig G, Eibel H, Le Gros G, Van der Putten H. Mice deficient in CD23 reveal its modulatory role in IgE production but no role in T and B cell development. J Immunol. 1994;152:3378–90.
75. Fujiwara H, Kikutani H, Suematsu S et al. The absence of IgE antibody-mediated augmentation of immune responses in CD23-deficient mice. Proc Natl Acad Sci USA. 1994; 91:6835–9.
76. Reljic R, Cosentino G, Gould HJ. Function of CD23 in the response of human B cells to antigen. Eur J Immunol. 1997;27:572–5.
77. Carter RH, Spycher MO, Ng Y, Hoffman R, Fearon DT. Synergistic interaction between complement receptor type 2 and membrane IgM on B lymphocytes. J Immunol. 1988;141: 457–63.
78. Davey EJ, Bartlett WC, Kikutani H et al. Homotypic aggregation of murine B lymphocytes is independent of CD23. Eur J Immunol. 1995;25:1224–9.
79. Gustavsson S, Hjulström S, Tianmin L, Heyman B. CD23/IgE-mediated regulation of the specific antibody response in vivo. J Immunol. 1994;152:4793–800.
80. Pirron U, Schlunck T, Prinz JC, Rieber EP. IgE-dependent antigen focusing by human B lymphocytes is mediated by the low-affinity receptor for IgE. Eur J Immunol. 1990;20:1547–51.
81. Kehry MR, Yamashita LC. Fc1996; receptor II (CD23) function on mouse B cells: role in IgE dependent antigen focusing. Proc Natl Acad Sci USA. 1989;86:7556–60.
82. Capron A, Dessaint JP, Capron M, Joseph M, Ameisen J, Tonnel AB. From parasites to allergy: a second receptor for IgE. Immunol Today. 1986;7:15–18.
83. Yokota A, Yukawa K, Yamamoto A et al. Two forms of the low-affinity Fc receptor for IgE differentially mediate endocytosis and phagocytosis: Identification of the critical cytoplasmic domains. Proc Natl Acad Sci USA. 1992;89:5030–4.
84. Dugas B, Mossalayi MD, Damais C, Kolb JP. Nitric oxide production by human monocytes: evidence for a role of CD23. Immunol Today. 1995;16:574–80.
85. Vouldoukis I, Riveros-Moreno V, Dugas B et al. The killing of Leishmania major by human macrophages is mediated by nitric oxide induced after ligation of the Fc epsilon RII/CD23 surface antigen. Proc Natl Acad Sci USA. 1995;92:7804–8.
86. Wang F, Gregory CD, Rowe M et al. Epstein-Barr virus nuclear antigen 2 specifically induces expression of the B-cell activation antigen CD23. Proc Natl Acad Sci USA. 1987;84:3452–6.
87. Rabatic S, Gagro A, Medar-Lasic M. CD21-CD23 ligand pair expression in children with allergic asthma. Clin Exp Immunol. 1993;94:337–40.
88. Pfeil T, Fischer A, Bujanowski-Weber J, Luther H, Altmeyer P, König W. Effect of cytokines on spontaneous and allergen-induced CD23 expression, sCD23 release and Ig(E,G) synthesis from peripheral blood lymphocytes. Immunology. 1989;68:37–44.
89. Chomarat P, Briolay J, Banchereau J, Miossec P. Increased production of soluble CD23 in rheumatoid arthritis, and its regulation by interleukin 4. Arthritis Rheum. 1993;36:234–42.
90. Bansal AS, MacGregor AJ, Pumphrey RS, Silman AJ, Ollier WE, Wilson PB. Increased levels of sCD23 in rheumatoid arthritis are related to disease status. Clin Exp Rheumatol. 1994;12: 281–5.
91. Plater-Zyberk C, Bonnefoy JY. Marked amelioration of established collagen-induced arthritis by treatment with antibodies to CD23 *in vivo*. Nature Med. 1995;1:781–5.
92. Pene J, Chretein I, Rousset F, Briere F, Bonnefoy J, DeVries J. Modulation of IL-4-induced human IgE production in vitro by IFN-τ and IL-5: the role of soluble CD23 (sCD23). J Cell Biochem. 1989;39:253–64.

93. Sarfati M, Rubio-Trujillo M, Wong K, Rector E, Sehon AH, Delespesse G. In vitro synthesis of IgE by human lymphocyte. I. The spontaneous secretion of IgE by B lymphocytes from allergic individuals: A model to investigate the regulation of human IgE synthesis. Immunology. 1984;53:187
94. Sarfati M, Bettler B, Letellier M et al. Native and recombinant soluble CD23 fragments with IgE suppressive activity. Immunology. 1992;76:662–7.
95. Liu YJ, Cairns JA, Holder MJ et al. Recombinant 25-kDa CD23 and interleukin 1 alpha promote the survival of germinal center B cells: evidence for bifurcation in the development of centrocytes rescued from apoptosis. Eur J Immunol. 1991;21:1107–14.
96. Mossalayi D, Dalloul A, Arock M, Debre P. Effect of CD23 on purified human hematopoietic cells. Bone Marrow Transplant. 1992;9 Suppl. 1:50–3.
97. Mossalayi MD, Lecron J-C, Dalloul AH et al. Soluble CD23 (FcεRII) and interleukin 1 synergistically induce early human thymocyte maturation. J Exp Med. 1990;171:959–64.
98. Herbelin A, Elhadad S, Ouaaz F, De Groote D, Descamps-Latscha B. Soluble CD23 potentiates interleukin-1-induced secretion of interleukin-6 and interleukin-1 receptor antagonist by human monocytes. Eur J Immunol. 1994;24:1869–73.
99. Bartlett WC, Conrad DH. Murine soluble FcεRII: a molecule in search of a function. Res Immunol. 1992;143:431–6.

19
Commentary on FcR regulation of development and function of the immune system

J. C. CAMBIER

While acknowledging the importance of receptors for immunoglobulin constant regions (FcR) in regulation of the development and function of the immune system, most immunologists, and more broadly most biologists, are somewhat intimidated by this subdiscipline. This is the result of a number of factors, including FcR isoform multiplicity, general low affinity, ubiquity and frequent expression of more than one FcR isoform by individual cells. All of these factors conspire to make FcR function appear extremely complex. The reviews in this section provide a refreshing summation of our current knowledge of immunoregulatory function of FcR, and do much to demystify the discipline. Provided in the following paragraphs is an effort to integrate some of the concepts presented in the individual reviews and provide some additional information to make the treatise more seamless.

EFFECTS OF IMMUNOGLOBULINS ON DEVELOPMENT OF THE IMMUNE SYSTEM.

Immunoglobulins have a significant effect on both the development and function of the immune system. Perhaps the best evidence for effects on development comes from studies in which the initial exposure of developing animals to immunoglobulins has been manipulated [for review see Reference 1]. Among the best animal models for such studies are the artiodactyls, in which the first exposure to immunoglobulin comes through consumption of colostrum. In

J.G.J. van de Winkel and P.M. Hogarth (eds.), The Immunoglobulin Receptors and their Physiological and Pathological Roles in Immunity. 207–211.

artificially reared swine, for example, ingestion of 3–5 g purified IgG at birth has been shown to suppress the later *de novo* synthesis of IgG and IgA to the same extent as colostrum [2]. Studies in other immunological models also provide insight regarding this issue. In a mouse model, suckling of pups on the SCID versus conventional dams resulted in the accelerated development of intestinal IgA [3]. Finally, immunity develops more rapidly in formula-fed infants than in those nursed by their mothers [4]. Obviously, multiple mechanisms may be operative in mediating these effects. Maternal antibody may simply reduce antigen load [5], removing immunogens that stimulate immunoglobulin production and, in some cases, repertoire diversification [6]. However, the profound and long lasting effect of IgG ingestion at birth on serum immunoglobulin concentration in swine tempts one to speculate that exposure of the developing immune system to immunoglobulin plays a role in establishing the set-point of levels of mature lymphoid cells poised to respond to immunogen. Exposure to immunoglobulins early in development may reduce the size of this compartment. Interesting in this context is the fact that injection of heavy chain-specific IgG antibodies into neonatal mice induced profound B cell deficiency and prevented subsequent antibody responses [7].

HOW MIGHT IgG IMMUNOGLOBULINS REGULATE DEVELOPMENT OF THE IMMUNE SYSTEM?

Insight regarding how IgG immunoglobulins may regulate development of the immune system is provided by analyses of FcR expression by lymphocytes and their progenitors. As reviewed in Chapter 16, both FcγRII (CD32) and FcγRIII (CD16) are expressed by murine thymic late pro-T cells that are the precursors of both T cells and NK cells. A counterpart of these cells can be found in the intestinal epithelium. The expansion of these cycling CD3, CD4, CD8 negative cells is regulated by various cytokines. When these cells mature into pre-T cells and begin to express IL-2Rα, rearrange TCRβ genes and express pre-TCR both FcR are down regulated. FcR apparently do not reappear in the αβ T cell lineage until mature T cells are activated through TCR. However, maturing γδ T cells in the thymus express FcR.

FcγRII and FcγRIII are expressed by precursors of B cells and macrophage precursors in fetal liver [8] and apparently also in the yolk sac (see Chapter 16). FcγRIII is down-regulated as pro-B cells become pre-B cells and express pre-BCR. However, FcγRII expression is retained and is expressed in high levels throughout the B lineage from the pre-B stage to the plasma cell stage.

Clearly precursors of B and T cells that do not express pre-BCR or pre-TCR, express receptors for IgG Fc regions that could regulate the expansion and differentiation of this pool. Obviously, this regulation could not be antigen specific. Paradoxically, these cells co-express both FcγRIII and FcγRIIB, which have opposing biological functions. As discussed in Chapters 15 and 16, FcγRIII transduces activating signals using associated transducers, FcRγ and perhaps FcRβ chains, that are structural and functional counterparts of antigen receptor transducer subunits CD79a (Igα), CD79b (Igβ), CD3 components, and ζ and η chains of TCR. These chains all contain ITAM motifs that act as the respective receptors interface with cytoplasmic effectors of signal transduction [9]. In the mouse, B1, B1′ and B2 isoforms of FcγRIIB, are single chain receptors whose cytoplasmic tails contain an ITIM motif that forms the receptors interface with phosphatases that function in an inhibitory fashion (see Chapter 15).

Although it might seem that co-expression of these receptors would be counterproductive, recent findings suggest an alternate interpretation. Hippen *et al.* [10] have shown in B cells that FcγRIIB co-ligation with B cell antigen receptors does not totally abort signalling. Rather the FcγRIIB effectors block PI3-kinase activation, phosphoinositide hyrolysis and Ca^{2+} mobilization while leaving BCR-mediated activation of Src- and Syk-family tyrosine kinases intact. Thus, the FcRIIB signal is a qualitative modifier rather than an inhibitor of signal transduction by ITAM containing receptors. In B cells, this qualitative modulation can result in cell death by apoptosis (co-aggregation of FcγRIIb and BCR) [11] rather than proliferation (aggregation of BCR alone).

Viewed in this light, co-expression of FcγRIIB and FcγRIII takes on new significance; relative expression of the two receptors becomes key. High FcγRIII and low FcγRIIB expression versus the inverse should lead to equally significant but qualitatively distinct biologic responses. Thus, the response of developing T and B cells to FcγR ligation may be determined by differentiation-dependent changes in relative expression of FcγRII and FcγRIII. No evidence is available regarding changes in relative surface expression of these receptors early B and T cells. By co-igation of FcγRIIB and FcγRIII, immune-complexed maternal IgG derived from, for example, colostrum in artiodactyls, may modify development of the immune system; reducing immunoglobulin production long term by reducing the size of lymphoid compartments generated from T and B cell progenitors.

The alternative possibility that FcγRIIB and/or FcγRIII may have ligands other than IgG should, however, not be disregarded. Sandor and Lynch (Chapter 16) discuss experiments which have shown that blockade of FcR-interaction in fetal thymus organ culture by FcR specific antibodies, or blockade of ligands with soluble FcR, inhibits T cell development. This soluble FcR may bind a non-IgG ligand on thymic stromal cells [12]. Maternal immunoglobulins could act by blocking signal delivery by this ligand.

Finally, while the studies discussed above suggest a role for FcγRIIB and FcγRIII in the development of the murine immune system, ablation of genes encoding FcγRIIB or the FcγRIIIγ chain has no obvious effect on lymphoid development (see Chapters 15 and 16). Teleology dictates, however, that when expressed on pro-B and pro-T cells, these receptors must have some biological function.

LYMPHOID FcR AND REGULATION OF THE IMMUNE RESPONSE

While mature murine B cells constitutively express FcγRIIB, T cells only express this receptor after activation. FcγRIIB has been shown by Daëron and colleagues to modify both BCR and TCR signaling transduction. A large body of evidence reviewed in Chapter 15 illustrates the ability of IgG immune complexes to block the humoral response via co-aggregation of FcγRIIB with BCR. These findings are confirmed by observations that responses to soluble and particulate T cell-dependent antigen are enhanced 3 to 5-fold in FcγRII knockout mice. It seems most likely that the immunoregulatory role of this receptor comes into play late in the antibody response when immunoglobulin class switching has occurred and the affinity of the response has matured. At this point, IgG antibodies bind to immunogen remaining in the system, and deviating to apoptosis or otherwise preventing the activation of newly matured B cells that recognize epitopes on the immunogen that are not sterically hindered by bound antibody. The activation of these cells would be counter-productive. Because they express receptors of low average affinity, they would only serve to reduce the affinity of the ongoing response. In addition, activation of these cells may increase the likelihood of generating autoantibodies. Finally, the antibody-mediated coligation of FcγRIIB and BCR may also function to prevent or terminate the activation of idiotype specific B cells.

The role of FcγRIIB in T cell biology is less clear. They could be involved in the deviation or inactivation of T cells by naturally arising clonotypic antibodies. Additionally, they may be brought into play when T cells recognize target cells that are coated with IgG antibodies. However, inhibition of T cell activation by this mechanism would seem to be counterproductive.

FcγRIII is not expressed by B cells or most T cells, but plays an important role in antibody-dependent cytotoxicity (ADCC) mediated by a subpopulation of NK cells – the so-called TNK cells. Many human peripheral blood γ/δ T cells express FcγRIII and mediate ADCC. In the mouse intestinal, CD8 α/α T cells express FcγRIII, as do epithelial dendritic T cells in skin. It seems likely that in these cells FcγRIII serves only an effector function.

FcεRII (CD23) AND REGULATION OF THE IMMUNE RESPONSE

FcεRII or CD23, the low affinity receptor for IgE, also has been implicated in immune regulation (see Chapter 15). This receptor is expressed predominantly on B cells and follicular dendritic cells. Ligation of this receptor on B cells inhibits IgE production but does not affect production of other isotypes. This inhibition appears to occur subsequent to isotype switching. FcεRII can participate in uptake of IgE immune complexes for antigen presentation enhancing IgG production. In the latter context, expression of the receptor is thought to exacerbate allergy in atopic patients.

Relatively little is known about FcεRII-mediated signal transduction. It has been suggested that the human CD23a FcεRII isoform and the single mouse

FcεRII isoforms both contain an ITIM motif [13]. However, neither the phosphorylation of the tyrosine in this motif nor its ability to interact with phosphatases implicated in FcγRIIB signalling, has been demonstrated.

CONCLUSIONS

It is clear that receptors for immunoglobulin Fc regions play multiple roles in the immune function. As reviewed here by Guyre and O'Shea (Chapter 17), they play an important role in antigen uptake by antigen-presenting cells, including macrophages, B cells and dendritic cells. Available evidence indicates that FcγRIIB and FcγRIII play roles in the development of both T and B lymphoid populations, and FcγRIIB and FcεRII play important roles in regulation of the immune response. Particularly for FcγRIIB regulation of B cell responses, underlying molecular mechanisms are yielding to intense scrutiny. However, much remains to be done before the dynamic and diverse functions of Fc receptors will be fully understood.

References

1. Butler JE. Immunoglobulins and immune cells in animals milks. In: Ogra PL, Mesteeky J, Lanum ME, Strober W, McGhee JR, Bienenstock J, eds, Mucosal Immunology. New York: Academic Press (in press).
2. Klobasa F, Werhahn E, Butler JE. Regulation of humoral immunity in the piglet by immunoglobulins of maternal origin. Res Vet Sci. 1981;31:195–206.
3. Kramer DR, Cebra JJ. Early appearance of 'natural' mucosal IgA responses and germinal centers in suckling mice developing in the absence of maternal antibodies. J Immunol. 1995; 154:2051–62.
4. Stephens S. Development of secretory immunity in breast fed and bottle fed infants. Arch Dis Childhood. 1986;61:263–9.
5. Van Maanen C, Bruin G, de Boer-Luijtze E, Smolders G, deBoer GF. Interference of maternal antibodies with the immune response of foals after vaccination against equine influenza. Vet Quarterly. 1992;14:13–7.
6. Knight KL, Winstead CR. Generation of antibody diversity in rabbits. Curr Opin Immunol. 1997;9(2):228–32.
7. Manning DD, Jutila JW. Immunosuppression of mice injected with heterologous anti-immunoglobulin heavy chain antisera. J Exp Med. 1972;135:1316–23.
8. Carlsson L, Candéias S, Staerz U, Keller G. Expression of FcγRIII defines distinct subpopulations of fetal liver B cell and myeloiod precursors. Eur J Immunol. 1995;25:2308–17.
9. Cambier JC. Antigen and Fc receptor signaling: the awesome power of the immunoreceptor tyrosine based activation motif (ITAM). J Immunol. 1995;155:3281–5.
10. Hippen KL, Buhl AM, D'Ambrosio D, Nakamura K, Persin C, Cambier JC. Inhibitory FcγRIIB1 signaling in B cells is integrated by CD19 dephosphorylation. Immunity. 1997;7: 49–58.
11. Ashman RF, Peckham D, Stunz LL. Fc receptor off-signal in the B cell involves apoptosis. J Immunol. 1996;157:5–11.
12. Lynch RG, Hagen M, Mueller A, Sandor M. Potential role of FcγR in early development of murine lymphoid cells: evidence for functional interaction between FcγR on pre-thymocytes and an alternative, non-Ig ligand on thymic stromal cells. Immunol Lett. 1995;44:105–9.
13. Vivier E, Daëron M. Immunoreceptor tyrosine-based inhibition motifs. Immunol Today. 1997;18:286–90.

Part B

CLINICAL ASPECTS OF Fc RECEPTORS

20 The role and use of recombinant receptors in the investigation and control of antibody-induced inflammation

M. S. POWELL and P. M. HOGARTH

The interaction of immune complexes with cell surface Fc receptors is a potent stimulus for the activation of inflammatory cells. In autoimmune diseases this activation process can generate widespread tissue destruction, such as vasculitis associated with rheumatoid arthritis or glomerulonephritis induced by autoantibody deposition in systemic lupus erythematosus. Whilst the cause of autoimmunity is unknown, it is clear that immune complexes play a key role in the development of the disease pathology, through the activation of cells by complement or Fc receptors. Thus, the development of recombinant glycoproteins targeted at the neutralization of the cytotoxic mediators of immune complex-induced inflammation has been the focus of many recent investigations. Glycoproteins targeted to different mediators of the inflammation cascade limit the secretion of pro-inflammatory cytokines, act as cytokine antagonists or prevent receptor–ligand interactions. These have been developed as potential therapeutic agents in the treatment of immune complex diseases. Here we review examples of these proteins which are used to intervene in models of immune complex- or antibody-mediated diseases. Such recombinant glycoproteins could intervene at different points in the inflammation cascade but FcR blockade would also act to prevent re-presentation of antigen after complexing with Ig, i.e. could act at multiple points of immune complex formation, immune complex precipitation or re-presentation of antigen which

J.G.J. van de Winkel and P.M. Hogarth (eds.), The Immunoglobulin Receptors and their Physiological and Pathological Roles in Immunity. 215–231.

may be responsible for perpetuation of the disease once initiated (Figure 20.1). The role of recombinant soluble Fc receptors in the regulation of immune complex-mediated models of autoimmunity is also discussed.

Fc RECEPTOR AND INFLAMMATION

The production of autoantibodies against specific target cell structures or the formation and subsequent deposition of immune complexes can induce severe tissue damage. Moreover, the damage has largely been thought to be mediated through the activation of complement as the primary event with a subsequent role for Fc receptor-mediated activation of cells [1]. The seminal work of Dixon and colleagues on the mechanism by which immune complexes initiate inflammation indicated a major role for complement. It was also clear that inflamma-

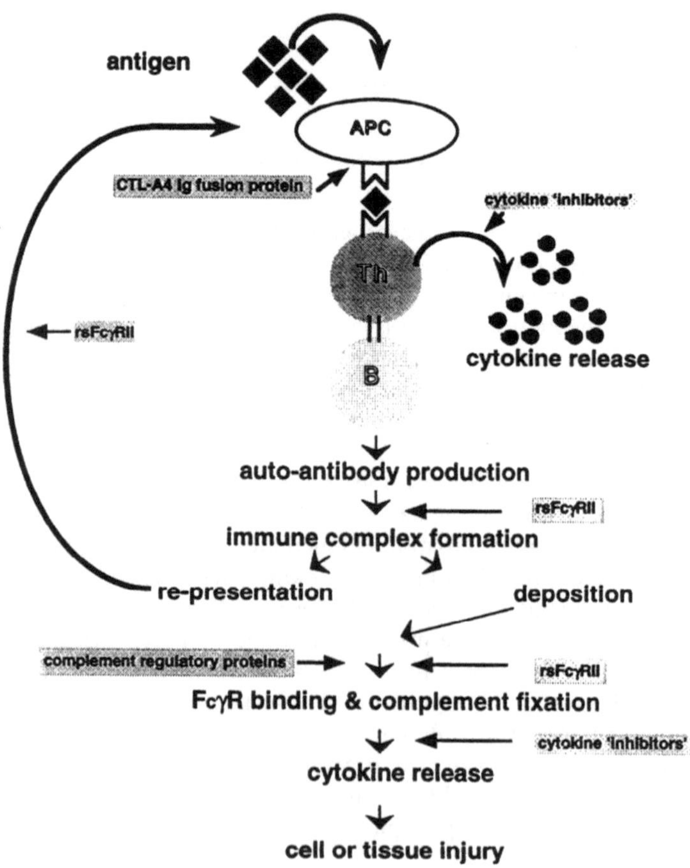

Figure 20.1 Sequence of events leading to antibody mediated tissue injury and possible perpetuation of disease. Sites of intervention by recombinant proteins are indicated in the boxes

tion was obtained following the inactivation of complement using a variety of agents, including cobra venom factor, implying that other mechanisms for the induction of inflammatory responses may also exist. The extent of Fc receptor involvement in this process is not well understood, but there is sufficient evidence available to show that blockade of Fc receptors is a potentially useful therapeutic approach in certain autoimmune conditions. Certainly the use of intravenous pooled human immunoglobulin (IVIg) to treat a variety of autoimmune conditions, including those with immune complex production, has implied a role for FcR [2]. Many explanations have been advanced to account for the 'therapeutic activity of IVIg including correction of idiotype anti-idiotype network imbalances, removal of pathogens, or FcR blockage [3]'. However, more direct evidence has come from the use of anti-human FcγRIII monoclonal antibodies to block uptake of immune complexes in a chimpanzee model of immune thrombocytopenia purpura [4,5] and in the treatment of experimental ITP [6]. Recent experiments using recombinant FcR or *in vivo* models of immune complex inflammation in gene deficient animals have pointed to a major role of FcR in the exacerbation of immune complex diseases. Moreover, Ierino *et al.* [7] demonstrated that Arthus reaction in rats was completely inhibited by using a recombinant soluble form of the human FcγRII. The treatment of these animals profoundly inhibited neutrophil localization in the blood vessels. Later analysis showed that mice deficient for the FcεRI γ-subunit, and therefore FcγRIII, failed to generate an Arthus lesion. Inactivation of FcγRIII also resulted in a diminished though not totally absent response, to immune complex deposition [8]. Taken together, these results clearly point to a major role for Fc receptors in the generation of severe immune complex-mediated inflammation and any strategy designed to prevent the interaction of immune complexes with Fc receptor bearing cells may have a profound influence on the outcome of the inflammatory process.

TREATMENT OF TYPE II HYPERSENSITIVITY REACTIONS

Type II (antibody-dependent cytotoxic) hypersensitivity reactions occur as a result of the inappropriate binding of antibody to the cell surface or tissue antigens (self or foreign), causing FcγR-mediated phagocytosis, or killer cell activity and/or complement mediated lysis of the target cell (Figure 20.1). Type II hypersensitivity reactions may contribute to the pathogenesis of a variety of disorders including Goodpasture's syndrome, myasthenia gravis, immune thrombocytopenia purpura (ITP), and autoimmune haemolytic anaemia (AIHA), as well as hyperacute graft rejection. These hypersensitivity reactions and treatment using recombinant glycoproteins are discussed each in turn and are summarized in Table 20.1.

Goodpasture's syndrome

Goodpasture's disease is an uncommon autoimmune disorder that typically

Table 20.1 Use of recombinant glycoproteins in the treatment of type II hypersensitivity reactions

Disease model	*Recombinant glycoprotein*	*References*
Goodpasture's syndrome	rshGoodpasture's antigen	11
	CTL-A4 Ig fusion protein (h)	13
Myasthenia gravis	CTL-A4 Ig fusion protein (h)	18, 19
	rTNF-β_1	20
	sCR1	21
Hyperacute xenograft rejection	rshCR1	40, 42, 44
	rshCD46	45
	Gal 1pep	46

rsh, recombinant soluble human; r. recombinant; rsm, recombinant soluble mouse; h, human

afflicts young men, is rapidly progressive and can cause glomerulonephritis and/or lung haemorrhage [9,10]. Patients produce autoantibodies to the Goodpasture antigen found on the NC1 domain of the α3 domain of type IV collagen, a minor component of the glomerular basement membrane (GBM) [11]. The development of a human recombinant Goodpasture antigen specific for the NC1 domain, can inhibit autoantibody binding to human GBM *in vitro* [11] and the development of a single chain antibody reactive for the Goodpasture antigen [12] has been reported. These may be useful as physical blockers of autoantibody binding to GBM *in vivo.*

The use of CTL-A4 Ig fusion protein has shown promising effects in the treatment of experimental Goodpasture's syndrome (Table 20.1). Animals treated with recombinant human CTL-A4 Ig fusion protein from day zero to 14 or to day 35 had reduced disease severity in a model following immunization with bovine GBM [13]. These beneficial effects were observed if treatment was initiated after the onset of glomerulonephritis, it appears that the CTL-A4 Ig fusion protein may limit generated B cell stimulation and hence autoantibody production. Clearly, this form of therapy acts to limit the down-stream events in the inflammation cascade (see Figure 20.1). It is also tempting to speculate that rsFcγRII may be useful in the treatment of Goodpasture's syndrome given this material can avidly inhibit immune complex induced tissue damage at sites of inflammation [7], and may limit any further damage by preventing effector cell activation.

Myasthenia gravis (MG)

Myasthenia gravis is an autoimmune disorder mediated by autoantibody against the acetylcholine receptors present on the post-snyaptic membrane of the neuromuscular junction [14]. Normally, transmission of impulses from the nerve to the muscle takes place through the release of acetylcholine from the nerve termini and diffusion across the nerve junction to the muscle fibre. However, in MG autoantibodies to the acetylcholine receptors and complement proteins C3 and C9, present at the postsynaptic folds, are thought to increase the rate of turnover of the acetylcholine receptors and act to block acetylcholine binding [14,15].

It is well established that T cells play a key role in the pathogenesis of MG. Strategies designed to limit T cell proliferation have included the use of peptide analogues to pathogenic epitopes of the acetylcholine receptor which can specifically interfere with myasthogenic specific T cell receptors without concomitant activation of other T cells [16,17]. In an experimental rodent model of autoimmune MG, CTL-A4 Ig fusion protein has been shown to interfere with T cell co-stimulatory signals. Moreover, the fusion protein results in a profound decrease in the production of IL-2, which is known to be elevated in active disease [18,19], and can significantly decrease lymphoproliferative and antibody responses by primed lymph node cells, when restimulated with acetylcholine receptor *in vitro*.

The inhibition of pro-inflammatory cytokine secretion can act as an alternative treatment in MG patients. Low concentrations of recombinant TGF-β_1 (rTGF-β_1), could suppress autoantigen up-regulation of pro-inflammatory cytokines IFN-γ, IL-4, IL-6, and TNF-α from cultured mononuclear cells of a MG patient, as well as the number of acetylcholine reactive IFN-γ and IL-4 secreting cells *in vitro*. Thus, given these results it would suggest that rTGF-β_1 has an immunosuppressive effect on the up-regulation of pro-inflammatory cytokine secretion – this recombinant cytokine may be an attractive alternative in the treatment of MG [20].

Finally, until recently little evidence has been provided linking complement and possibly FcR to the pathogenesis of MG. In a recent paper it was elegantly demonstrated that the use of recombinant soluble complement receptor 1 (sCR1) could significantly alter the clinical severity of experimental MG [21].

Immune thrombocytopenia purpura (ITP)

Immune thrombocytopenia purpura (ITP) is a disorder characterized by autoantibody induced destructive thrombocytopenia [22,23]. Autoantibodies from patients have been found to predominantly bind to glycoprotein complexes IIb/IIIa and Ib/IX, causing FcγR mediated uptake of sensitized platelets in the spleen and liver [22,24,25]. The treatment of ITP in patients has primarily focused on the administration of IVIg causing both Fc receptor blockage [23,26] and on anti-idiotypic suppression of the autoreactive B cell clone [27]. Moreover, from the vast amount of literature available concerning the pathogenesis and treatment of ITP, it is evident that FcγRs play a crucial role in the pathogenesis of this disease. Indeed, based on the observations that treatment of chimpanzees and humans with anti-FcγRIII monoclonal antibody protected from platelet destruction led investigators to examine further the role of the FcγRs in the pathogenesis of ITP [4,5]. Clinical studies using *in vivo* immune complexes generated by anti-human red blood cells [27], IgG Fc fragments [28], and an antibody against FcγRIII [4,5,29] have all shown that blockage of cell surface FcγR inhibits the depletion of sensitized cells or platelets. Interestingly, as an alternative to blockade of FcγRs, the use of an extracorporeal protein A immunoabsorption of patient serum caused a decline in circulating immune complex levels and anti-platelet IgG [30]. Also, it has been demonstrated that mice deficient for the FcεRI γ-chain and therefore the cell surface expression of FcγRIII protein [31], were protected from platelet depletion following administration of a cytotoxic platelet mAb [31,32]. To date recombinant soluble FcγR glycoprotein has not been used in therapy but, in light of the role of FcγR in ITP and AIHA, use of rsFcγRII glycoprotein in therapy would be of interest.

Hyperacute graft rejection

Xenograft transplantation is being considered as an alternative to allograft transplantation. However, the rejection of xenografts largely involves antibody and complement, although some cell machinated processes are also involved in the absence of antibody as indeed they are in allograft rejection [32–42]. In experimental models of xenograft rejection several agents have been used to prolong graft survival – cobra venom factor [41], sCRI with other anti-inflammatory agents [40,42,43]. In addition, other regulators of complement action e.g. CD46, DAF have been shown to be potentially useful [45–47] (Table 20.1).

An alternative approach for the prolongation of xenograft survival is the development of inhibitors that can specifically block naturally occurring anti-Gal α(1,3)Gal, low titred IgM antibodies from binding to cell surface antigens present on the surface of the transplanted organ. These antibodies initiate both complement and coagulation cascades [48–50], but synthetic octapeptide, Gal pep1 (DAHWESWL) that mimics the Gal epitope can specifically inhibit the interaction between human anti-Gal α(1,3) Gal antibodies and the carbohydrate Gal α(1,3) Gal *in vitro* [50].

TREATMENT OF TYPE III HYPERSENSITIVITY REACTIONS

Type III hypersensitivity reactions are mediated principally by immune complex formation and precipitation. Many of the down-stream processes elicited in type III hypersensitivity reactions are identical to those by Type II reactions (Figure 20.1), where the persistent stimulation of FcγRs and complement are crucial parameters in the pathogenesis of these diseases. Indeed, the involvement of complement in the modulation of Type III hypersensitivity reactions is well documented. Treatment of rats with cobra venom factor inhibits the Arthus reaction, a classic type III hypersensitivity reaction [51,52]. Moreover, the recombinant soluble complement regulatory proteins, CR1, decay accelerating factor and CD46 can inhibit the Arthus reaction [46,47,53]. Recent evidence also indicates that immune complex deposition and down-stream activation of inflammatory cells via Fc receptor binding is a major stimulus of Type III hypersensitivity reactions, resulting in damage to the tissue or organ, depending on site of immune complex deposition [7,54,55]. The administration of rsFcγRII not only blocks effector function but also alters the behaviour of immune complexes [55]. Moreover, the precipitation of immune complexes is profoundly delayed by the presence of rsFcγRII, making it an attractive therapeutic modality. Other examples of Type III hypersensitivity reactions that have been extensively studied using experimental models and treated using recombinant soluble glycoproteins include; systemic lupus erythematosus (SLE), experimental allergic encephalomyelitis (EAE) and rheumatoid arthritis (RA). Each model will be discussed in turn in the following sections and are summarized in Table 20.2.

Systemic lupus erythematosus (SLE)

Immune complex production and deposition is a usual feature of SLE. This disease is predominantly seen in women and the clinical manifestations may include arthritis, vasculitis, and/or glomerulonephritis. Autoimmune thrombocytopenia and haemolytic anaemia are also common complications of SLE [56]. Several animal models that resemble the human disease have been described: the (NZW × NZB) F_1; MRL/*lpr* and BXSB mouse models are clinically and immunopathologically similar although there are a number of distinct differences [57]. However, each has been utilized to study disease pathology as well as possible use of recombinant proteins in treatment.

Cytokines are known to play a significant but somewhat imprecise role in the pathogenesis of SLE in mouse models.. Moreover, the development of either Th_1 or Th_2 responses has profound immunological consequences, although it remains unclear which subset of Th cells plays a more critical role in lupus pathogenesis. To date the treatment of mice or *in vitro* analysis with anti-IFN-γ monoclonal antibody [58], mouse soluble IFN-γ receptor [59], recombinant TNF-α [60,61], anti-IL-10 monoclonal antibody [62,63], anti-IL-4 mAb [64,65], IL-4 receptor [65], IL-1 receptor [66], anti IL-6 mAb and anti-IL-6 receptor mAb [67] have all delayed the onset of autoimmunity in mice strains prone to the development of lupus, or shown promising effects by limiting autoantibody production. It is interesting to note that from the recombinant glycoproteins used thus far in the treatment of mouse lupus models, there are some surprising results. For example, several activities of IFN-γ and TNF-α are synergistic [59], making it unexpected that recombinant TNF-α should inhibit disease progression in mice prone to the development of lupus. In a recent study, an IL-4 transgene in B cells completely prevented the development of lupus in (NZW × C57BL/6.Yaa)F_1 mice. It was reported that the constitutive expression of IL-4 resulted in a down-regulation of IgG subclasses of autoantibodies such

Table 20.2 Use of recombinant glycoproteins in the treatment of Type III hypersensitivity reactions

Disease/model	*Recombinant glycoprotein*	*References*
Reverse passive Arthus reaction	rshFcγRII	7
	rshCD46	46
	rshDAF (CD55)	47
	rshCR1 (CD35)	53
Systemic lupus erythematosus (SLE)	anti-IFN γ mAb	58
	rsmIFN-γR	59
	rTNF-α	60, 61
	anti-IL-10 mAb	62, 63
	anti-IL-4 mAb	64, 65
	IL-4 receptor	65
	IL-1 receptor	66
	anti-IL-6 mAb	67
	anti-IL-6 receptor mAb	67
	CTL-A4 Ig fusion protein (m)	71
	rshCR1	72
Experimental allergic encephalomyelitis (EAE)	rshCR1	77
	rshIL-1 receptor antagonist	78, 79
	rhIL-10	80, 81
	rhIL-13	82
	rIL-2	80
	rhIL-6	83
	rIFN-τ (o)	84
	TGF-β_1	20
	rhIFN-β_{1b}	81, 85–89
	anti-mouse CD4 mAb	90–94
	mouse CD4 analogue	95
	long-lived MHC peptides	75, 96, 97
Rheumatoid arthritis (RA)	anti-human TNF mAb	98–104
	rshIL-1 receptor antagonist	

rsh, recombinant soluble human; h, human; r, recombinant; rsm, recombinant soluble mouse; (m), mouse; (o), ovine

as the IgG_3, which are believed to be especially nephritogenic [68]. This result is surprising given the earlier studies that specifically show that the interference with IL-4 secretion limits the disease pathogenesis (see above). However, as it recognized that a number of factors contribute to disease pathogenesis, it is suggested that the usefulness of a particular cytokine may well depend on the progression of the disease state, and/or indeed the age of the animal oat the onset of therapy [59,69,70] (see also Table 20.2), which may clearly bear significance in the treatment of patients.

Other alternatives studied in the treatment of mouse SLE have focused on prevention of T cell stimulation by inhibition of the B7–CD28 interaction using the CTL–A4 Ig fusion protein. Indeed it has been demonstrated that the murine CTL–A4 If fusion protein can selectively retard the progress of SLE [71] (Table 20.2). Mice treated with the fusion protein did not make autoantibodies to dsDNA during treatment or three months following treatment withdrawal. Suppression of autoantibody production was also accompanied by reduced severity of nephritic damage. There was no depletion of T and B lymphocytes *in vivo* and, importantly CTL-A4 could suppress an established immune response [71] making this recombinant glycoprotein an attractive alternative for the treatment of Ig mediated hypersensitivity reactions that are already established.

Finally, as it is the inappropriate immune complex formation and deposition that causes the tissue damage, it remains to be seen whether rsFcγRII can effectively limit the immune dysfunction seen in the experimental models of SLE. However sCR1, or anti-C5 monoclonal antibody, significantly reduced both morphologic and functional consequences of complement-mediated glomerulonephritis [72,73] (Table 20.2).

Experimental allergic encephalomyelitis (EAE)

Experimental allergic encephalomyelitis (EAE) is an acute autoimmune disease elicited in rodents by immunization with myelin basic protein (MBP) [74]. The disease phenotype is characterised by the invasion of the central nervous system by $CD4^+$ T cell lymphocytes, demyelination and acute, chronic, or chronic relapsing paralysis [75]. It is used as a model for the analysis of the human disease multiple sclerosis (MS), although it is recognized that the human disease MS is probably a more complex disease than EAE. Given the pathological correlations between MS in humans and EAE in rodents, however, it makes this model an important experimental tool for further study of the human disease [76].

The pathogenesis of EAE is exacerbated by complement activation and/or inappropriate cytokine production. The treatment of antibody-mediated demyelinating EAE, (a variant of the EAE model) using human recombinant soluble CR1 protein was successful [77] (Table 20.2). The recombinant glycoprotein was shown to have significant effect on the disease progression in treated animals, significantly reducing the clinical severity of disease in these animals. Other glycoproteins that limit immune complex activation of target cells have not been reported to date.

Numerous studies that have evaluated the use of cytokines or cytokine antagonist as effective agents in the treatment of EAE. The known recombinant cytokines or cytokine receptor antagonists demonstrated to significantly alleviate clinical signs of EAE include: recombinant human IL-1 receptor antagonist [78,79], recombinant human IL-10 [80,81], human recombinant IL-13 [82], recombinant IL-2 [80], human recombinant IL-6 in a murine viral model of MS, Theiler's murine encephalomyelitis virus (TMEV), [83], ovine recombinant IFN-τ [84], TGF β_1 [20], and human recombinant IFN-β_{1b} (Betaseron), [81,85–89]. Currently recombinant human IFN-β_{1b} (rhIFN-β_{1b}) is the approved drug in the management of MS [89],although mechanism of action is still unclear (Table 20.2).

The pathogenesis of EAE is attributable to $CD4^+$ T lymphocytes as anti-CD4 monoclonal antibody and human CTLA-4 Ig fusion protein can both inhibit the development of EAE in rodents [90–94]. A small, synthetic, analogue of murine CD4 was demonstrated to have a potent capacity to inhibit EAE [95] (Table 20.2). Created by rational drug design, the cyclic peptide is composed of D-amino acids and restrained by introduced disulphide bridges to mimic the surface of CD4. The CD4 analogue has biological activity *in vivo*, is capable of inhibiting EAE, presumably by blocking T cell activation, and does not deplete the $CD4^+$ T cell subset and has no inherent immunogenecity [95]. As yet the use of this analogue for therapy has not been reported, but poses an exciting prospect for future treatment.

In keeping with strategies designed to inhibit T cell mediated activation, others have utilised myelin basic peptides to form long life peptide–MHC complexes for reversal of acute EAE [75,96]. These soluble peptide variants are thought to operate by MHC blockade [97], or by altering IFN-γ and TNF-α production [73]. Currently it is not known how these soluble peptides will affect normal MHC presentation or exactly how long they will elicit their immunoregulatory functions.

Rheumatoid arthritis (RA)

Rheumatoid arthritis (RA) is a systemic and chronic autoimmune inflammatory disorder that can have several effects on joints with severe extra-articular complications, including uveitis and vasculitis. The development of autoantibodies to the Fc portion of IgG (rheumatoid factor), type II collagen, or heat shock proteins (hsp) may also result in the precipitation of immune complexes in the synovial space and surrounding blood vessels initiating the release of inflammatory cytokines [14]. Analysis of mRNA and protein in RA tissues found that pro-inflammatory cytokines TNF-α, IL-6, IL-1, GMCSF and IL-8 are abundant. Moreover, the production of these was partly regulated *in vivo* by high levels of anti-inflammatory cytokines IL-10, TGF-β and cytokine inhibitors such as IL-1 receptor antagonist (IL-IRA), and soluble TNF receptor [98]. However, these were not present in sufficient quantities to neutralize all the pro-inflammatory cytokines produced. *In vitro* neutralization of TNF-α results in down-regulation of pro-inflammatory cytokine production implying that TNF-α is the 'dominant' cytokine in the pathogenesis of RA. Indeed a chimeric anti-TNF-α monoclonal antibody decreased damage in mouse models of destructive arthritis [99] (Table 20.2). Clinical trials using this mAb have demonstrated a dose-dependant efficacy and decrease in disease activity with acute-phase responses lasting weeks [98–103].

An alternative recombinant glycoprotein trialed as a possible treatment of RA has been the IL-1 receptor antagonist, (IL-IRA). this glycoprotein can also effectively ameliorate high levels of pro-inflammatory cytokines in models of RA, but it has been demonstrated to be less effective than treatment with the anti-TNF-α mAb [104]. Recently, it was reported that gene therapy using a retrovirus construct containing the human IL-IRA protein (IRAP) gene can effectively inhibit the influx of leukocytes into the joint space of rabbits for a period of time and the concentration of rabbit IL-1 was reduced by the presence of the IRAP gene [105]. This would suggest that the introduction of the IRAP gene and subsequent secretion of IL-IRA can inhibit the production of pro-inflammatory cytokines and effectively change the course of arthritic disease in rabbit knees. Studies examining the systemic expression of human IL-IRA in mice showed significant levels of IL-IRA protein were present in sera for at least 15 months after transplantation of retroviral transduced haematopoietic stem cells [106], suggesting that treatment can be potentially maintained for long periods of time. There has been some speculation in the literature that a trial in human RA will also use this approach [107,108].

It remains to be seen whether other recombinant glycoproteins that are known to effectively inhibit the inflammation cascade (Figure 20.1) would be useful in the amelioration of RA (see above). RsFcγRII has a promising role in the treatment of this disorder as it is effective in reduction of Arthus reaction lesions [7], which show characteristics of the extra-articular complications in RA including vasculitis, and interestingly rsFcγRII can inhibit the binding of rheumatoid factor to FcR-bearing cells (B. Wines, unpublished). It is important to note that the binding of immune complexes to FcR^+ cells is a potent inducer of TNF-α release [109–111]. Thus, by limiting the interaction of immune complexes with Fc receptor bearing cells by using rsFcγRII, could make this particular recombinant glycoprotein a crucial part of the treatment for RA.

CONCLUSIONS

Many strategies are now emerging in attempts to neutralize key components in immune complex-mediated inflammatory processes. The extent to which such approaches will be effective remains to be seen but there are encouraging signs in clinical trials that are in progress. However, recombinant proteins have inherent problems: they are expensive to make and may have glycosylation or other post-translational differences as a result of production in non-human production systems; genetic polymorphisms also exist which result in amino acid differences between individuals. Such facts raise the issue of immunogenicity of the therapeutic agent either limiting its life span or possibly inducing autoimmunity [112–144]. Nonetheless, whatever the commercial or scientific limitations of recombinant protein use, the observation that inhibition of Fc–FcR or other ligand–receptor interactions may prevent inflammation, encourages the pursuit of other strategies with the same aim. To this end, the use of modern high throughput screening technologies in conjunction with combinatorial libraries of low molecular weight organic molecules promises to yield many potential new 'traditional-type' drugs which may have anti-inflammatory activity. Combinatorial libraries have the potential to generate huge diversity in the compounds synthesized in these libraries. Indeed, there have been recent reports of the isolation of such compounds as potential inhibitors of the IgE/FcεRI interaction [115], and the CD4/MHC class II interaction [116]. However, putting aside the applied and commercial considerations it is clear that new technologies and new approaches have enabled a more thorough dissection of the inflammatory processes and the role of Fc and other receptors in the induction of these.

References

1. Dixon FJ. Antigen-antibody complexes and autoimmunity. Ann NY Acad Sci. 1965;124:162–6.
2. Bussel JB, Kimberly RP, Inman RD et al. Intravenous gamma globulin treatment of chronic idiopathic thrombocytopenic purpura. Blood. 1983;62:480–6.
3. Bussel JB. Modulation of Fc receptor clearance and antiplatelet antibodies as a consequence of intravenous immune globulin infusion with immune thrombocytopenic purpura. J Allergy Clin Immunol. 1989;84:566–71.
4. Clarkson SB, Kimberly KP, Valinsky JE et al. Blockade of clearance of immune complexes by an anti-Fc gamma receptor monoclonal antibody. J Exp Med. 1986;164:474–89.
5. Kimberly RP, Edberg JC, Merriam LT, Clarkson SB, Unkeless JC, Taylor RP. *In vivo* handling of soluble complement fixing AB/dsDNA immune complexes in chimpanzees. J Clin Invest. 1989;84:962–70
6. Clarkson SB, Bussel JB, Kimberly RP, Valinsky JE, Nachman RL, Unkeless JC. Treatment of refractory immune thrombocytopenic purpura with an anti-Fcγ-receptor antibody. N Engl J Med. 1986b;314:1236–9.
7. Ierino FL, Powell MS, McKenzie IFC, Hogarth PM. Recombinant soluble FcγRII; production, characterisation and inhibition of the Arthus reaction. J Exp Med. 1993;178:1617–28.
8. Hazenbos WL, Gessner JE, Hofhuis FM et al. Impaired IgG-dependent anaphylaxis and Arthus in Fc gamma RIII (CD16) deficient mice. Immunity. 1996;5:181–8.
9. Kelly PT, Haponik EF. Goodpasture syndrome molecular and clinical advances. Medicine (Baltimore). 1994;73:171–85.
10. Bosch T. Current status in extracorporeal immunomodulation: immune disorders. Artif Organs. 1996;20:902–5.
11. Turner N, Forstová J, Rees A, Pusey CD, Mason PJ. Production and characterisation of recombinant Goodpasture's antigen in insect cells. J Biol Chem. 1994;269:17141–5.
12. Ross CN, Turner N, Savage P, Cashman SJ, Spooner RA, Pusey CD. A single-chain Fv reactive with the Goodpasture antigen. Lab Invest. 1996;74:1051–9.
13. Nishikawa K, Linsey PS, Collins AB, Stamenkovic I, McClusky RT, Andres G. Effect of CTL-A4 chimeric protein on rat autoimmune anti-glomerular basement membrane glomerulonephritis. Eur J Immunol. 1994;24:1249–54.
14. Roitt I, Brostoff J, Male D. Immunology, 3rd edn. London: Mosby; 1993.
15. Kernich CA, Kaminski HJ. Myasthenia gravis: pathophysiology, diagnosis and collaborative care. J Neurosci Nurs. 1995;27:207–15.
16. Zisman E, Katz-Levy Y, Dayan M et al. Peptide analogs to pathogenic epitopes of the human acetylcholine receptor a subunit as potential modulators of myasthenia gravis. Proc Natl Acad Sci USA. 1996;93:4492–7.
17. Wauben MH, Hoedemaekers AC, Graus YM, Wagenaar JP, van Eden W, de Baets MH. Inhibition of experimental autoimmune myasthenia gravis by major histocompatibility complex class II competitor peptides results not only in a suppressed but also in an altered immune response. Eur J Immunol. 1996;26:2866–75.

18. McIntosh KR, Linsey PS, Drachman DB. Immunosuppression and induction of energy by CTL-A4 Ig *in vitro*: effects on cellular and antibody responses of lymphocytes from rats with experimental autoimmune myasthenia gravis. Cell Immunol. 1995;166:103–12.
19. Yi Q, Ahlberg R, Pirskanen R, Lefvert AK. Acetylcholine receptor-reactive T cells in myasthenia gravis: evidence for the involvement of different subpopulations of T helper cells. J Neuroimmunol. 1994;50:177–86.
20. Link J, He B, Navikas V et al. Transforming growth factor-beta 1 suppresses autoantigen-induced expression of pro-inflammatory cytokines but not of interleukin-10 in multiple sclerosis and myasthenia gravis. J Neuroimmunol. 1995;58:21–35.
21. Piddleston SJ, Jiang S, Levin JL, Vincent A, Morgan BP. Soluble complement receptor 1 (sCR1) protects against experimental autoimmune myasthenia gravis. J Neuroimmunol. 1996;71:173–7.
22. McMillan R. Chronic idiopathic thrombocytopenic purpura. New Engl J Med. 1981;304: 1135–91.
23. Bussel JB. Autoimmune thrombocytopenic purpura. Hematol Oncol Clin N Am. 1990;4: 179–91.
24. Kiefel V, Santoso S, Kaufmann E, Mueller-Eckhardt C. Autoantibodies against platelet glycoprotein Ib/IX: a frequent finding in autoimmune thrombocytopenic purpura. Br J Haematol. 1991;79:256–62.
25. Kiefel V, Freitag E, Kroll H, Santoso S, Mueller-Eckhardt C. Platelet autoantibodies (IgG, IgM, IgA) against glycoproteins IIb/IIIa and Ib/IX in patients with thrombocytopenia. Ann Hematol. 1996;72:280–5.
26. Imbach P, Barandun S, d'Apuzzo V et al. High-dose intravenous gamma-globulin for idipathic thrombocytopenic purpura in childhood. Lancet. 1981;1:1228–31.
27. Salama A, Mueller-Eckhardt C, Kiefel V. Effect of intravenous immunoglobulin in immune thrombocytopenia. Lancet. 1983;2:193–5.
28. Debre M, Bonnetr MC, Fridman WH et al. Infusion of Fc gamma fragments for treatment of children with acute immune thrombocytopenic purpura. Lancet. 1993;342:945–9.
29. Soubrane C, Tourani JM, Andrieu JM et al. Biological response to anti CD-16 monoclonal antibody therapy in a human immunodeficiency virus-related immune thrombocytopenic purpura patient. Blood. 1993;81:15–19.
30. Balint JP, Cochran SK, Jones FR. Modulation of idiotypic and anti-isotypic immunoglobulin G responses in an immune thrombocytopenic purpura patient as a consequence of extracorporeal protein A immunoadsorption. Arfit Organs. 1995;19:496–509.
31. Takai T, Li M, Sylevestre D, Clynes R, Ravetch JV. Fcγ chain deletion results in pleiotrophic effector cell defects. Cell. 1994;76:519–29.
32. Clynes R, Ravetch JV. Cytotoxic antibodies trigger inflammation through Fc receptors. Immunity. 1995;3:21–6.
33. Linsley PS, Wallace PM, Johnson J et al. Immunosuppression in vivo by a soluble form of the CTLA4 T cell activation molecule. Science. 1992;257:792–5.
34. Lenschow DJ, Zeng Y, Thistlethwaite JR et al. Long-term survival of xenogeneic pancreatic islet grafts induced by CTLA4Ig. Science. 1992;257:789–92.
35. Turka LA, Linsley PS, Lin H et al. T-cell activation by the CD28 ligand B7 is required for cardiac allograft rejection in vivo. Proc Natl Acad Sci USA. 11992;89:11102–5.
36. Lin H, Bolling SF, Linsley PS et al. Long-term acceptance of major histocompatibility complex mismatched cardiac allografts induced by CTLA4Ig plus donor-specific transfusion. J Exp Med. 1993;178:1801–6.
37. Pearson TC, Alexander DZ, Winn KJ, Linsley PS, Lowry RP, Larsen CP. Transplantation tolerance by CTLA4-Ig. Transplantation. 1994;57:1701–6.
38. Wallace PM, Johnson JS, MacMaster JF, Kennedy KA, Gladstone P, Linsley PS. CTLA4Ig treatment ameliorates the lethality of murine graft-versus-host disease across major histocompatibility complex barriers. Transplantation. 1994;58:602–10.
39. Zhow X-J, Niesin N, Pawlowski I et al. Prolongation of survival of discordant kidney xenografts by C6 deficiency. Transplantation. 1990;50:896–8.
40. Pruitt SK, Baldwin WM, Marsh HC, Lin SS, Yeh CG, Bollinger RR. The effect of soluble complement receptor type 1 on hyperactive xenograft rejection. Transplantation. 1991;52: 868–73.
41. Leventhal JR, Dalmasso AP, Cromwell JW et al. Prolongation of cardiac xenograft survival by depletion of complement. Transplantation. 1993;55:857–66.

42. Zehr KJ, Herskowitz A, Lee PC, Kumar P, Gillinov AM, Baumgartner WA. Neutrophil adhesion and complement inhibition prolongs survival of cardiac xenografts in discordant species. Transplantation. 1994;57:900–6.
43. Burch RM, Weitzberg M, Blok N et al. N-(fluorenyl-9-methoxycarbonyl) amino acids, a class of antiinflammatory agents with a different mechanism of action. Proc Natl Acad Sci USA. 1991;88:355–9.
44. Pruitt SK, Kirk AD, Bollinger RR et al. The effect of soluble complement receptor type 1 on hyperactive rejection of porcine xenografts. Transplantation. 1994;57:363–70.
45. Christiansen D, Milland J, Thorley BR et al. Engineering of recombinant soluble CD46: an inhibitor of complement activation. Immunology. 1996;87:348–54.
46. Christiansen D, Milland J, Thorley BR, McKenzie IFC, Loveland BE. A functional analysis of recombinant soluble CD46 *in vivo* and a comparison with recombinant soluble forms of CD55 and CD35 *in vitro*. Eur J Immunol. 1996;26:578–85.
47. Moran P, Beasley H, Gorrell A et al. Human recombinant soluble decay accelerating factor inhibits complement activation *in vitro* and *in vivo*. J Immunol. 1992;149:1736–43.
48. Sandrin MS, Vaughan HA, McKenzie IFC. Identification of Galα(1,3)Gal as a major epitope for pig-to-human vascularised xenografts. Transplantation Rev. 1994;128:134–45.
49. Sandrin MS, McKenzie IFC. Galα(1,3)Gal, the major xenoantigen(s) recognised in pigs by human natural antibodies. Immunol Rev. 1994;141:169–75.
50. Vaughan HA, Oldenburg KR, Gallop MA, Atkin JD, McKenzie IFC, Sandrin MS. Recognition of an octapeptide sequence by multiple Galα(1,3)Gal-binding proteins. Xenotransplantation. 1996;3:18–23.
51. Ward PA, Cochrane CG. Bound complement and immunologic injury of blood vessels. J Exp Med. 1965;121:215–31.
52. Cochrane CG, Janoff A. In: Zweifach BW, Grant L, McCluskey RT, eds, The Inflammatory Process. Academic Press; 1974.
53. Yeh CG, Marsh HC Jr, Carson GR et al. Recombinant soluble human complement receptor type I inhibits inflammation in the reversed passive Arthus reaction in rats. J Immunol. 1991; 146:250–6.
54. Sylvestre DL, Ravetch JV. Fc receptors initiate the Arthus reaction: redefining the inflammatory cascade. Science. 1994;265:1095–8.
55. Gavin AL, Wines BD, Powell MS, Hogarth PM. Recombinant soluble FcγRII inhibits immune complex precipitation. Clin Exp Immunol. 1995a;102:620–5.
56. Mills JA. Systemic lupus erythematosus. N Engl J Med. 1994;330:1871–9.
57. Andrews BS, Eisenberg RA, Theofilopoulos AN et al. Spontaneous murine lupus-like syndromes. J Exp Med. 1978;148:1198–215.
58. Jacob CO, van der Meide PH, McDevitt HO. *In vivo* treatment of (NZB × NZW)F_1 lupus-like nephritis with monoclonal antibody to γ-interferon. J Exp Med. 1987;166:798–803.
59. Ozman L, Roman D, Fountoulakis M, Schmid G, Ryffel B, Garotta G. Experimental therapy of systemic lupus erythematosus: the treatment of NZB/W mice with mouse soluble interferon-γ receptor inhibits the onset of glomerulonephritis. Eur J Immunol. 1995;25:6–12.
60. Jacob CO, McDevitt MO. Tumor necrosis factor-α in murine autoimmune 'lupus' nephritis. Nature (London). 1988;331:356–8.
61. Gordon C, Ranges GE, Greenspan JS, Wofsy D. Chronic therapy with recombinant tumor necrosis factor-alpha in autoimmune NZB/NZW f1 mice. Clin Immunol Immunopathol. 1989;52:421–34.
62. Llorente L, Zou W, Levy Y et al. Role of interleukin 10 in the B lymphocyte hyperactivity and autoantibody production of human systemic lupus erythematosus. J Exp Med. 1995;181:839–44.
63. Ishida H, Muchamuel T, Sakaguchi S, Andrade S, Menon S, Howard M. Continuous administration of anti-interleukin 10 antibodies delays onset of autoimmunity in NZB/W F_1 mice. J Exp Med. 1994;179:305–10.
64. Nakajima A, Hirose S, Yagita H, Okumura K. Roles of IL-4 and IL-12 in the development of lupus in NZB/W F1 mice. J Immunol. 1997;158:1466–72.
65. Schorlemmer HU, Dickneite G, Kanzy EJ, Enssle KH. Modulation of the immunoglobulin dysregulation in GvH- and SLE-like diseases by the murine IL-4 receptor (IL-4-R). Inflamm Res. 1995;44:S194–6.
66. Schorlemmer HU, Kanzy EJ, Langner KD, Kurrle R. Immunoregulation of SLE-like disease by the IL-1 receptor: disease modifying activity on BDF1 hybrid mice and MRL autoimmune mice. Agents Actions. 1993;39:C117–20.

67. Kitani A, Hara M, Hirose T et al. Autostimulatory effects of IL-6 on excessive B cell differentiation in patients with systemic lupus erythematosus: analysis of IL-6 production and IL-6R expression. Clin Exp Immunol. 1992;88:75–83.
68. Santiago ML, Fossati L, Jacquet C, Muller W, Izui S, Reiniger L. Interleukin-4 protects against a genetically linked lupus-like autoimmune syndrome. J Exp Med. 1997;185:65–70.
69. Jacob CO, Holoshitz J, van der Meide P, Strober S, McDevitt HO. Heterogeneous effects of IFN-γ in adjuvant arthritis. J Immunol. 1989;142:1500–5.
70. Brennan DC, Yui MA, Wuthrichg RP, Kelley VE. Tumor necrosis factor and IL-1 in New Zealand black/white mice. Enhanced gene expression and acceleration of renal injury. J Immunol. 1989;143:3470–5.
71. Finck BF, Linsey PS, Wofsy D. Treatment of murine lupus with CTLA-4 Ig. Science. 1994; 265:1225–7.
72. Couser WG, Johnson RJ, Young BA, Yeh CG, Toth CA, Rudolph AR. The effects of soluble recombinant complement receptor 1 on complement-mediated experimental glomerulonephritis. J Am Soc Nephrol. 1995;5:1888–94.
73. Wang Y, Hu Q, Madri JA, Rollins SA, Chodera A, Matis LA. Amelioration of lupus-like autoimmune disease in NZB/W F1 mice after treatment with a blocking monoclonal antibody specific for complement component C5. Proc Natl Acad Sci USA. 1996;93:8563–8.
74. Karin N, Mitchell DJ, Brocke S, Ling N, Steinman L. Reversal of experimental encephalomyelitis by a soluble peptide variant of a myelin basic protein epitope: T cell receptor antagonism and reduction of interferon γ and tumor necrosis factor α production. J Exp Med. 1994;180:2227–37.
75. Samson MF, Smilek DE. Reversal of acute experimental autoimmune encephalomyelitis and prevention of relapses by treatment with a myelin basic protein peptide analogue modified to form long-live peptide-MHC complexes. J Immunol. 1995;155:2737–46.
76. Wraith DC, Smilek DE, Mitchell DJ, Steinman L, McDevitt HO. Antigen recognition in autoimmune encephalomyelitis and the potential for peptide-mediated immunotherapy. Cell. 1994;59:247–55.
77. Piddlesden SJ, Storch MK, Hibbs M, Freeman AM, Lassmann H, Morgan BP. Soluble recombinant complement receptor 1 inhibits inflammation and demyelination in antibody-mediated demyelinating experimental allergic encephalomyelitis. J Immunol. 1994;152: 5477–84.
78. Thompson RC, Dripps DJ, Eisenberg SP. Interleukin-1 receptor antagonist (IL-1ra) as a probe and as a treatment for IL-1 mediated disease. Int J Immunopharmacol. 1992;14:475–80.
79. Martin D, Near SL. Protective effects of the interleukin-1 antagonist (IL-1ra) on experimental allergic encephalomyelitis in rats. J Neuroimmunol. 1995;61:241–5.
80. Willenborg DO, Fordham SA, Cowden WB, Rawshaw IA. Cytokines and murine autoimmune encephalomyelitis: inhibition or enhancement of disease with antibodies to select cytokines, or by delivery of exogenous cytokines using a recombinant vaccinia virus system. Scand J Immunol. 1995;41:31–41.
81. Porrini AM, Gambi D, Reder AT. Interferon effects on interleukin-10 secretion. Mononuclear cell response to interleukin-10 is normal in multiple sclerosis patients. J Neuroimmunol. 1995;61:27–34.
82. Cash E, Minty A, Ferrara P, Caput D, Fradelizi D, Rott O. Macrophage-inactivating IL-13 suppresses experimental autoimmune encephalomyelitis in rats. J Immunol. 1994;153:4258–67.
83. Rodriguez M, Pavelko KD, McKinney CW, Leibowitz JL. Recombinant human IL-6 suppresses demyelination in a viral model of multiple sclerosis. J Immunol. 1994;153:3811–21.
84. Soos JM, Subramaniam PS, Hobeika AC, Schiffenbauer J, Johnson HM. The IFN pregnancy recognition hormone IFN-τ blocks both development and superantigen reactivation of experimental allergic encephalomyelitis without associated toxicity. J Immunol. 1995;155: 2747–53.
85. Panitch HS. Interferons in multiple sclerosis. A review of the evidence. Drugs. 1992;44:946–62.
86. Goodkin DE. Role of steroids and immunosuppression and effects of interferon β-1b in multiple sclerosis. West J Med. 1994;161:292–8.
87. Goodkin DE. Interferon β-1b. Lancet. 1994;344:1511.
88. Connelly JF. Interferon beta for multiple sclerosis. Am Pharmacother. 1994;28:610–16.

89. Weinstock-Guttman B, Ransohoff RM, Kinkel P, Rudick RA. The interferons: biological effects, mechanisms of action and use in multiple sclerosis. Ann Neurol. 1995;37:7–15.
90. Brostoff SW, Mason DWJ. Experimental allergic encephalomyelitis: successful treatment *in vivo* with a monoclonal that recognises T helper cells. J Immunol. 1984;133:1938–42.
91. Waldor MK. Reversal of experimental allergic encephalomyelitis with monoclonal antibody to a T-cell subset marker. Science. 1985;227:415–17.
92. Sriram S, Roberts CA. Treatment of established chronic relapsing experimental allergic encephalomyeliti with anti-L3T4 antibodies. J Immunol. 1986;136:4464–9.
93. O'Neill JK, Baker D, Davison AN et al. Control of immune-mediated disease of the central nervous system with monoclonal (CD4-specific) antibody. J Neuroimmunol. 1993;45:1–14.
94. Arima T, Rehman A, Hickey WF, Fle MW. Inhibition of CTLA4Ig of experimental allergic encephalomyelitis. J Immunol. 1996;156:4916–24.
95. Jameson BA, McDonnell JM, Marini JC, Krongold R. A rationally designed CD4 analogue inhibits experimental allergic encephalomyelitis. Nature. 1994;368:744–6.
96. Smilek DE, Wraith DC. Hodgkinson S, Dwivedy S, Steinman L, McDevitt HO. A single amino acid change in a myelin basic protein peptide confers the capacity to prevent rather than induce experimental autoimmune encephalomyelitis. Proc Natl Acad Sci USA. 1993;88: 9633–7.
97. Gautam AM, Pearson CI, Smilek DE, Steinman L, McDevitt HO. A polyanine peptide with only five native myelin basic protein residues induces autoimmune encephalomyelitis. J Exp Med. 1992;176:605–9.
98. Feldmann M, Brennan FM, Maini RN. Role of cytokines in rheumatoid arthritis. Annu Rev Immunol. 1996;14:397–440.
99. Maini RN, Elliott M, Brennan FM, Williams RO, Feldmann M. Targeting TNF alpha for the therapy of rheumatoid arthritis. Clin Exp Rheumatol. 1994;12:S63–6.
100. Feldmann M, Brennan FM, Elliott MJ, Williams RO, Maini RN. TNF alpha is an effective therapeutic target for rheumatoid arthritis. Ann NY Acad Sci. 1995;766:272–8.
101. Maini RN, Elliott M, Brennan FM, Feldmann M. Beneficial effects of tumour necrosis factor-alpha (TNF-alpha) blockade in rheumatoid arthritis (RA). Clin Exp Immunol. 1995; 101:207 12.
102. Elliott MJ, Maini RN, Feldmann M et al. Repeated therapy with monoclonal antibody to tumour necrosis factor alpha (cA2) in patients with rheumatoid arthritis. Lancet. 1994;344: 1125–7.
103. Elliott MJ, Maini RN, Feldmann M et al. Randomised double-blind comparison of chimeric monoclonal antibody to tumor necrosis factor alpha (cA2) versus placebo in rheumatoid arthritis. Lancet. 1994;344:1105 10.
104. Butler DM, Maini RN, Feldmann M, Brennan FM. Modulation of proinflammatory cytokine release in rheumatoid synovial membrane cell cultures. Comparison of monoclonal anti TNG-alpha antibody with interleukin-1 receptor antagonist. Eur Cytokine Netw. 1995; 6:225–30.
105. Otani K, Nita I, Macauley W, Georgescu HI, Robbins PD, Evans CH. Suppression of antigen-induced arthritis in rabbits by ex vivo gene therapy. J Immunol. 1996;156:3558–62.
106. Boggs SS, Patrene KD, Mueller GM, Evans CH, Doughty LA, Robbins PD. Prolonged systemic expression of human IL-1 receptor antagonist (hIL-1ra) in mice reconstituted with hematopoietic cells transduced with a retrovirus carrying the hIL-1ra cDNA. Gene Ther. 1995;2:632–8.
107. Evans CH, Robbins PD. The promise of a new clinical trial – intra-articular IL-1 receptor antagonist. Proc Assoc Am Physicians. 1996;108:1–5.
108. Evans CH, Robbins PD. The interleukin-1 receptor antagonist and its delivery by gene transfer. Receptor. 1994;4:9–15.
109. Mulligan MS, Jones ML, Vaporciyan AA, Howard MC, Ward PA. Protective effects of IL-4 and IL-10 against immune complex-induced lung injury. J Immunol. 1993;151:5666–74.
110. Plat GL, Laufer J, Fabian I, Passwell JH. Cross-linking plasma membrane Fc alpha, Fc gamma or mannose receptors induces TNF production. Immunology. 1993;80:287–92.
111. Mulligan MS, Ward PA. Immune complex-induced lung and dermal vascular injury. Differing requirements for tumor necrosis factor-alpha and IL-1. J Immunol. 1992;149:331–9.
112. Genian CP, Abel K, Belmar N et al. Late complications of immune deviation therapy in a nonhuman primate. Science. 1996;274:2054–6.

113. MacFarland HF. Complexities in the treatment of autoimmune disease. Science. 1996;274: 2037–8.
114. Vial T, Descotes J. Immune-mediated side-effects of cytokines in humans. Toxicology. 1995; 105:31–57.
115. Weigand TW, Williams PB, Dreskin SC, Jouvin M-H, Kinet J-P, Tasset D. High-affinity oligonucleotide ligands to human IgE inhibit binding to Fcε receptor I. J Immunol. 1996;157: 221–30.
116. Li S, Gao J, Satoh T et al. A computer screening approach to immunoglobulin superfamily structures and interactions. Discovery of small non-peoptidic CD4 inhibitors as novel immunotherapeutics. Proc Natl Acad Sci USA. 1997;94:73–8.

21
FcR and autoimmunity

R. REPP and J. G. J. van de WINKEL

INTRODUCTION

Human immunoglobulin receptors (FcR) provide an important link between the humoral and cellular branches of the immune response. FcR engagement in autoimmune diseases may result in many biological responses including phagocytosis, endocytosis, antibody-dependent cellular cytotoxicity (ADCC), release of inflammatory mediators, facilitation of antigen presentation and clearance of immune complexes (IC). Most FcR belong to the family of multi-chain immune recognition receptors (MIRR), consisting of distinct ligand binding α-chains, which covalently associate with shared signalling components (γ-, β- or ζ-chains).

Three classes of leukocyte FcγR are currently distinguished (FcγRI (CD64), FcγRII (CD32), FcγRIII (CD16), encoding 12 receptor isoforms. These molecules differ in affinity and specificity for IgG isotypes, cell distribution patterns, capacity to trigger intracellular signals, and molecular weights. In addition to the FcγR heterogeneity in each person, genetic polymorpisms introduce further variations between individuals [1].

One α-gene is known for the IgA receptor (FcαRI, CD89), which encodes several alternatively spliced transmembrane isoforms [2]. Two classes of IgE receptors are distinguished, which can bind IgE with either high (FcεRI), or low affinity (FcεRII, CD23).

ABNORMALITIES IN Fc RECEPTOR FUNCTION IN AUTOIMMUNE DISEASES

Altered expression of Fc receptors

Alterations in expression of Fc receptors on monocytes/macrophages have been described in various autoimmune diseases (Table 2.1).

J.G.J. van de Winkel and P.M. Hogarth (eds.), The Immunoglobulin Receptors and their Physiological and Pathological Roles in Immunity. 233–248.

Table 21.1 Altered expression of FcR in SLE and RA

	FcγRI	*FcγRII*	*FcγRIII*	*FcεRII*
SLE				
Peripheral blood				
Monocytes	↑	↓	–/↓	↑
Neutrophils	–/↑	↓	↓	NE
Rheumatoid arthritis				
Peripheral blood				
Monocytes	↑	↑	NR	↑
Neutrophils	–	–	–	NE
Synovial fluid				
Macrophages	↑/↑↑	NR	↑	↑
Neutrophils	↑	↓	–/↓	NE

Expression of FcR on mononuclear phagocytes and neutrophils from patients with SLE and RA is altered compared to healthy controls. Increased (↑) and decreased (↓) expression is indicated (NE, not expressed; NR, not reported)

FcγRI is constitutively expressed on monocytes and macrophages and upregulated by IFN-γ or IL-10. Enhanced expression of FcγRI has been reported on peripheral blood monocytes in systemic lupus erythematosus (SLE) [3] and rheumatoid arthritis (RA) [4,5], but also on RA macrophages from synovial fluid (SF) [4], synovial tissue, and rheumatoid nodules [6] and on liver macrophages in autoimmune hepatitis [7]. Interestingly, only low levels of IFN-γ can be found in synovial fluid [8], whereas high levels of IL-4 and IL-13 are present in SF [9], which are both known to downregulate FcγRI expression *in vitro*. FcγRII levels are increased on circulating monocytes of RA patients [5], but decreased in SLE [3]. The low affinity FcεRII (CD23) is expressed on subsets of monocytes/macrophages, and is induced by IL-4 *in vitro*. Increased CD23 expression on circulating monocytes, was found in patients with progressive systemic sclerosis (PSS), RA, and in SLE patients [10].

Peripheral blood neutrophils constitutively express FcγRIIa and FcγRIIIb, as well as FcαRI. G-CSF and IFN-γ are known to induce FcγRI expression on PMN. FcγRI expression on circulation PMN is found in some SLE patients [11], but not in RA [12]. Synovial fluid PMN express FcγRI in most RA patients [13,14] but also in patients with reactive arthritis [14].

FcγRII is diminished on circulating PMN in SLE patients, correlating with high levels of circulating immune complex level [11], and in synovial fluid PMN in RA [14]. Decreased expression of FcγRIII is also found on circulating PMN from SLE patients [11] and patients with spondylarthropathies [15] and in synovial fluid neutrophils [13]. Interestingly, addition of SLE serum to granulocytes from healthy donors inhibits the detection of FcγRII and FcγRIII [11], suggesting that saturation of Fcγ receptors by serum immune complexes and their subsequent internalization may result in a depletion of detectable FcγR.

FcεRII (CD23) is an early activation antigen of B-cells and an increased proportion of FcεRII positive B-cells is found in patients with RA [16], but not in SLE patients [17], whereas levels of soluble sCD23 are enhanced in both diseases [16,18]. The increase in CD23 expression is not disease-specific and seems to reflect increased production of IL-4 [9] and enhanced susceptibility of B-cells to IL-4 [16]. In synovial tissue, high proportions of CD23-positive lymphocytes were found not only in RA patients, but also in osteoarthritis and even in chronic inflamed non-articular tissues. This indicates, that CD23 expression may be a characteristic feature of any chronic inflammatory response.

Fc receptors on bovine chondrocytes have been described by Saura *et al.* These showed a macrophage like pattern of IgG binding and response to immune complex stimulation, including production of superoxide anion [19].

Function of Fc receptors *in vitro*

Immune complex uptake plays a central role in autoimmune diseases. All three classes of FcγR on mononuclear phagocytes can mediate phagocytosis of IgG opsonized erythrocytes [20], depending on the degree of target cell sensitization and IgG isotype [21]. Low levels of target sensitization, achieved with anti-(Rh)D antibodies, mainly involve FcγRI, whereas FcγRII becomes more important at higher levels of opsonisation [22]. In cultured monocytes, used as a model for tissue macrophages, FcγRIII becomes the predominant FcγR and is involved in binding and internalization of opsonized erythrocytes [23].

Circulating monocytes from SLE patients show an impaired phagocytosis of IgG-opsonized bovine erythrocytes despite enhanced binding [24]. Uptake of small complexes such as heat-aggregated IgG and BSA–anti-BSA complexes is impaired [25], whereas binding and uptake of ovalbumin–anti-ovalbumin complexes proved enhanced in SLE patients with active disease [26] (Table 21.2). Fc receptor-triggered oxidative burst was decreased upon stimulation with IgG latex particles [27]. Altered FcγR function and diminished B7-1 expression were found to contribute to an abnormal antigen presentation capability of SLE monocytes [28].

Monocytes from patients with active RA patients showed enhanced erythrocyte rosette formation and phagocytosis [29] and normal uptake of aggregated IgG [30], despite increased numbers of binding sites for labelled IgG aggregates [30]. This enhanced binding could reflect an increased number of Fc receptors [4,5], but also the interaction of cell bound IgG- and IgM- rheumatoid factor (RF) with immune complexes [31].

The phagocytic capacity of neutrophils was documented as defective in SLE patients [32], in RA and Felty's syndrome [33], contributing to an enhanced susceptibility to infections. Oxidative burst proved normal in SLE [32] and RA [33]. Reduced chemiluminescence upon stimulation with aggregated IgG,

Table 21.2 Altered FcR function *in vitro* in SLE and RA

	IgG binding	*Phagocytosis*	*Oxidative burst*
SLE			
Monocytes	EA rosetting ↑ IC binding ↓/–/↑	EA uptake ↓ IC uptake ↓/–/↑	↓/–
Neutrophils		↓	–
Rheumatoid arthritis			
Monocytes	EA rosetting ↑ IC binding ↑	EA uptake ↑ IC uptake ↓/–	–
Neutrophils	EA rosetting ↓	S. aureus ↓	↓/–

In vitro function of circulating neutrophils and monocytes from patients with SLE and RA is compared to healthy donors. Altered binding of complexed IgG, phagocytosis of IgG-coated particles, and oxidative burst triggered via Fcγ receptor has been reported (EA: IgG sensitized bovine erythrocytes; IC: soluble immune complexes)

despite normal response to non-IgG stimuli and enhanced binding of aggregated IgG, was reported in a patient with primary Sjögren's syndrome [34]. Neutrophils also play a critical role in leukocytoclastic vasculitis, where immune complexes bound to endothelial cells trigger their activation. Moser *et al.* showed FcγRII and FcγRIII on circulating PMN to cooperate in initiation of oxidative burst by immune complex-bearing endothelial cells [35]. Concerning FcγR signalling, Goulding *et al.* found depressed calcium fluxes after specific stimulation of FcγRII or FcγRIII [12], although a recent report failed to confirm this [36].

Acquired and genetic factors influencing Fc receptor function

Soluble immune complexes

These are found in a variety of autoimmune diseases and proposed to play a critical role in disease pathogenesis. Saturation of Fcγ receptors by serum immune complexes and their subsequent internalization may result in diminished numbers of available low affinity FcγR on SLE granulocytes [11] and monocytes [3,37], as well as specific defects in granule exocytosis [33]. Indirect hints for a role of FcγR-blockade (leading to altered immune complex clearance) were obtained by Lockwood *et al.*, who documented reversal of a clearance defect in patients with nephritis or vasculitis after plasma exchange [38].

Cryoglobulin complexes

In type I cryoglobulinemia these increase surface expression of FcγRI and down-regulates FcγRII and FcγRIII in neutrophils [39]. In addition, cryoglobulin complexes trigger H_2O_2 generation and calcium mobilization in neutrophils which could be inhibited by blocking Ab to FcγRIII, but not FcγRII [39], supporting a role for FcγRIII in the pathogenesis of leukocytoclastic vasculitis.

Polyclonal hypergammaglobulinemia

This is a feature of various autoimmune diseases, including SLE and RA. Halma *et al.* reported that elevated IgG serum levels were the most important factor predicting the alterations of immune complex clearance [40].

Characteristics of IgG molecules

FcγR are known to cooperate with complement receptors (see Chapter 13) and fixation of C3-derived ligands to immune complexes alters phagocytosis and cellular responses to IC stimulation [20]. IgG molecules lacking the terminal galactose from the oligosaccharide chains in the CH2 domain (Gal(O)-IgG) were found in patients with rheumatoid arthritis, correlating with disease activity, and are believed involved in the pathogenesis of RA. Delayed clearance of agalacto-mIgG_{2a} but not of agalacto-mIgG_1 was found in mice, suggesting impaired binding to FcγRI [41]. In addition, impaired FcγRIII-dependent K-cell mediated haemolysis of red cells coated with agalactosylated autoantibodies was described [42]. No alteration in FcγRIIb1-mediated inhibition of B-cells were found using agalactosylated IgG [43].

Mediators of inflammation

Many of these are found in autoimmune diseases, and are known to influence Fc receptor expression and function or are produced after triggering cells via Fc receptors. A detailed discussion of cytokines and FcR can be found in Chapter 2.

Genetic factors

In addition to acquired factors, genetic differences between Fc receptors are described in autoimmune disease patients [44]. In the non-obese diabetic (NOD) mouse the FcγRI allele on chromosome 3 is defective, resulting in a deletion in the cytoplasmic tail, with altered turnover of receptor/antibody complexes. A NOD mouse strain that contains a chromosome segment encoding FcγRI proved less susceptible to autoimmune diabetes [45]. Four members of a single family lacking FcγRI were reported, but they were not suffering from any disease [46].

FcγRII-deficient mice display elevated immunoglobulin levels after antigen challenge, probably due to perturbations in immune complex-mediated feed-back inhibition of antibody production. These data are compatible with a possible contribution of FcγRIIb defects in the development of autoimmunity

[47]. At present, no null alleles for FcγRIIA or FcγRIIB molecules are known in humans.

A point mutation in the second Ig-like domain of FcγRIIa results in an arginine (R^{131}) or histidine (H^{131}) at amino acid position 131 and has been shown critical for the binding of human IgG_2 and IgG_3 (described in detail in Chapter 23). A skewed distribution towards IIa-R^{131} was described in African–American SLE patients, particular in lupus nephritis [48]. Duits *et al.* documented an association of FcγRIIa-R^{131} with lupus nephritis in an Caucasian population [49], whereas no association of the FcγRIIa polymorphism with SLE and lupus nephritis was found in another study [50]. Besides the possible role of FcγRIIa allotypes as genetic risk factor predisposing to the development of SLE, an association between FcγRIIa-R^{131} and several clinical parameters, including earlier disease onset, occurrence of autoimmune haemolytic anaemia, Sm and nRNP autoantibodies, and hypocomplementemia was found in a series of 108 caucasian SLE patients [Manger *et al.* manuscript in preparation]. Since FcγRIIa polymorphism is critical for IgG_2 binding, and IgG_2 represents the major humoral response to *S. pneumoniae*, Yee *et al.* analysed five SLE patients with atypical or complicated pneumoccocal infections: four were homozygeous for the IIa-R^{131} allele and one was heterozygeous [51].

FcγRIIIb deficiency has been described in a series of 21 individuals [52]. Most of them had no signs of infection or autoimmunity but two suffered from autoimmune thyroiditis [52]. Patients with paroxysmal nocturnal haemoglobinuria, who are deficient in FcγRIIIb (due to an acquired defect of phosphatidylinositol anchor synthesis) show a decreased clearance of immune complexes [53]. FcγRIIIb deficiency has been found in two SLE patients, although it remains unclear whether this defect is linked to the development or progression of SLE [54].

The neutrophilic FcγRIIIb bears a co-dominant, bi-allelic polymorphism (IIIb-NA1 and IIIb-NA2 allele), with an association of the IIIb-NA2 with lower levels of phagocytosis of IgG-opsonized erythrocytes [55]. In SLE patients, no skewing for this polymorphism has been found so far [49]. A significant skewing towards NA1 and an increased rate of renal dysfunction was found in Wegener's granulomatosis patients [56].

The γ subunit of immunoglobulin receptors is critical for signalling capacity and stable surface expression of FcγRI and FcγRIII. FcRγ chain-deficient mice are resistant to the development of experimental immune haemolytic anaemia, and immune thrombocytopenia, due to a failure to phagocytose opsonized cells [57]. This animal model clearly demonstrated a crucial role for Fcγ receptors in immune cytopenias.

Fc receptor-mediated clearance of immune complexes

Immune complexes are found in most autoimmune diseases including SLE and RA, and play a critical role in their pathophysiology. Inadequate clearance of immune complexes leads to enhanced tissue deposition and inflammation. The major site of immune complex clearance is the mononuclear phagocyte system of liver and spleen. Naked immune complexes are primarily removed via Fcγ receptor-dependend mechanisms from spleen phagocytes. In contrast, C3b-opsonized immune complexes are primarily bound to erythrocyte CR1 receptors and transported to liver and spleen phagocytes, which bind to CR1-associated immune complexes via Fcγ and complement receptors.

The major role of Fcγ receptors is supported by animal studies showing impaired clearance of immune complexes after FcγR blockade with aggregated IgG and infusion of FcγRIII antibodies in primates [58], and absent alteration of clearance in mice depleted of complement by cobra venom factor [59]. Mice that are genetically deficient in the expression of Fc receptors showed strongly reduced clearance of IgG-coated erythrocytes and platelets and diminished reactions to soluble immune complexes, whereas no impairment was found in C3 and C4 deficient mice [60].

Studies in patients with active SLE show a profound defect in Fc receptor-specific clearance of anti-Rhesus(D)-opsonized autologous erythrocytes [24] [reviewed in Reference 61] that correlates with disease activity, lupus nephritis and IC levels [61]. Removal of autologous erythrocytes opsonized with low amounts of anti-(Rh)D antibodies occurs mainly by spleen phagocytes. More densely opsonized erythrocytes, however, may be cleared by hepatic cells (involving complement receptors). The role of impaired splenic FcR function in SLE is underlined by the fact that prolongation of clearance of erythrocytes is more marked at low levels of sensitization [62].

Several mechanisms for the impaired clearance have been proposed. First, the impaired clearance despite enhanced binding of opsonized erythrocytes [24], suggests a dissociation of Fc receptor-ligand binding and internalization. Second, high levels of immune complexes and polyclonal hypergammaglobulinemia could contribute to Fc receptor blockade [3,37,40], although van der Woude *et al.* could find no relationship between serum IgG levels and reduced clearance [63]. Third, genetic HLA-linked phagocytosis defects are known for individuals with HLA DR3 or DR2 [64], HLA haplotypes associated with increased incidence of autoimmune diseases. Such defects may contribute to the predisposition and pathogenesis of SLE but do not represent a major factor. Clearance defects did not correlate with presence of HLA DR3 or DR2 haplotypes [65]. Finally, an association with humoral immune response parameters is suggested by the fact that Fc receptor-mediated immune clearance correlated remarkably well with a decrease of antigen-specific IgG after immunization with a primary antigen, and spontaneous IgG release *in vitro* of B cells obtained from peripheral blood [63].

Interestingly, in patients with drug induced lupus, with and without clinical symptoms, a normal clearance of Ig-coated autologous RBC was reported, and this may represent an explanation for protection against renal and central nervous system disease in these patients [66].

In patients with rheumatoid arthritis, clearance of anti-(Rh)D opsonized RBC was not significantly impaired in most studies [67,68]. Fields *et al.* reported prolonged clearance in RA patients, which was not correlated with immune complex levels or disease activity [69]. Kimberly *et al.* showed that only the first fast phase of elimination, which mainly reflects complement-dependent trapping of erythrocytes in the liver was impaired [70]. A qualitative alteration of erythrocyte clearance was reported by Malaise *et al.* who found a decreased spleen to liver uptake ratio despite normal overall clearance in RA patients [67].

Reduced clearance rates of IgG-coated erythrocytes were also found in patients with Sjögren's syndrome [68], dermatitis herpetiformis [71], and autoimmune hepatitis [72].

SPECIFIC ACTIONS OF AUTOANTIBODIES MEDIATED BY Fcγ RECEPTORS

Anti-neutrophil cytoplasmic antibodies (ANCA) are a hallmark of Wegener's granulomatosis. Neutrophils from patients with Wegener's granulomatosis often express ANCA target antigens (myeloperoxidase and proteinase 3) on their surface, and ANCA can bind and activate neutrophils via FcγRIIa [73] and FcγRIIIb [74] (Table 21.3). Fc interaction of ANCA and anti-nuclear antibodies also plays a role in the induction of CD11b on neutrophils which has been proposed to play a role in the development of vasculitis [75].

Anticardiolipin antibodies (ACA) are found in various autoimmune diseases and are associated with thrombosis, recurrent abortion and thrombocytopenia. ACA, purified from serum of SLE patients, were shown to inhibit phagocytosis

Table 21.3 FcγR-mediated action of autoantibodies

	PMN	*Monocyte*	*NK cell*	*Platelet*	*B cell*
ANCA	FcγRIIa, FcγRIIIb				
Anticardiolipin antibodies	FcγRIIa, FcγRIIIb				
Anti-β_2, GPI Ab	FcγRIIa			FcγRIIa	
Anti-epithelial BM Ab	FcγRIIIb				
Rheumatoid factors			FcγRIIIa		FcγRIIb
AIHA		FcγRI, FcγRIIIa			
ITP		FcγRI, FcγRIIIa		FcγRIIa	

Triggering of different effector cell populations via Fcγ receptors is described for various autoantibodies. Crosslinking of Fcγ receptors occurs via soluble immune complexes, autoantibody-coated target cells (e.g. erythrocytes) or binding of autoantibodies to specific antigens of FcγR-bearing cells of neutrophils and to reduce the ability to bind aggregated IgG [76]. Plasma with high ACA levels induces respiratory burst in neutrophils via interaction with FcγRIIa and FcγRIIIb [77] (Table 21.3).

β_2 glycoprotein I (β_2 GPI) is a plasma protein required for the binding of some antiphospholipid antibodies. Antibodies to β_2 GPI have been shown to possess lupus anticoagulant properties, and to activate platelets via FcγRIIa crosslinking. Anti-β_2 GPI also stimulate neutrophil oxidative burst and exocytosis via binding to β_2 GPI and FcγRIIa [78].

Anti-squamous epithelial basement membrane antibodies are characteristic for some autoimmune blistering skin diseases. Neutrophils can bind to opsonized epithelia via FcγRIIIb, and this immune adherence is stimulated mainly by rTNF-α and rGM-CSF [79].

Autoantibodies to the Fc part of IgG (rheumatoid factors, RF), can inhibit immunoglobulin production of B cells via crosslinking of FcγR with the B cell antigen receptor [80] or co-ligation of aggregated B cell receptors with FcγRIIb-fixed immune complexes [81]. IgG-containing RF complexes from RA patients induce decreased FcγRIIIa expression on NK cells, a decrease in NK activity, and a complete loss of ADCC [82].

Fcγ receptor-mediated erythrophagocytosis and sequestration of agglutinated erythrocytes in spleen and liver are the major pathogenic mechanisms of autoimmune haemolytic anemia (AIHA). The pathogenicity of autoantibodies to erythrocytes is determined by their subclass [83], affinity, ability to cause hemagglutination [84], and by the degree of glycosylation [42]. The role of FcγRI and FcγRIII is underlined by the fact that in FcRγ chain-deficient mice, immune haemolytic anaemia and immune thrombocytopenia cannot be induced experimentally [57] (Table 21.3).

Cross-linking FcγRIIa on platelets by anti-platelet antibodies or immune complexes triggering platelet activation [85]. Anti-platelet antibodies can interact with the platelet target antigen by means of their Fab portion, and with the platelet FcγRII via their Fc region [86]. An ITP patient was described with activating platelet autoantibody probably recognizing the CD9 antigen, and inhibitory platelet autoantibody, recognizing FcγRII, which inhibited CD9 antibody-induced platelet activation mediated via this receptor [87] (Table 21.3).

PATHOPYSIOLOGICAL ROLE OF Fc RECEPTOR AUTOANTIBODIES

Autoantibodies to Fc receptors can lead to destruction or sequestration of target cells, alter binding of complexed or monomeric IgG, and trigger signals in FcR-positive cells by crosslinking Fcγ receptors. FcγR autoantibodies may also play a role in downodulation of FcγR, thereby contributing to a reduced clearance of immune complexes.

In rheumatoid arthritis, IgM auto-antibodies reacting with FcγRII and FcγRIII [88] and IgG anti-FcγRIII autoantibodies [89] were found. IgG and IgM autoantibodies directed against all three types of FcγR were found in patients with Sjögren's syndrome, with a predominance of anti-FcγRIII antibodies [90] and in 30–50% of patients with progressive systemic sclerosis (PSS) [88]. IgM autoantibodies against FcγRI, FcγRII, and FcγRIII are found in SLE patients [88], wheras IgG autoantibodies binding to FcγRIII were found predominantly in sera from patients with Raynaud's syndrome [88]. Interestingly, many patients diagnosed with degenerative osteoarthritis also had IgG autoantibodies, directed primarily against FcγRII with lesser reactivity toward FcγRIII [88].

Autoimmune neutropenia (AIN) is characterized by the immune-mediated destruction of neutrophils (most anti-FcγR antibodies do not mediate cell destruction) by antibodies directed against neutrophil antigens. Autoantibodies against the NA1 allele of FcγRIIIb occur mainly in children with primary immmune neutropenia, but rarely in primary AIN of the adult, or in secondary forms of AIN [91].

Autoantibodies with specificity for the α-subunit of FcεRI (FcεRIa) were found in about 30% of patients with severe chronic idiopathic urticaria [92,93]. These autoantibodies mediate histamine release from skin mast cells *in vitro* (with purified IgG) [93] and after intradermal injection of autologous serum, supporting that anti-FcεRIa autoantibodies are relevant to the pathogenesis of severe chronic urticaria.

SOLUBLE FcR IN AUTOIMMUNE DISEASE

Soluble FcγR are described in detail in Chapter 24. Briefly, in autoimmune diseases soluble Fcγ receptors can modulate immune complex formation and solubility, thereby delaying immune precipitation [94]. Increased levels of soluble FcγRIII were found in synovial fluid from patients with various forms of arthritis including RA [89]. Plasma sFcγRIII was shown to be mainly derived from neutrophils (sFcγRIIIb) and to a lesser degree from NK cells (sFcγRIIIa) [95], although high levels of sFcγRIIIa were found in plasma of two patients with rheumatoid arthritis [95]. Increased serum sFcγRIII has been detected only in a minority of patients with SLE [96] but the fact that IgG was found complexed with sFcγR suggests a potential role in immune complex clearance [97]. sFcγRIII was detectable in one third of patients with primary Sjögren's syndrome and paralleled diminished adherence and chemotaxis of neutrophils [98].

Soluble FcεRII (sCD23] is closely related to B-cell activation and elevated serum levels of sCD23 have been reported in several autoimmune disorders, including rheumatoid arthritis [16,99], SLE [18,100], Sjögren's syndrome [18], autoimmune thyroiditis [99], Graves' disease [101], bullous pemphigoid [102], autoimmune chronic active hepatitis [100], and myasthenia gravis [99]. IL-4 levels are increased in classic autoimmune conditions, reflecting enhanced Th2 activity, and increases production and release of sCD23.

FcR-BASED STRATEGIES IN TREATMENT OF AUTOIMMUNOPATHIES

Many therapeutic strategies in autoimmune diseases have an direct or indirect influence on Fc receptor function (a detailed discussion of Fc receptor based therapies can be found elsewhere in Chapter 25 and in Reference 1). In immune thrombocytopenic purpura, anti-FcγRIII antibodies [103] were used to block FcγR, which resulted in a reduced uptake of autoantibody-coated platelets. An anti-FcγRI antibody [104] likewise resulted in clinical improvement of an ITP patient, via a marked down-regulation of FcγRI on monocytes, although the platelet count remained unchanged. Functional blockade of Fc receptors on splenic macrophages is also one of the proposed mechanisms of action of intravenous immunoglobulin (IVIg) in immune cytopenias [105,83]. In addition, IVIg suppresses immunoglobulin productions by altering signal transduction through FcγRII in B lymphocytes [106]. In ITP, intravenous anti-(Rh)D is thought to act via FcγR-blockade by substitution of antibody-coated platelets for autoantibody-coated RBC, although additional mechanisms of action are likely [107]. Danazol and vinblastine are also used in ITP and their clinical effect may be mediated in part by decreasing FcγR on monocytes [108].

Soluble FcεRII is increased in RA patients. The involvement of CD23 and the role of CD23 as a potential therapeutic target in RA is suggested by the finding that antibodies to FcεRII lead to a marked amelioration of arthritis in mice with established collagen-induced arthritis [109].

Fusion proteins of FcγRI antibodies and specific peptides dramatically enhance antigen presentation [110]. Such fusion proteins may be of therapeutic interest for the inhibition of autoreactive T cell clones by FcγRI-targeted presentation of antagonistic peptides.

Effects of glucocorticoids on monocyte Fcγ receptor expression and phagocytic function vary with the level of monocyte activation and the dose and schedule of drug administration. Whereas regular doses of steroids are inhibitory in normal donors, high dose, pulse methylprednisolone in lupus nephritis enhances mononcyte FcγR expression and phagocytosis and decreases levels of circulating immune complexes [111].

CONCLUSIONS

Fc receptors play an important role in autoimmune diseases by three major mechanisms. First, triggering of Fc receptors on cytotoxic effector cells can lead to the destruction of autoantibody-opsonized cells, such as erythrocytes in AIHA or platelets in ITP. Second, tissue-deposited immune complexes can crosslink FcRs causing the release of (pro)inflammatory molecules mainly from neutrophils and macrophages. Third, FcR dysfunction can cause inadequate clearance of immune complexes, leading to IC-mediated diseases such as lupus nephritis or some vasculitic disorders. FcγR polymorphisms and various acquired factors seem to play a relevant role in autoimmune diseases, influencing the interaction of IgG isotypes with select Fc receptors. Autoantibodies against Fc receptors, found in many autoimmune disorders, can block FcR function (e.g. IC clearance) or trigger FcR-positive cells, leading to enhanced tissue destruction. Soluble Fc receptors may ameliorate the clinical course of autoimmune diseases by influencing IC clearance and inhibition of type III hypersensitivity reactions. Therapeutic strategies, influencing Fc receptor function in autoimmune diseases, include steroids as well as IVIg and anti-(Rh)D antibodies, and newer approaches using FcγR antibodies are currently tested in clinical trials.

References

1. Deo YM, Graziano RF, Repp R, van de Winkel JGJ. Clinical significance of IgG Fc receptors and FcγR-directed immunotherapies. Immunol Today. 1997;18(3):127–35.
2. Morton HC, van Egmont M, van de Winkel JGJ. Structure and function of human IgA Fc receptors (FcαR). Crit Rev Immunol. 1996;16:423–40.
3. Szucs G, Kavai M, Suranyi P, Kiss E, Csipo I, Szegedi G. Correlation of monocyte phagocytic receptor expressions with serum immune complex level in systemic lupus erythematosus. Scand J Immunol. 1994;40(5):481–4.
4. Highton J, Carlisle B, Palmer DG. Changes in the phenotype of monocytes/macrophages and expression of cytokine mRNA in peripheral blood and synovial fluid of patients with rheumatoid arthritis. Clin Exp Immunol. 1995;102(3):541–6.
5. Shinohara S, Hirohata S, Inoue T, Ito K. Phenotypic analysis of peripheral blood monocytes isolated from patients with rheumatoid arthritis. J Rheumatol. 1992;19(2):211–5.
6. Broker BM, Edwards JC, Fanger MW, Lydyard PM. The prevalence and distribution of macrophages bearing FcγRI, FcγRII, and FcγRIII in synovium. Scand J Rheumatol. 1990; 19(2):123–35.
7. Yamamoto K, Ohmoto M, Matsumoto S et al. Activated liver macrophages in human liver diseases. J Gastroenterol Hepatol. 1995;10(Suppl 1):S72–6.
8. Hessian PA, Highton J, Palmer DG. Quantification of macrophage cell surface molecules in rheumatoid arthritis. Clin Exp Immunol. 1989;77(1):47–51.
9. Isomaki P, Luukkainen R, Toivanen P, Punnonen J. The presence of interleukin-13 in rheumatoid synovium and its antiinflammatory effects on synovial fluid macrophages from patients with rheumatoid arthritis. Arthritis Rheum. 1996;39(10):1693–702.
10. Becker H, Potyka P, Weber C, Federlin K. Detection of circulating FcεRII/CD23+ monocytes in patients with rheumatic diseases. Clin Exp Immunol. 1991;85(1):61–5.
11. Szucs G, Kavai M, Kiss E, Csipo I, Szegedi G. Correlation of IgG Fc receptors on granulocytes with serum immune complex level in systemic lupus erythematosus. Scand J Immunol. 1995;42(5):577–80.
12. Goulding NJ, Guyre PM. Impairment of neutrophil Fcγ receptor mediated transmembrane signalling in active rheumatoid arthritis. Ann Rheum Dis. 1992;51(5):594–9.
13. Watson F, Robinson JJ, Phelan M, Bucknall RC, Edwards SW. Receptor expression in synovial fluid neutrophils from patients with rheumatoid arthritis. Ann Rheum Dis. 1993; 52(5):354–9.
14. Felzmann T, Gadd S, Majdic O et al. Analysis of function-associated receptor molecules on peripheral blood and synovial fluid granulocytes from patients with rheumatoid and reactive arthritis. J Clin Immunol. 1991;11(4):205–12.
15. Kahan A, Collantes Estevez E, Barel M, Frade R, Amor B. Expression of complement receptors (CR1 and CR3) and FcγRIII on polymorphonuclears from patients with spondylarthropathies. Clin Exp Rheumatol. 1993;11(1):27–34.
16. Chomarat P, Briolay J, Banchereau J, Miossec P. Increased production of soluble CD23 in rheumatoid arthritis and its regulation by interleukin-4. Arthritis Rheum. 1993;36(2):234–42.
17. Kumagai S, Ishida H, Iwai K et al. Possible different mechanisms of B cell activation in systemic lupus erythematosus and rheumatoid arthritis: opposite expression of low-affinity receptors for IgE (CD23) on their peripheral B cells. Clin Exp Immunol. 1989;78(3):348–53.
18. Bansal A, Roberts T, Hay EM, Kay R, Pumphrey RS, Wilson PB. Soluble CD23 levels are elevated in the serum of patients with primary Sjögren's syndrome and systemic lupus erythematosus. Clin Exp Immunol. 1992;89(3):452–5.
19. Saura R, Uno K, Satsuma S, Kurz EU, Scudamore RA, Cooke TDV. Mechanisms of cartilage degradation in inflammatory arthritis interaction between chondrocytes and immunoglobulin G. J Rheumatol. 1993;20(2):336–43.

20. Shen L, Graziano RF, Fanger MW. The functional properties of FcγRI, II and III on myeloid cells: a comparative study of killing of erythrocytes and tumor cells mediated through the different Fc receptors. Mol Immunol. 1989;26(10):959–69.
21. van de Winkel JG, Boonen GJ, Janssen PL, Vlug A, Hogg N, Tax WJ. Activity of two types of Fc receptors, FcγRI and FcγRII, in human monocyte cytotoxicity to sensitized erythrocytes. Scand J Immunol. 1989;29(1):23–31.
22. Indik ZK, Hunter S, Huang MM et al. The high affinity Fc gamma receptor (CD64) induces phagocytosis in the absence of its cytoplasmic domain: the gamma subunit of FcγRIIIA imparts phagocytic function to FcγRI. Exp Hematol. 1994;22(7):599–606.
23. Clarkson SB, Ory PA. CD16. Developmentally regulated IgG Fc receptors on cultured human monocytes. J Exp Med. 1988;167(2):408–20.
24. Salmon JE, Kimberly RP, Gibofsky A, Fotino M. Defective mononuclear phagocyte function in systemic lupus erythematosus: dissociation of Fc receptor-ligand binding and internalization. J Immunol. 1984;133(5):2525–31.
25. Kavai M, Csipo I, Sonkoly I, Csongor J, Szegedi GY. Defective immune complex degradation by monocytes in patients with systemic lupus erythematosus. Scand J Immunol. 1986;24(5):527–32.
26. Kabashima T, Sakurai T, Yamane K, Kono I, Kashiwagi H. Enhanced Fc receptor function of monocytes from patients with clinically active systemic lupus erythematosus: binding and degradation of soluble immune complexes in vitro. Jpn J Med. 1986;25(3):263–9.
27. Gyimesi E, Kavai M, Kiss E, Csipo I, Szucs G, Szegedi G. Triggering of respiratory burst by phagocytosis in monocytes of patients with systemic lupus erythematosus. Clin Exp Immunol. 1993;94(1):140–4.
28. Tsokos GC, Kovacs B, Sfikakis PP, Theocharis S, Vogelgesang S, Via CS. Defective antigen-presenting cell function in patients with systemic lupus erythematosus. Arthritis Rheum. 1996;39(4):600–9.
29. Hoch S, Schur PH. Monocyte receptor function in patients with rheumatoid arthritis. Arthritis Rheum. 1981;24(10):1268–7.
30. Heurkens AH, Westedt ML, Breedveld FC, Jonges E, Cats A, Stijnen T, Daha MR. Uptake and degradation of soluble aggregates of IgG by motion. Scand J Rheumatol. 1989;18:97–105.
33. Minty CA, Hall ND, Bacon PA. Depressed exocytosis by rheumatoid neutrophils in vitro. Rheumatol Int. 1983;3(3):139–42.
34. Schopf RE, Rehder M, Laux B, Korting GW. Functional Fc-receptor defect of polymorphonuclear leukocytes in a patient with Sjögren's syndrome. Klin Wochenschr. 1987;65(7):342–4.
35. Moser R, Etter H, Oligati L, Fehr J. Neutrophil activation in response to immune complex-bearing endothelial cells depends on the functional cooperation of FcγRII (CD32) and FcγRIII (CD16). J Lab Clin Med. 1995;126(6):588–96.
36. Jones J, Laffafian I, Lawson T, Williams BD, Morgan BP. Signalling through neutrophil FcγRIII, FcγRII, and CD59 is not impaired in active rheumatoid arthritis. Ann Rheum Dis. 1996;55(5):294–7.
37. Katayama S, Chia D, Knutson DW, Barnett EV. Decreased Fc receptor avidity and degradative function of monocytes from patients with systemic lupus erythematosus. J Immunol. 1983;131(1):217–22.
38. Lockwood CM, Worlledge S, Nicholas A, Cotton C, Peters DK. Reversal of impaired splenic function in patients with nephritis or vasculitis (or both) by plasma exchange. N Engl J Med. 1979;300(10):524–30.
39. Hundt M, Zielinska Skowronek M, Schmidt RE. Fc-gamma receptor activation of neutrophils in cryoglobulin-induced leukocytoclastic vasculitis. Arthritis Rheum. 1993; 36(7):974–82.
40. Halma C, Breedveld FC, Daha MR et al. Elimination of soluble 123-labelled aggregates of IgG in patients with systemic lupus erythematosus. The effect of serum IgG and numbers of erythrocyte complement receptor Type 1. Arthritis Rheum. 1991;34:442–52.
41. Newkirk MM, Novick J, Stevenson MM, Fournier MJ, Apostolakos P. Differential clearance of glycoforms of IgG in normal and autoimmune-prone mice. Clin Exp Immunol. 1996; 106(2):259–64.
42. Hadley AG, Zupanska B, Kumpel BM et al. The glycosylation of red cell autoantibodies affects their functional activity in vitro. Br J Haematol. 1995;91(3):587–94.
43. Groenink J, Spijker J, Van Den Herik Oudijk IE et al. On the interaction between agalactosyl IgG and Fc-gamma receptors. Eur J Immunol. 1996;26(6):1404–7.

44. Rascu A, Repp R, Westerdaal NAC, Kalden JR, van de Winkel JGJ. Clinical relevance of Fcγ receptor polymorphism. Ann NY Acad Sci. 1997;815:282–95.
45. Prins JB, Todd JA, Rodrigues NR et al. Linkage on chromosome 3 of autoimmune diabetes and defective Fc receptor for IgG in nod mice. Science. 1993;260(5108):695–7.
46. van de Winkel JG, de Wit TP, Ernst LK, Capel PJ, Ceuppens JL. Molecular basis for a familial defect in phagocyte expression of IgG receptor I (CD64). J Immunol. 1995;154(6): 2896–903.
47. Takai T, Ono M, Hikida M, Ohmori H, Ravetch JV. Augmented humoral and anaphylactic responses in FcγRII- deficient mice. Nature. 1996; 379.
48. Salmon JE, Millard S, Schachter LA et al. FcγRIIa alleles are heritable risk factors for lupus nephritis in african americans. J Clin Invest. 1996;97(5):1348–54.
49. Duits AJ, Bootsma H, Derksen RHWM et al. Skewed distribution of IgG Fc receptor IIa (CD32) polymorphism is associated with renal disease in systemic lupus erythematosus patients. Arthritis Rheum. 1995;38(12):1832–6.
50. Botto M, Theodoridis E, Thompson EM et al. FcγRIIa polymorphism in systemic lupus erythematosus (SLE): no association with disease. Clin Exp Immunol. 1996;104(2):264–8.
51. Yee AMF, Ng SC, Sobel RE, Salmon JE. FcγRIIa polymorphism and atypical pneumococcal infections in SLE. Arthritis Rheum. 1996;39(9 Suppl):S190.
52. De Haas M, Kleijer M, Van Zwieten R, Roos D, Von Dem Borne AEGK. Neutrophil FcγRIIIb deficiency, nature, and clinical consequences: a study of 21 individuals from 14 families. Blood. 1995;86(6):2403–13.
53. Selvaraj P, Rosse WF, Silber R, Springer TA. The major Fc receptor in blood has a phosphatidylinositol anchor and is deficient in paroxysmal nocturnal haemoglobinuria. Nature. 1988;333(6173):565–7.
54. Clark MR, Liu L, Clarkson SB, Ory PA, Goldstein IM. An abnormality of the gene that encodes neutrophil Fc receptor III in a patient with systemic lupus erythematosus. J Clin Invest. 1990;86(1):341–6.
55. Bredius RG, Fijen CA, De Haas M et al. Role of neutrophil FcγRIIa (CD32) and FcγRIIIb (CD16) polymorphic forms in phagocytosis of human IgG_1- and IgG_3-opsonized bacteria and erythrocytes. Immunology. 1994;83(4):624–30.
56. Wainstein E, Edberg J, Csernok E et al. FcγRIIIb alleles predict renal dysfunction in Wegener's Granulomatosis (WG). Arthritis Rheum. 1996;39(9 Suppl):S210.
57. Clynes R, Ravetch JV. Cytotoxic antibodies trigger inflammation through Fc receptors. Immunity. 1995;3(1):21–6.
58. Kimberly RP, Edberg JC, Merriam LT, Clarkson SB, Unkeless JC, Taylor RR. In vivo handling of soluble immune complement fixing Ab/dsDNA immune complexes in chimpanzees. J Clin Invest. 1989;84:962–70.
59. Palermo MS, Minnucci FS, Isturiz MA. Impairment of the murine mononuclear phagocyte system function by antigenic stimulation. Clin Immunol Immunopathol. 1994;73(1):103–8.
60. Sylvestre D, Clynes R, Ma M, Warren H, Carroll MC, Ravetch JV. Immunoglobulin G-mediated inflammatory responses develop normally in complement-deficient mice. J Exp Med. 1996;184(6):2385–92.
61. Salmon JE. Wallace DJ, Hahn BH, eds. Dubois' Lupus Erythematosus, 4th edn. Abnormalities in Immune Complex Clearance and Fc Receptor Function. Baltimore: Williams & Wilkins; 1993:108–19.
62. Kabbash L, Esdaile J, Shenker S, Decary F, Danoff D, Fuks A, Shuster J. Reticuloendothelial system Fc receptor function in systemic lupus erythematosus: effect of decreased sensitization on clearance of autologous erythrocytes. J Rheumatol. 1987;14(3):487–9.
63. van der Woude FJ, Kallenberg CG, Limburg PC et al. Reticuloendothelial Fc receptor function in SLE patients. II. Associations with humoral immune response parameters in vivo and in vitro. Clin Exp Immunol. 1984;55(3):481–6.
64. Kimberly RP, Gibofsky A, Salmon JE, Fotino M. Impaired Fc-mediated mononuclear phagocyte system clearance in HLA-DR2 and MT1-positive healthy young adults. J Exp Med. 1983;157(5):1698–703.
65. van der Woude FJ, van der Giessen M, Kallenberg CG et al. Reticuloendothelial Fc receptor function in SLE patients. I. Primary HLA linked defect or acquired dysfunction secondary to disease activity? Clin Exp Immunol. 1984;55(3):473–80.
66. Fields TR, Zarrabi MH, Gerardi EN, Bennett RS, Zucker S, Hamburger MI. Reticuloendothelial system Fc receptor function in the drug induced lupus erythematosus syndrome. J Rheumatol. 1986;13(4):726–31.

67. Malaise MG, Foidart JB, Hauwaert C, Mahieu P, Franchimont P. In vivo studies on the mononuclear phagocyte system Fc receptor function in rheumatoid arthritis. Correlations with clinical and immunological variables. J Rheumatol. 1985;12(1):33–42.
68. O'Sullivan MM, Walker DM, Williams BD. Reticuloendothelial Fc receptor function in patients with Sjögren's syndrome. Clin Exp Immunol. 1985;61(3):483–8.
69. Fields TR, Gerardi EN, Ghebrehiwet B et al. Reticuloendothelial system Fc receptor function in rheumatoid arthritis. J Rheumatol. 1983;10(4):550–7.
70. Kimberly RP, Meryhew NL, Runquist OA. Mononuclear phagocyte system complement receptor dysfunction in rheumatoid arthritis. J Immunol. 1987;138(12):4165–8.
71. Lawley TJ, Hall RP, Fauci AS, Katz SI, Hamburger MI, Frank MM. Defective Fc-receptor functions associated with the HLA-B8/DRw3 haplotype: studies in patients with dermatitis herpetiformis and normal subjects. N Engl J Med. 1981;304(4):185–92.
72. Lin RY, Green LJ, Winny D, Ramaswamy G. Immunological abnormalities in autoimmune chronic active hepatitis. J Rheumatol. 1989;16(11):1489–93.
73. Porges AJ, Redecha PB, Kimberly WT, Csernok E, Gross WL, Kimberly RP. Anti-neutrophil cytoplasmic antibodies engage and activate human neutrophils via FcγRIIa. J Immunol. 1994;153(3):1271–80.
74. Kocher M, Edberg J, Fleit H, Kimberly R. Anti-neutrophil cytoplasmic antibody (ANCA) engagement of FcγRIIIb on neutrophils (PMN) is blocked by soluble Fcγ receptor. Arthritis Rheum. 1996;39(9 Suppl):S210.
75. Johnson PA, Alexander HD, McMillan SA, Maxwell AP. Up-regulation of the granulocyte adhesion molecule Mac-1 by autoantibodies in autoimmune vasculitis. Clin Exp Immunol. 1997;107:513–9.
76. Yu CL, Sun KH, Tsai CY, Wang SR. Inhibitory effects of anticardiolipin antibodies on lymphocyte proliferation and neutrophil phagocytosis. Ann Rheum Dis. 1991;50(12):903–8.
77. Perniok A, Gaubitz M, Domschke W, Schneider M. Anticardiolipin antibodies induce respiratory burst in neutrophils via interaction with FcγII and -III receptors. Arthritis Rheum. 1994;37:S212.
78. Arvieux J, Jacob MC, Roussel B, Bensa JC, Colomb MG. Neutrophil activation by anti-beta-2 glycoprotein I monoclonal antibodies via Fc-gamma receptor II. J Leukoc Biol. 1995;57(3):387–94.
79. Gammon WR, Hendrix JD, Mangum K, Jeffes EWB. Recombinant human cytokines stimulate neutrophil adherence to IgG autoantibody-treated epithelial basement membranes. J Invest Dermatol. 1990;95(2):164–71.
80. Terness P, Berteli A, Susal C, Opelz G. Regulation of antibody response by an IgG-anti-Ig autoantibody occurring during alloimmunization II. Selective inactivation of antigen receptor-occupied B cells. Transplantation. 1992;54(1):92–6.
81. Fazekas G, Palfi G, Wolff Winiski B et al. IgG isotype-specific auto-antibodies bind preferentially to cross-linked membrane Ig. Int Immunol. 1995;7(7):1125–34.
82. Hendrich C, Kuipers JG, Kolanus W, Hammer M, Schmidt RE. Activation of CD16+ effector cells by rheumatoid factor complex. Role of natural killer cells in rheumatoid arthritis (see comments). Arthritis Rheum. 1991;34(4):423–31.
83. Pottier Y, Pierard I, Barclay A, Masson PL, Coutelier JP. The mode of action of treatment by IgG of haemolytic anaemia induced by an anti-erythrocyte monoclonal antibody. Clin Exp Immunol. 1996;106(1):103–7.
84. Izui S, Berney T, Shibata T, Fulpius T, Fossati L, Merino R. Molecular and cellular basis for pathogenicity of autoantibodies. Tohoku J Exp Med. 1994;173(1):15–30.
85. Clark WF, Tevaarwerk GJ, Reid BD. Human platelet-immune complex interaction in plasma. J Lab Clin Med. 1982;100(6):917–31.
86. Perutelli P, Mori PG. Activation of human platelets by monoclonal antibodies. Haematologica. 1993;78(3):172–7.
87. Yanabu M, Nomura S, Fukuori T et al. Platelet activation induced by an antiplatelet autoantibody against CD9 antigen and its inhibition by another autoantibody in immune thrombocytopenic purpura. Br J Haematol. 1993;84(4):694–701.
88. Boros P, Odin JA, Chen J, Unkeless JC. Specificity and class distribution of FcγR-specific autoantibodies in patients with autoimmune disease. J Immunol. 1994;152(1):302–6.
89. Lamour A, Baron D, Soubrane C et al. Anti-Fc gamma receptor III autoantibody is associated with soluble receptor in rheumatoid arthritis serum and synovial fluid. J Autoimmun. 1995;8(2):249–65.

90. Lamour A, Le Corre R, Soubrane C, Khayat D, Youinou P. Anti-Fc gamma receptor autoantibodies from patients with Sjögren's syndrome do not react with native receptor on human polymorphonuclear leukocytes. J Autoimmun. 1996;9(2):181–91.
91. Shastri K, Logue GL. Autoimmune neutropenia. Blood. 1993;81(8):1984–95.
92. Fiebiger E, Maurer D, Holub H et al. Serum IgG autoantibodies directed against the alpha chain of FcεRI: a selective marker and pathogenetic factor for a distinct subset of chronic urticaria patients? J Clin Invest. 1995;96(6):2606–12.
93. Niimi N, Francis DM, Dermani F et al. Dermal mast cell activation by autoantibodies against the high affinity IgE receptor in chronic urticaria. J Invest Dermatol. 1996;106(5): 1001–6.
94. Gavin AL, Wines BD, Powell MS, Hogarth PM. Recombinant soluble FcγRII inhibits immune complex precipitation. Clin Exp Immunol. 1995;102(3):620–5.
95. De Haas M, Kleijer M, Michinton RM, Roos D, Von Dem Borne AEGK. Soluble FcγRIIIa is present in plasma and is derived from natural killer cells. J Immunol. 1994;152(2):900–7.
96. Hutin P, Lamour A, Pennec YL, Soubrane C, Dien G, Khayat D, Youinou P. Cell-free FcγRIII in sera from patients with systemic lupus erythematosus: correlation with clinical and biological features. Int Arch Allergy Immunol. 1994;103(1):23–7.
97. Szucs G, Kavai M, Kiss E et al. Soluble form of FcγRII and III in sera of patients with SLE. (Abstract) Proceedings of the 9th International Congress of Immunology; 1995:533.
98. Lamour A, Soubrane C, Ichen M, Pennec YL, Khayat D, Youinou P. FcγRIII shedding by polymorphonuclear cells in primary Sjögren's syndrome. Eur J Clin Invest. 1993;23(2):97–101.
99. Bansal AS, Ollier W, Marsh MN, Pumphrey RSH, Wilson PB. Variations in serum sCD23 in conditions with either enhanced humoral or cell-mediated immunity. Immunology. 1993; 79(2):285–9.
100. Al Janadi M, Al Wabel A, Raziuddin S. Soluble CD23 and interleukin-4 levels in autoimmune chronic active hepatitis and systemic lupus erythematosus. Clin Immunol Immunopathol. 1994;71(1):33–7.
101. Sayinalp S, Akalin S, Sayinalp N et al. Serum immunoglobulin E and soluble CD23 in patients with graves' disease. Hormon Metabol Res. 1996;28(3):133–7.
102. Maekawa N, Hosokawa H, Soh H et al. Serum levels of soluble CD23 in patients with bullous pemphigoid. J Dermatol. 1995;22:315.
103. Clarkson SB, Bussel JB, Kimberly RP, Valinsky JE, Nachman RL, Unkeless JC. Treatment of refractory immune thrombocytopenic purpura with an anti-Fcγ-receptor antibody. N Engl J Med. 1986;314(19):1236–9.
104. Ericson SG, Coleman KD, Wardwell K et al. Monoclonal antibody 197 (anti-FcγRI) infusion in a patient with immune thrombocytopenia purpura (ITP) results in down-modulation of FcγRI on circulating monocytes. Br J Haematol. 1996;92(3):718–24.
105. Mouthon S, Kaveri SV, Spalter SH et al. Mechanisms of action of intravenous immune globulin in immune-mediated diseases. Clin Exp Immunol. 1996;104:3–9.
106. Kondo N, Kasahara K, Kameyama T et al. Intravenous immunoglobulins suppress immunoglobulin productions by suppressing Ca^{++}-dependent signal transduction through Fcγ receptors in B lymphocytes. Scand J Immunol. 1994;40(1):37–42.
107. Scaradavou A, Woo B, Woloski BMR et al. Intravenous anti-D treatment of immune thrombocytopenic purpura: experience in 272 patients. Blood. 1997;89(8):2689–700.
108. Schreiber AD, Chien P, Tomaski A, Cines DB. Effect of danazol in immune thrombocytopenic purpura. N Engl J Med. 1987;316(9):503–8.
109. Plater Zyberk C, Bonnefoy JY. Marked amelioration of established collagen-induced arthritis by treatment with antibodies to CD23 in vivo. Nature Med. 1995;1(8):781–5.
110. Liu C, Goldstein J, Graziano RF et al. FcγRI-targeted fusion proteins results in efficient presentation by human monocytes of antigenic and antagonist T cell epitopes. J Clin Invest. 1996;98(9):2001–7.
111. Salmon JE, Kapur S, Meryhew NL, Runquist OA, Kimberly RP. High-dose, pulse intravenous methylprednisolone enhances Fcγ receptor-mediated mononuclear phagocyte function in systemic lupus erythematosus. Arthritis Rheum. 1989;32(6):717–25.

22 Heparin-induced thrombocytopenia as a model for FcγRII-mediated disease

R. BAKER and B. DALE

THE FcγRII-DEPENDENT NATURE OF HEPARIN-INDUCED THROMBOCYTOPENIA

The clinical association of thrombocytopenia and often thrombosis has been recognized for over 20 years as a common and serious complication of heparin therapy. Heparin is widely used as an anticoagulant for the prevention and treatment of thromboembolic disorders, so when this adverse reaction to heparin occurs it can not only exacerbate the existing thrombosis for which heparin was originally indicated, but may initiate new episodes of life-threatening arterial and/or venous thrombosis. These events include peripheral arterial occlusion, ischaemic stroke, acute myocardial infarction, pulmonary embolism and lower limb venous gangrene. Each complication can cause substantial morbidity resulting in limb amputation, hemiplegia, cardiac failure and not infrequently death [1,2].

Recent evidence implicates the development of heparin-platelet factor 4–IgG immune complexes in the pathogenesis of this disorder which results in FcγRII dependent platelet activation [3–6]. On platelets, FcγRII is the pivotal molecule that mediates rapid signal transduction in response to complexed IgG [7,8]. Stimulation of this pathway causes intense irreversible platelet aggregation, release of granule contents, thromboxane A_2 production and the generation of platelet derived microvesicles producing excess thrombin and platelet-rich clot formation [7–11]. These changes can profoundly disturb the anticoagulant/procoagulant haemostatic balance as clinically demonstrated by the substantially increased risk of deep venous thrombosis (27-fold) and pulmonary

J.G.J. van de Winkel and P.M. Hogarth (eds.), The Immunoglobulin Receptors and their Physiological and Pathological Roles in Immunity. 249–266.

embolism (93-fold) in patients with heparin-induced thrombocytopenia (HIT) compared with those without the disorder following lower limb orthopaedic surgery [12]. FcγRII-dependent platelet activation can be replicated in the laboratory by adding either serum or purified IgG from patients with HIT and therapeutic concentrations of heparin to normal platelets. This reaction is the basis for the diagnostic bioassays widely used in clinical practice [7,13–16].

As FcγRII is the only IgG Fc receptor found on platelets [17], the understanding of the pathogenesis of HIT provides a prototype for the study of the impact of genetic and acquired factors that modulate the ligand–receptor coupling of complexed IgG and cellular FcγRII. This interaction could be important in other poorly understood but common immune-mediated disorders where FcγRII-mediated cellular activation may play an important pathophysiological role. Directly analogous clinical examples where this process may be relevant include other conditions associated with thrombocytopenia and/or thrombosis and include reports of FcγRII-dependent platelet activation in patients with immune thrombocytopenic purpura [18], anti-streptokinase antibodies [19] and the antiphospholipid syndrome [20] and up-regulation in the expression number of platelet FcγRII receptors in those with systemic illness [21] or myeloproliferative disorders [22]. Other applications of the understanding of the IgG–FcγRII interaction in HIT include non-haematological immune mediated diseases where the impaired handling of complexed IgG by the genetic polymorphism of FcγRII may be important to predispose patients with systemic lupus erythematosis to glomerulonephritis [23,24] and childhood susceptibility to infection with encapsulated bacteria [25]. However conflicting results regarding the impact of the FcγRII polymorphism have been obtained in both these disease groups [26,27] which may reflect either ethnicity or referral bias of the patient cohorts or other unrecognized factors that modulate the disease phenotype.

CLINICAL ASPECTS OF HEPARIN-INDUCED THROMBOCYTOPENIA

Immune-mediated heparin-dependent thrombocytopenia is usually of delayed onset occurring after at least 4 and up to 14 days after the initiation of heparin therapy, but sooner if there has been previous exposure to the drug [1]. The platelet count usually declines to less than 100×10^9/L. Despite severe thrombocytopenia of less than 40×10^9/L, haemorrhage is rare and thrombosis common [1,2]. Infrequently the platelet count may be normal after falling from a higher level [28]. A retrospective study of 127 medical and surgical inpatients with serologically confirmed HIT, showed that over 80% of new thrombotic events were venous complications, whereas arterial events occurred in those patients with previous underlying cardiovascular risk factors [29,30].

It is now apparent that the clinical expression of thrombosis with HIT underestimates the development of antibodies that cause FcγRII platelet activation. A recent prospective study of heparin thromboprophylaxis in 665 patients following orthopaedic surgery found that 8% of patients on standard heparin developed heparin-dependent IgG platelet activating antibodies [12]. These antibodies were detected by the FcγRII-dependent platelet serotonin release assay and strongly predicted the development of thrombocytopenia but not thrombosis. Delayed thrombocytopenia characteristic of HIT occurred in approximately one-third of these patients (2.7%), most of whom developed thrombosis (2.3%). The incidence of heparin-dependent IgG antibodies was lower with the use of a low molecular weight heparin (2.2%) and no patient in this arm developed thrombocytopenia, raising this feature as a safety advantage of low molecular weight heparin over standard heparin. The development of HIT is probably dose- and duration-dependent, occurring more frequently with bovine than porcine heparin (estimated incidence 2.9% and 1.1%, respectively [31]. Rare cases of HIT are still described with small doses of heparin given as flushes to maintain the patency of indwelling catheters [32] or even as coating for intra-arterial catheters [33].

Treatment of HIT requires immediate cessation of heparin and substitution of alternative antithrombotic drugs whilst awaiting the 2–5 day delayed therapeutic anticoagulant activity of oral anticoagulants. The most recent experience suggest the preferred treatment is the low molecular weight heparinoid, Orgaran [34]. This can be effectively used in the majority of cases for short term anticoagulation provided it does not cross-react with the HIT antibody which occurs in about 10–15% of cases [35]. Other agents that have been used successfully are low molecular weight heparins, which cross-react with the heparin-dependent antibody in up to 20% of cases [36]; ancrod, a defibrinogenating snake venom [37], a stable prostaglandin analogue, iloprost [38]; thrombin inhibitors such as Hiridin [39]; the cyclooxygenase inhibitor, aspirin [1]; the plasma volume expander dextran [1]; and to correct the platelet count, intravenous IgG [40,41]. Apart from the latter, the target of therapy is not immune modulation but anticoagulation which does not interrupt the continued immune-induced injury caused by the antibodies binding to heparin-like molecules (glycosaminoglycans) on the endothelial surface [3,42]. After heparin withdrawal the platelet count generally returns to normal over the next 5–7 days although this can be delayed by several weeks [1,2]. If heparin is inadvertently reinstituted, severe and rapid thrombocytopenia recurs. Most patients become antibody negative 1–2 months after ceasing heparin therapy [2].

NEW OBSERVATIONS ON THE PATHOGENESIS OF HEPARIN-INDUCED THROMBOCYTOPENIA

Despite convincing evidence that antibodies in patients with HIT are specific for heparin and that heparin–IgG immune complexes cause FcγRII-mediated platelet activation, attempts to demonstrate the direct binding of the patient IgG to heparin generally yielded negative or equivocal results. The crucial recent observation is that the antibodies in HIT are not specific for heparin alone but for complexes of heparin and platelet factor 4 (PF4), a heparin binding protein, present at high concentration in platelet α granules which is released from activated platelets [3–6]. Heparin binds to lysine-rich domains near the COOH terminus of each PF4 monomer and it is likely that the HIT antibodies bind to combinatorial epitopes in the PF4 tetramer–heparin macromolecule [43]. It has become clear that many other glycosaminoglycans can substitute for heparin in this reaction but require a certain degree of sulphation (between 0.64 and 1.31) and optimal monosaccharide chain length (molecular weight of 4.8 kDa and a 16 chain polysaccharide) [44]. This observation is supported by the finding that HIT antibodies only cause FcγRII platelet activation in the presence of other polysulphated glucans (low molecular weight heparin and dextran sulphate; degree of sulphation 1.25). No response occurs with either the partially desulphated heparin and with the low sulphated heparinoid, Orgaran (degree of sulphation less than 0.6) or the shortest pentasaccharide heparin fragment with antithrombin III binding activity [44–46]. This explains the clinical efficacy of Orgaran in treatment of HIT [34] and provides data for the design of safer new drugs for anticoagulation.

Different antibodies have been found in about 15% of patients with HIT which cause platelet aggregation but do not bind to the heparin–PF4 complex [47,48]. These antibodies have recently been identified to be directed against homologous molecules to PF4 that are also released by platelet activation or expressed as a chemokine of the inflammatory response [47]. A recent study found that 9 of the 15 patients who had positive heparin-dependent platelet aggregation but lacked heparin–PF4 antibodies, had significant antibodies to interleukin 8 (IL-8) and to the cleavage cytokine of the platelet derived basic protein, neutrophil-activating peptide 2 (NAP-2) [47]. These significant antibody titres were not found in patients with HIT who had heparin–PF4 antibodies or control subjects and were similarly associated with the development of clinical thrombosis. It is of interest that the binding of these antibodies in the ELISA assay was not dependent on heparin. Like PF4, these molecules are released into the circulation and their concentration and antibody titre is significantly increased in inflammatory disorders, malignancy, infection and following major surgery [49,50]. Both chemokines bind to heparin and it is likely that the known direct binding of heparin to platelets may target the IgG–IL-8 and NAP-2 immune complexes to the platelet surface explaining the heparin-dependent FcγRII-mediated platelet activation and thrombosis in these patients.

These findings support the following model as shown in Figure 22.1 for the pathogenesis of HIT and thrombosis and the importance of the platelet FcγRII receptor in modulating this process.

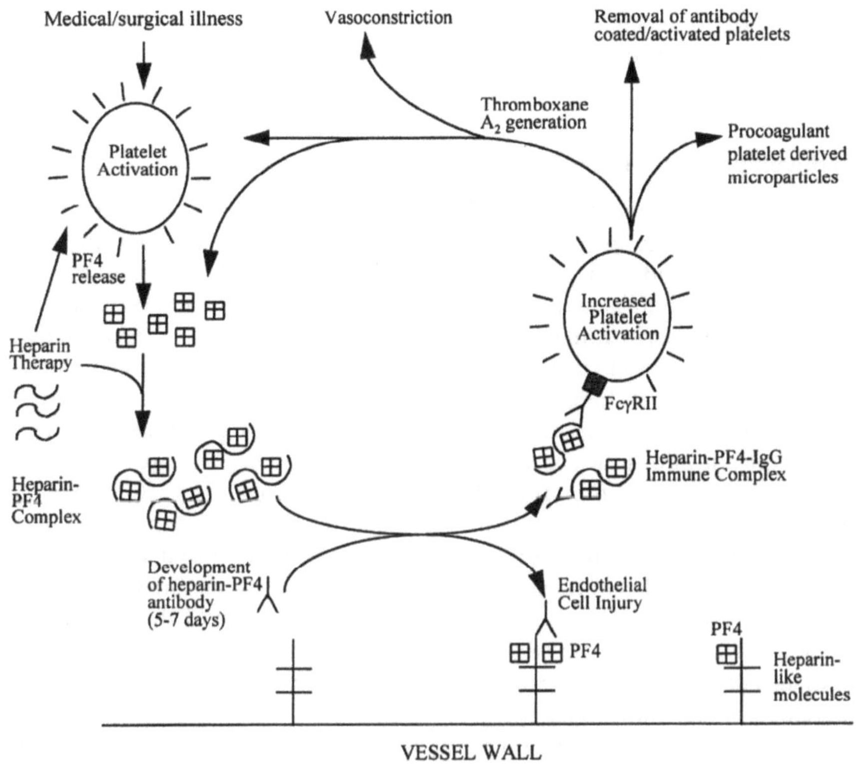

Figure 22.1 Model of the pathogenesis of HIT (see text for details)

Pre-existing conditions and PF4 release

HIT often occurs in previously unwell patients particularly from the intensive care unit and during the postoperative course [1,29]. It has been shown that subaggregating concentrations of adrenaline or thrombin can not only activate platelets to release PF4, but can replace heparin to promote platelet aggregation in response to HIT sera [51]. It is likely in hospitalized patients predisposed to HIT that traces of thrombin are generated, causing slight platelet activation that both primes the platelets for enhanced stimulation by the HIT antibodies and increases the plasma concentrations of platelet α granule proteins, including PF4. Heparin therapy also contributes to substantial increases in circulating levels of PF4 by both directly enhancing platelet activation, causing further platelet PF4 release and displacement of PF4 by heparin from the glycosamino-glycans from the endothelial surface [52,53]. After bolus heparin injection the plasma levels of PF4 rise by 15- to 30-fold and persist at that level for several hours [52,53]. The requirement for PF4 is confirmed by studies in a patient with the grey platelet syndrome whose platelets lack α granules and their contents [54]. No heparin-dependent platelet aggregation in response to HIT plasma occurred but the abnormality could be corrected by the addition of purified PF4.

The development of heparin–PF4 immune complexes

The ratio of the concentration of PF4 and heparin appear crucial for the development of the immunogenic epitope in the PF4–heparin macromolecule for the HIT antibody. In solid phase assays the apparent optimal ratio of heparin to PF4 binding is approximately 1:2 [3,44] with the concentration of unfractionated heparin ranging from 25 to 40 IU/mg native or recombinant PF4 [44]. In the presence of the low heparin concentrations, several PF4 tetramers bind to each heparin chain leading to the formation of the macromolecular complexes which is the main target for the HIT antibody. At high heparin concentrations the complexes between PF4 and heparin are smaller and there is little binding of the HIT IgG antibody [44]. This observation explains why excess heparin (100 U/ml) inhibits heparin FcγRII-dependent platelet activation from HIT plasma. It is an interesting observation that some batches of heparin with similar anticoagulant properties have a very different optimal PF4–heparin ratio [3]. The variation was related to the age of the batches but this finding highlights the possibility that there may be unrecognized variations in standard unfractionated heparin that may influence the propensity of patients to develop the HIT antibody.

Surface glycosaminoglycans on endothelial cells can substitute for heparin and bind to PF4, making this a target for the HIT antibody [3,42]. Serum from HIT patients containing IgG antibodies bind to human umbilical vein endothelial cells, resulting in the prothrombotic surface expression of tissue factor that may predispose to the thrombotic tendency of HIT patients [3,42]. Later it was found that this binding was dependent on PF4 and inhibited by high concentrations of heparin and prior heparinase treatment of the endothelial cells but not with the anti-FcγRII monoclonal antibody IV.3 [3,42]. IgM and IgA heparin–PF4 antibodies have also been described as the only abnormality in HIT patients with thrombosis [55,56]. However most studies often find these antibodies in association with IgG heparin–PF4 complexes and it has been suggested that the IgM antibodies occur early after exposure to heparin [55]. The mechanism for thrombosis and thrombocytopenia of the IgM and IgA antibodies in HIT is open to speculation, but it has been shown that they bind to the heparin–PF4 complexes on endothelial cells and platelets [42,56]. Conceivably other immunocompetent cells with FcμR or FcαR receptors may interact with bound IgM and/or IgA complexes causing cellular activation or destruction of platelets.

FcγRII-mediated platelet activation

The macromolecule of heparin–PF4–IgG can assemble and then bind to the platelet FcγRII in at least two ways. One mechanism is the formation of soluble immune complexes that bind directly to platelet FcγRII, initiating receptor clustering and thereby platelet aggregation [3]. This pathway of FcγRII-dependent platelet aggregation is analogous to using heat-aggregated IgG as the stimulus for platelet aggregation. The second mechanism is a two step process involving the initial assembly of the heparin–PF4 complex on the platelet surface followed by a secondary process of antibody attachment causing Fc-dependent platelet activation [57]. The antigen combining site rather than the Fc portion of the HIT IgG antibody binds to the platelet by recognition of the PF4–heparin macromolecule [57]. The assembly of the PF4–heparin complex can form after initial platelet activation induced PF4 release providing a coupling site [4] or heparin molecules may bind to the platelet surface first, releasing PF4 which subsequently complexes with the membrane bound heparin [5]. In either process the second step to platelet activation is achieved by the Fc portion of the antibody associating with the FcγRII on the same platelets (intraplatelet) or on adjacent platelets (interplatelet) [58]. All anti-platelet monoclonal antibodies that mediate Fc-dependent platelet activation and purified IgG from a patient with immune thrombocytopenia [18] appear to act by this mechanism. The HIT antibody also binds to endothelial cells through recognition of the heparin–PF4 epitope and this complex will localize platelets through the platelet FcγRII receptor close to the endothelial cells surface [3,42]. This can lead to FcγRII-dependent platelet activation which would result in further prothrombotic immune injury.

Both mechanisms of FcγRII-dependent platelet activation (Fc immune complex and direct $F(ab')_2$ binding) probably operate simultaneously in patients with HIT. The factors that influence the preference for one mechanism over another are presently unclear, but may be influenced by the degree of baseline platelet activation prior to the development of the HIT antibody or the relative ratio of the concentration of the PF4 and heparin in the plasma and near to the surface of the platelet. It is also uncertain whether priming of the platelets for aggregation after the initial PF4 or heparin binding may lower the antibody stimulus threshold for FcγRII-mediated platelet activation.

High levels of platelet activation plasma markers (soluble P selectin and β thromboglobulin) are found in patients with HIT [59], reflecting platelet FcγRII-dependent release of procoagulant microparticles [11] and platelet activating substances such as thromboxane A_2 and ADP [8–10]. The platelet-derived microparticles contain activated clotting factors and their membranes act as a source of phospholipid. This catalyses activation of the clotting cascade and the formation of thrombin that leads to further platelet activation and deposition of cross-linked fibrin [29]. These procoagulant conditions may predispose to clinical thrombosis in some individuals [30].

IMPLICATIONS OF THESE RECENT FINDINGS

Diagnosis of heparin-induced thrombocytopenia

The clinical diagnosis of HIT is difficult and currently there is no diagnostic gold standard test. Careful diagnostic criteria have been proposed and include thrombocytopenia occurring during heparin administration, exclusion of other causes of thrombocytopenia such as infection, other drugs or autoimmune thrombocytopenia, resolution of thrombocytopenia after cessation of heparin and the demonstration of a heparin-dependent platelet antibody by a diagnostic test [1,13]. Despite these helpful guidelines there is often clinical uncertainty, either because patients at highest risk for developing HIT generally have other possible causes for thrombocytopenia (the most common being infection), or they have the clinical expression of HIT with negative diagnostic tests. These comments highlight the confounding factors in studies examining the role and specificity and sensitivity of the diagnostic assays for HIT because each group uses its own 'in house' diagnostic methods and each have a different population of possible HIT patients depending on their definition and referral pattern. This may partly explain the described heterogeneity in the reports describing the frequency and significance of HIT-dependent antibodies.

In response to the recent discovery that the heparin–PF4 complex is the target antigen for the HIT antibody, ELISA methods have been developed utilizing this principal. In a recent study comparing the FcγRII functional bioassay for HIT, the [^{14}C] serotonin release assay (SRA), the heparin PF4 ELISA was positive in approximately 90% of cases with both the clinical diagnosis of HIT and positive SRA [48]. The level of sensitivity of the ELISA is in agreement with another similar bioassay, the heparin-induced platelet activation assay [60]. The false negatives in the heparin–PF4 assay are probably caused by platelet activating antibodies directed against other heparin binding proteins such as IL-8 and NAP-2, as previously described [47]. Antibodies to PF4–heparin were also detected in 8% of possible HIT patients with negative SRA and up to 8% of patients who were receiving heparin and who were not thrombocytopenic [48]. These results suggest either the ELISA is more sensitive than the SRA for detecting pathological antibodies or the positive result is a consequence of heparin therapy with no clinical relevance. Like functional bioassays the titre of the antibody did not appear to distinguish the clinical severity of HIT. Other studies confirm that the development of heparin–PF4 antibodies is common after exposure to heparin, with a prevalence ranging from 4–23% of patients depending upon the clinical situation and method of assay [48,61–63]. It has been shown that if the platelet count is normal when HIT antibodies develop there is no apparent adverse clinical outcome [12]. Isolated and multiple immunoglobulin types (IgG, IgM and IgA) directed against the heparin–PF4 molecule have also been detected after administration of heparin. Only occasionally can the IgM and IgA antibodies be associated with thrombocytopenia and only when this occurs may the patient develop thrombosis [55,56].

The thrombocytopenia may be the result of non-FcγRII-mediated platelet destruction and the thrombosis in susceptible people is caused by endothelial perturbation upon binding of the IgM antibodies.

The relative role of functional FcγRII-dependent bioassays compared to the heparin–PF4 ELISA assay in predicting the *in vitro* cross-reactivity of the heparin antibodies to low molecular weight heparin or heparinoids remains uncertain. There is greater experience using Fc-dependent functional bioassays such as platelet aggregation and SRA to predict the safety of the proposed substitute heparin-like anticoagulant, and it is generally accepted that it is probably unsafe clinically to use a cross-reacting anticoagulant detected by these methods [34–36]. No study has addressed this problem with the heparin–PF4 ELISA but a recent report describes successful use in a patient with HIT of a positive cross reactive LMWH that was retrospectively found in the heparin–PF4 ELISA [64]. Unfortunately a functional bioassay was not performed simultaneously for direct comparison between these two methods.

Due to these problems with sensitivity and specificity, the role of the heparin PF4 ELISA remains uncertain in clinical practice but it may be a useful adjunct combined with the bioassay in the clinical management of these patients. The practical role appears to be as a secondary test after a negative functional bioassay overcoming the sensitivity problems of the ELISA when platelet activation and HIT occurs by non–heparin–PF4 antibodies [47]. A subsequent negative ELISA result would be a strong predictor that the patient does not have HIT. A positive ELISA may not be as helpful, although in the setting of thrombocytopenia the clinician could cease heparin if it is thought that the ELISA assay is a more sensitive test and not a clinically irrelevant result arising during heparin therapy. Numerous questions remain concerning the clinical role of the diagnostic tests and HIT. Are patients on heparin with a positive ELISA and negative bioassay more likely to develop complications if heparin is continued? Is the heparin-dependent platelet activation in patients with a negative ELISA always FcγRII-dependent? A prospective study is required in strictly selected HIT patients comparing simultaneous rigorous bioassays of FcγRII-dependent platelet activation using both high concentrations of heparin and blockade of response by an anti-FcγRII monoclonal antibody compared with a standardized heparin platelet–PF4 ELISA assay. These results need to be correlated to platelet count and clinical outcomes.

Modulation of HIT antibody binding to the platelet Fc receptor

As HIT antibodies activate platelets by their FcγRII receptors, several genetic and acquired variables have been described that would influence complexed IgG and platelet FcγRII interaction. Such changes as outlined in Table 22.1, may protect or predispose the patient to the development of HIT and could explain the discrepancy between the common development of the heparin–PF4 antibodies and the infrequent detection of FcγRII platelet activation leading to the clinical manifestations of HIT.

Table 22.1 Factors that modulate the interaction of complexed IgG and platelet FcγRII in patients with heparin induced thrombocytopenia

Hereditary	*Acquired*
– Variation in FcγRII number	– Increased FcγRII number
– FcγRII-131 polymorphism	– HIT IgG
– Plasma IgG level	Titre
	Epitope specificity
	IgG subclass
	– Soluble FcγRII level

Surface expression of platelet FcγRII

The number of platelet FcγRII is stable among individuals over time and ranges between 1500 and 4500 receptors per platelet [65,66]. The number of receptors strongly correlates with platelet aggregation in response to heat-aggregated IgG and it has been proposed that subjects with the highest number of Fc receptors are more likely to develop immune complex-mediated disease [66]. Platelet activation *in vitro* by thrombin and PMA can increase the number of platelet FcγRII by 50% [67]. This expression may be important in medical/surgical patients who are at highest risk of developing HIT. As a result of their underlying illness these patients already have mild platelet activation which results in an increase in the platelet FcγRII number [21]. When the heparin–PF4 complexes arise during heparin therapy the binding of these complexes to the FcγRII will initiate further platelet activation, predisposing to thrombosis. This concept is supported by the observation that the number of platelet Fc receptors substantially rises in patients with HIT and in acute inflammatory illnesses, producing increased platelet responsiveness to the HIT antibody or aggregated IgG [21]. Those patients with HIT who had the highest FcγRII number had the most serious thrombosis and a poorer prognosis. In those surviving patients, the platelet FcγRII number returned to baseline by 3 months and the level was not different from patients who did not develop HIT [21]. These results suggest that the baseline individual level of the resting platelet FcγRII number did not in itself predispose to HIT. As platelets are non-nucleated cells, the increased expression of platelet FcγRII probably is a result of either release of the soluble FcγRII from platelet granules or up-regulation of the receptor by inflammatory cytokines at the megakaryocyte or earlier bone marrow progenitor cells [17,21,67].

Platelet FcγRII polymorphism

Three genes, FcγRIIA, FcγRIIB and FcγRIIC, encode for multiple transcripts which produce a 40 kDa protein with similar extracellular and transmembrane domains but different cytoplasmic tails (FcγRIIa,b,c) [68]. Platelets and megakaryocytes predominantly contain only the receptor from the FcγRIIA gene [17,68] and a common functional polymorphism at amino acid position 131 in the extracellular domain has been described [69,70]. This results in either arginine (R) or histidine (H) expression, altering the binding affinity for human IgG_2 and mouse IgG_1 [69,71]. Paradoxically FcγRIIa H/H^{131} genotype has high affinity for human IgG_2 but low affinity for mouse IgG_1 resulting in no platelet activation by the IgG_1 anti-platelet monoclonal antibodies; the so called non-responder phenotype [65,72,73]. In the presence of human IgG_2 complexes the opposite response occurs leading to significant antibody binding and rapid FcγRII-dependent platelet activation [69,71]. It is of interest that the predominant humoral immune response to other polysaccharide antigens is of the human IgG_2 subclass [74]. As the HIT antibody is directed against the heparin–PF4 complex and the exact antigenic role of the mucopolysaccharide component is still unclear, it was speculated that in HIT patients, the antibodies were the IgG_2 subclass and the FcγRIIa H/H^{131} genotype would be predisposed to FcγRII-mediated platelet activation.

Several groups have demonstrated an over-representation of the FcγRIIa H^{131} allele in HIT patients [75,76], but this finding has not been supported by another recent study [77]. They also could not confirm the antibody subclass specificity in HIT patients because the predominant IgG heparin–PF4 antibodies were IgG_1 (88%) rather than the IgG_2 subclass (6%) [77]. However associated with the major IgG_1 response, IgG_2 antibodies were also significantly raised in 62% of patient sera along with IgG_3 antibodies (15%) but not IgG_4 antibodies. No co-segregation was observed between the development of IgG_2 subclass heparin–PF4 antibodies and any of the FcγRII genotypes. There was a trend towards a lower concentration of IgG_2 antibodies in those patients with the FcγRIIa H/H^{131} genotype compared with the FcγRIIa R/R^{131} genotype [77]. This suggests that the level may be apparently lower because of FcγRII-dependent removal of IgG_2 heparin–PF4 complexes that may lead to platelet activation associated with HIT. Alternatively it has been postulated that the FcγRIIa H/H^{131} genotype may regulate immunoglobulin subclass production which results in lower IgG_2 levels [69].

The hypothesis of the importance of the IgG_2 subclass HIT antibodies and FcγRII polymorphism is in direct conflict with the report that the FcγRIIa H/H^{131} genotype is unresponsive to HIT plasma, whereas the FcγRIIa R/R^{131} genotype responded well [75]. These contrasting clinical and laboratory results highlight the heterogenous nature of the clinical syndrome and the antibody response in particular cohorts of HIT patients which may be only partially explained by the FcγRIIa 131 polymorphism.

Plasma IgG level

High concentrations of IgG suppress the interaction of the HIT antibody with the platelet FcγRII, and this property is the basis for intravenous IgG therapy of

HIT [13,40,41]. High levels of baseline plasma monomeric IgG can produce platelets unresponsive to FcγRII-mediated activation, making these individuals unsuitable for HIT testing in platelet-rich plasma [13]. Washed platelets show normal responses to immune complexes, suggesting that regulation of IgG production in health and in clinical situations may be important determinants of immune complex–FcγRII interaction. It is uncertain whether this *in vitro* finding has clinical relevance but it suggests that the development of hypergammaglobulinemia in response to immune stimulation may inhibit the immune complex–platelet FcγRII interaction. The exact mechanism of the IgG inhibition is uncertain because the platelet Fc receptor has low affinity for monomeric compared with high affinity for complexed IgG.

Plasma soluble FcγRII levels

Two alternatively spliced FcγRIIA transcripts have been found in platelets and megakaryocytes that either do (FcγRIIa1), or do not (FcγRIIa2), contain the transmembrane encoding exon for FcγRIIa [68]. Soluble FcγRIIa2 has been found in culture supernatant of malignant cell lines and it is released upon thrombin activation of platelets and megakaryocytes [68,78]. Recombinant soluble FcγRIIa2 was found to be a potent inhibitor of Fc-dependent anti-CD9 antibody-induced platelet activation, suggesting that this molecule may be an important regulator of complexed IgG and platelet FcγRII interaction [78]. We have studied the inhibition of HIT antibody-induced platelet activation by another recombinant soluble FcγRII (rsFcγRII) molecule, as previously described [79]. Briefly, the truncated form of the receptor was produced by inserting a translation termination codon into the FcγRIIa cDNA segment encoding the 5′ region of the transmembrane domain. The HFc3.0 cDNA encodes the FcγRIIa allelic variant expressing glutamine and histamine at amino acid positions 27 and 131 respectively [79]. This molecule will bind with high affinity to the human IgG_2 subclass of the HIT antibody. Heat-inactivated plasma from 11 patients with HIT were examined in platelet-rich plasma from the same donor for response by platelet aggregation, thromboxane B_2 production and mepacrine release from platelet dense granules. Donor platelet rich plasma (250 μl) was preincubated with 50 μl rsFcγRII (final concentration 50 μl) or buffer and HIT plasma (200 μl) for 5 minutes prior to the addition of heparin (0.5 U/ml). Inhibition of platelet activation was variable but reduced in eight of 11 patients, in four by more than 50% of all baseline levels. Analogous to the dose response finding with the platelet activating monoclonal antibody ALB6 [78], weaker and slower responding HIT antibodies were more sensitive to inhibition by rsFcγRII (Figure 22.2). It is likely that the high antibody titre in some HIT patients overwhelms the binding capacity of the rsFcγRII allowing the initiation of FcγRII-mediated platelet activation.

Plasma levels of soluble FcγRII in normal subjects have been reported to range between 0 and 30 ng/ml [80,81]. We examined whether plasma levels of soluble

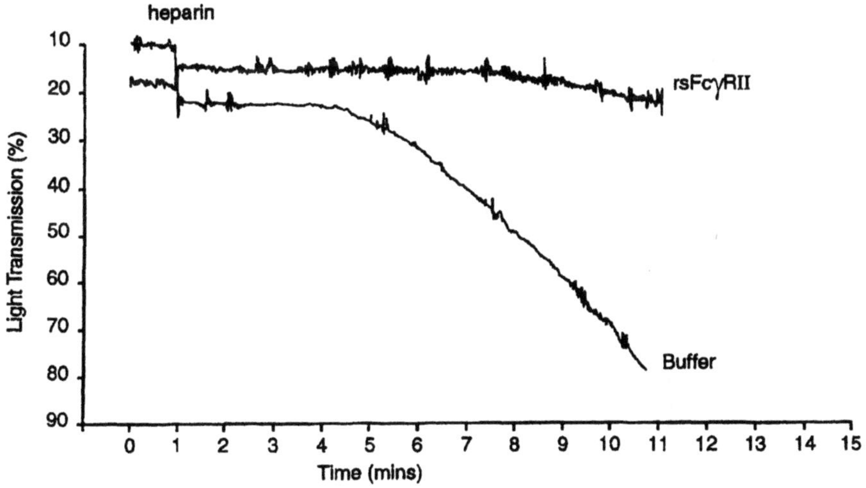

Figure 22.2 Inhibition of heparin-dependent antibody induced platelet aggregation from a patient with heparin induced thrombocytopenia by recombinant soluble FcγRII

FcγRII are raised in patients with HIT which may modulate the platelet FcγRII response to the HIT antibody. A capture:tag ELISA assay for human soluble FcγRII was developed using two non-competing anti-FcγRII mAb, 8.26 and 8.7 and recombinant soluble FcγRII was used as the standard [78]. The assay was sensitive to 1 ng/ml, reproducible (intra-assay coefficient of variation 5%, inter-assay coefficient of variation 12%) and not caused by a cellular remnant. Specificity of the ELISA for plasma FcγRII could be demonstrated by abolishing detection after preincubation of plasma with Affigel beads coated with the anti-FcγRII mAb IV.3 but not control mouse IgG. Compared with control levels (mean 36 ng/ml) we found a 10- to 30-fold elevation in the level of soluble FcγRII in patients with HIT and septicaemia, with only a modest increase in patients with immune thrombocytopenia at presentation (see Figure 22.3).

These findings suggest that the inflammatory response rather than platelet activation or consumption releases plasma soluble FcγRII. Increased levels could protect cells from FcγRII-dependent activation in response to immune complexes. It is uncertain whether the increased plasma FcγRII levels would be sufficient to completely inhibit platelet FcγRII-mediated platelet activation by the HIT antibody *in vivo* because much higher concentrations of recombinant soluble FcγRII were required to completely abolish *in vitro* HIT-induced platelet activation. Nevertheless the HIT assay conditions are contrived in favour of a sudden overwhelming immune complex load which requires only a small percentage of platelets to activate to produce irreversible generalized aggregation mediated by other potent platelet release substances such as thromboxane A_2 and ADP. It is likely that the concentration of plasma soluble FcγRII may

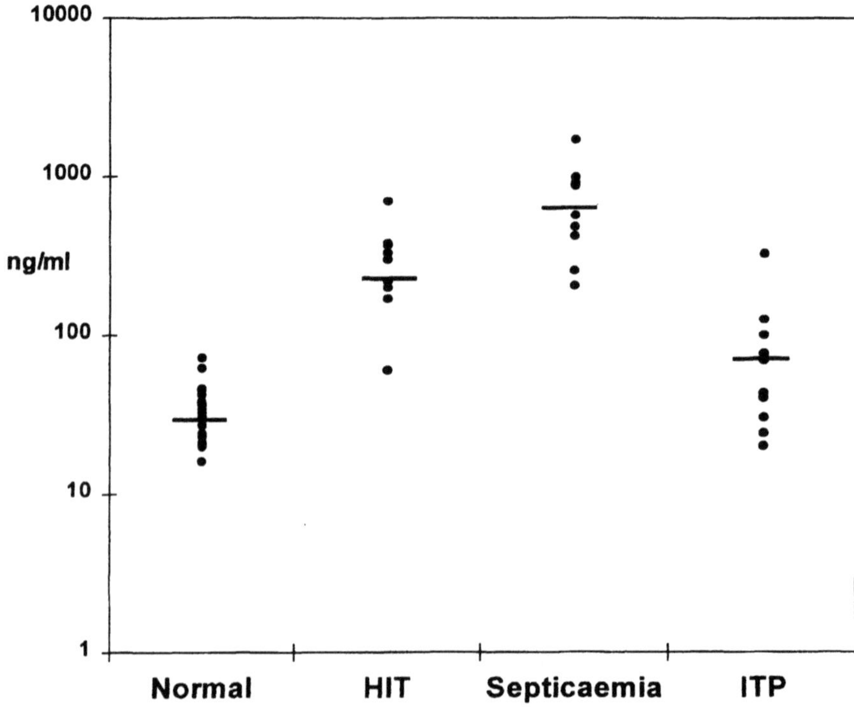

Figure 22.3 Plasma soluble FcγRII levels in patients with heparin-induced thrombocytopenia (HIT, $n = 10$), septicaemia ($n = 10$), immune thrombocytopenia purpura (ITP, $n = 10$) and 19 normal controls (logarithmic scale). The horizontal bars represent the mean value for each group. Soluble FcγRII levels were significantly raised in HIT (mean 293 ng/ml, $p < 0.0001$), ITP (mean 86 ng/ml, $p < 0.007$) and septicaemia (mean 738 ng/ml, $p < 0.0001$) when compared to normal controls (mean 35 ng/ml) (Student *t*-test)

play an important role in the initial development of HIT because raised levels may inhibit the low concentrations of the heparin–PF4–IgG immune complex stimulating the platelet FcγRII, breaking the cycle of increased FcγRII platelet activation, PF4 release and platelet microparticle development. It is probable that soluble FcγRII will have other protective indirect immunomodulating roles in HIT, such as its role in the recently described inhibition of immune complex precipitation [82]. Our finding of high plasma FcγRII levels in HIT protecting platelets from activation could be one explanation for the observation of the common development of HIT antibodies without thrombocytopenia or clinical evidence of disease.

CONCLUSION

HIT is the best example of a well described disease where the FcγRII receptor probably has an important role in disease pathogenesis. Modulation of the IgG–Fc interaction will result in better preventative strategies and treatment of HIT that may be applied to other poorly understood immune-mediated diseases. Further knowledge of the determinants of the platelet FcγRII–IgG interaction has direct practical relevance because it will improve the reliability of the functional Fc-dependent bioassays to diagnose HIT in the clinical laboratory.

References

1. Chong BH. Heparin-induced thrombocytopenia. Br J Haematol. 1995;89:431–9.
2. Visentin GP, Aster RH. Heparin-induced thrombocytopenia and thrombosis. Curr Opin Hematol. 1995;2:351–7.
3. Visentin GP, Ford SE, Scott P, Aster RH. Antibodies from patients with heparin-induced thrombocytopenia/thrombosis are specific for platelet factor 4 complexed with heparin or bound to endothelial cells. J Clin Invest. 1994;93:81–8.
4. Kelton JG, Smith JW, Warkentin TE, Hayward CP, Denomme GA, Horsewood P. Immunoglobulin G from patients with heparin-induced thrombocytopenia binds to a complex of heparin and platelet factor 4. Blood. 1994;83:3232–9.
5. Greinacher A, Pötzsch B, Amiral J, Dummel V, Eichner A, Mueller-Eckhardt C. Heparin-associated thrombocytopenia: isolation of the antibody and characterization of a multi-molecular PF4–heparin complex as the major antigen. Thromb Haemost. 1994;71:247–51.
6. Amiral J, Bridey F, Wolf M. Antibodies to macro molecular platelet factor 4-heparin complexes in heparin-induced thrombocytopenia: a study of 44 cases. Thromb Haemost. 1995;73:21–8.
7. Kelton JG, Sheridan D, Santos A et al. Heparin-induced thrombocytopenia: laboratory studies. Blood. 1988;72:925–30.
8. Anderson GP, Anderson CL. Signal transduction by the platelet Fc receptor. Blood. 1990;76:1165–72.
9. Chong BH, Fawaz I, Chesterman CN, Berndt MC. Heparin-induced thrombocytopenia: mechanism of interaction of the heparin-dependent antibody with platelets. Br J Haematol. 1989;73:235–40.
10. Chong BH, Pitney WE, Castaldi PA. Heparin-induced thrombocytopenia: association of thrombotic complications with heparin-dependent IgG antibody that induced thromboxane synthesis and platelet aggregation. Lancet. 1982;ii:1246–9.
11. Warkentin TE, Hayward CP, Boshkov LK et al. Sera from patients with heparin-induced thrombocytopenia generate platelet-derived microparticles with procoagulant activity: an explanation for the thrombotic complications of heparin-induced thrombocytopenia. Blood. 1994;84:3691–9.
12. Warkentin TE, Levine MN, Hirsh J et al. Heparin-induced thrombocytopenia in patients treated with low-molecular-weight heparin or unfractionated heparin. N Engl J Med. 1995; 332:1330–5.
13. Chong B.H, Burgess J, Ismail F. The clinical usefulness of the platelet aggregation test for the diagnosis of heparin-induced thrombocytopenia. Thromb Haemost. 1993a;69:344–50.
14. Greinacher A, Michels I, Kiefel V, Mueller-Eckhardt C. A rapid and sensitive test for diagnosing heparin-associated thrombocytopenia. Thromb Haemost. 1991;66:734–6.
15. Stewart MW, Etches WS, Boshkov LK, Gordon PA. Heparin-induced thrombocytopenia: an improved method of detection based on lumi-aggregometry. Br J Haematol. 1995;91:173–7.
16. Teitel JM, Gross P, Blake P, Garvey MB. A bioluminescent adenosine nucleotide release assay for the diagnosis of heparin-induced thrombocytopenia. Thromb Haemost. 1996;73:479.
17. Markovic B, Wu Z, Chesterman CN, Chong BH. Quantitation of soluble and membrane-bound Fc gamma RIIA (CD32) mRNA in platelets and megakaryoblastic cell line (Meg-01). Br J Haematol. 1995;91:37–42.

18. Jackson SP, Jane S.M, Mitchell CA et al. Arterial thrombosis associated with immune thrombocytopenia: Presence of a platelet aggregating IgG synergistic with thrombin and adrenalin. Thromb Haemost. 1989;62:846–9.
19. Lebrazi J, Helft G, Abdelouahed M et al. Human anti-streptokinase antibodies induce platelet aggregation in an Fc receptor (CD32) dependent manner. Thromb Haemost. 1995; 74:938–42.
20. Arnout J. The pathogenesis of the antiphospholipid syndrome: a hypothesis based on parallelisms with heparin-induced thrombocytopenia. Thromb Haemost. 1996;75:536–41.
21. Chong BH, Pilgrim RL, Cooley MA, Chesterman CN. Increased expression of platelet IgG Fc receptors in immune heparin-induced thrombocytopenia. Blood. 1993;81:988–93.
22. Moore A, Nachman RL. Platelet Fc receptor: increased expression in myeloproliferative disease. J Clin Invest. 1981; 67:1064–71.
23. Duits AJ, Bootsma H, Derksen RH et al. Skewed distribution of IgG Fc receptor IIa (CD32) polymorphism is associated with renal disease in SLE patients. Arthritis Rheumat. 1996;39: 1–5.
24. Salmon JE, Millard S, Schachter LA et al. Fc gamma RIIA alleles are heritable risk factors for lupus nephritis in African Americans. J Clin Invest. 1996;97:1348–54.
25. Sanders LA, van de Winkel JG, Rijkers GT et al. Fcγ receptor IIa (CD32) heterogeneity in patients with recurrent bacterial respiratory tract infections. J Infect Dis. 1994;170:854–61.
26. Botto M, Theodoridis E, Thompson EM et al. Fc gamma RIIa polymorphism in systemic lupus erythematosus (SLE): No association with disease. Clin Exp Immunol. 1996;104:264–8.
27. Norris CF, Surrey S, Bunin GR, Schwartz E, Buchanan GR, McKenzie SE. Relationship between Fc receptor IIA polymorphism and infection in children with sickle cell disease. J Paed. 1996;128:813–19.
28. Phelan BK. Heparin-associated thrombosis without thrombocytopenia. Ann Intern Med. 1983;99:637–8.
29. Warkentin TE. Heparin-induced thrombocytopenia: IgG-mediated platelet activation, platelet microparticle generation, and altered procoagulant/anticoagulant balance in the pathogenesis of thrombosis and venous limb gangrene complicating heparin-induced thrombocytopenia. Trans Med Reviews. 1996;10:249–58.
30. Boshkov LK, Warkentin TE, Hayward CP, Andrew M, Kelton JG. Heparin-induced thrombocytopenia and thrombosis: clinical and laboratory studies. Br J Haematol. 1993; 84:322–8.
31. Schmitt BP, Adelman B. Heparin-associated thrombocytopenia: a critical review and a pooled analysis. Am J Med Sci. 1993;305:208–15.
32. Doty JR, Alving BM, McDonnell DE, Ondra SL. Heparin-associated thrombocytopenia in the neurosurgical patient. Neurosurgery. 1986;19:69–72.
33. Laster J, Elfrink R, Silver D. Re-exposure to heparin of patients with heparin-associated antibodies. J Vasc Surg 1989;9:677–81.
34. Chong BH, Magnani HN. Orgaran in heparin-induced thrombocytopenia. Haemostasis. 1992;22:85–91.
35. Chong BH, Ismail F, Cade J, Gallus AS, Gordon S, Chesterman CN. Heparin-induced thrombocytopenia: Studies with a new low molecular weight heparinoid, Org 10172. Blood. 1989;73:1592–6.
36. Ramakrashna R, Manoharan A, Kwan YL, Kyle PW. Heparin-induced thrombocytopenia: cross-reactivity between standard heparin, low molecular weight heparin, dalteparin (Fragmin) and heparinoid, danaparoid (Orgaran). Br J Haem. 1995;91:736–8.
37. Demers C, Ginsberg JS, Brill-Edwards P et al. Rapid anticoagulation using ancrod for heparin-induced thrombocytopenia. Blood. 1991;78:2194–7.
38. Kappa JR, Fisher CA, Todd B et al. Intraoperative management of patients with heparin-induced thrombocytopenia. Ann Thoracic Surgery. 1990;49:714–22.
39. Nand S. Hirudin therapy for heparin-associated thrombocytopenia and deep venous thrombosis. Am J Haematol. 1993;43:310–11.
40. Grau E, Linares M, Olaso A, Ruvira J, Sauchris J. Heparin-induced thrombocytopenia-response to intravenous immunoglobulins in vivo and in vitro. Am J Hematol. 1992;39:312–13.
41. Greinacher A, Liebenhoff U, Kiefel V, Presek P, Mueller-Eckhardt C. Heparin-associated thrombocytopenia: the effects of various intravenous IgG preparations on antibody mediated

platelet activation – a possible new indication for high dose IV IgG. Thromb Haemost. 1994;71:641–5.
42. Cines DB, Tomaski A, Tannenbaum S. Immune endothelial-cell injury in heparin-associated thrombocytopenia. N Eng J Med. 1987;316:581–9.
43. Zycker MB, Katz IR. Platelet factor 4: production, structure, and physiologic and immunologic action. Proc Soc Exp Biol Med .1991;198:693–702.
44. Greinacher A, Alban S, Dummel V, Franz G, Mueller-Eckhardt C. Characterization of the structural requirements for a carbohydrate based anticoagulant with a reduced risk of inducing the immunological type of heparin-associated thrombocytopenia. Thromb Haemost. 1995;74:886–92.
45. Greinacher A, Michels I, Liebenhoff U, Presek P, Mueller-Eckhardt C. Heparin-associated thrombocytopenia: immune complexes are attached to the platelet membrane by the negative charge of highly sulfated oligosaccharides. Br J Haematol. 1993;84:711–16.
46. Greinacher A, Michels I, Mueller-Eckhardt C. Heparin-associated thrombocytopenia: the antibody is not heparin specific. Thromb Haemost. 1992;67:545–9.
47. Amiral J, Marfaing-Koka A, Wolf M et al. Presence of autoantibodies to interleukin-8 or neutrophil-activating peptide-2 in patients with heparin-associated thrombocytopenia. Blood. 1996;88:410–16.
48. Arepally G, Reynolds C, Tomaski A et al. Comparison of PF4/heparin ELISA assay with the 14C-serotonin release assay in the diagnosis of heparin-induced thrombocytopenia. Am J Clin Path. 1995;104:648–53.
49. Sylvester L, Yoshimura T, Sticherling M et al. Neutrophil attractant protein-1-immunoglobulin G immune complexes and free anti-NAP-1 antibody in normal human serum. J Clin Invest. 1992;90:471–81.
50. Bendtzen K, Hansen MB, Ross C, Poulsen LK, Svenson M. Cytokines and autoantibodies to cytokines. Stem Cells. 1995;13:206–22.
51. Pfueller SL, David R. Different platelet specificities of heparin-dependent platelet aggregating factors in heparin-associated immune thrombocytopenia. Br J Haem. 1986;64:149 159.
52. Dawes J, Pumphrey C, McLaren KM, Prowse CV, Pepper DS. The in vivo release of human platelet factor 4 by heparin. Thromb Res. 1982;27:65 76.
53. O'Brien JR, Etherington MD, Pashley M. Intra-platelet platelet factor 4 (IP.PF4) and the heparin-mobilizable pool of PF4 in health and atherosclerosis. Thromb Haemost. 1984;51: 354 7.
54. Horne MK, Alkins BR. Importance of PF4 in heparin-induced thrombocytopenia: confirmation with Gray platelets. Blood. 1995;85(5):1408–9.
55. Suh JS, Malik MI, Aster RH, Visentin GP. Characterization of the humoral immune response in heparin-induced thrombocytopenia/thrombosis (HITP) Blood. 1995;1763:444a.
56. Amiral J, Wolf M, Fischer A, Boyer-Neumann C, Vissac A, Meyer D. Pathogenicity of IgA and/or IgM antibodies to heparin-PF4 complexes in patients with heparin-induced thrombocytopenia. Br J Haem. 1996;92:954–9.
57. Horne MK, Alkins BR. Platelet binding of IgG from patients with heparin-induced thrombocytopenia. J Lab Clin Med. 1996;127:435–42.
58. Horsewood P, Hayward CP, Warkentin TE, Kelton JG. Investigation of the mechanisms of monoclonal antibody-induced platelet activation. Blood. 1991;78:1019–26.
59. Chong BH, Murray B, Berndt MC, Dunlop C, Brighton T, Chesterman CN. Plasma P-selectin is increased in thrombotic consumptive platelet disorders. Blood. 1994;83:1535–4.
60. Greinacher A, Amiral J, Dummel V, Vissac A, Kiefel V, Mueller-Eckhardt C. Laboratory diagnosis of heparin-associated thrombocytopenia and comparison of platelet aggregation test, heparin-induced platelet activation test, and platelet factor 4/heparin enzyme-linked immunosorbent assay. Transfusion. 1994;34:381–5.
61. Kappers-Klunne MC, Boon DMS, Hop WS, Michiels JJ, Stibbe J, van der Zwaan C. Heparin-induced thrombocytopenia and thrombosis: a prospective analysis of the incidence in patients with heart and cerebrovascular diseases. Br J Haem. 1997;96:442–6.
62. Yamamoto S, KoideM, Matsuo M, Suzuki S, Ohtaka M, Saika S et al. Heparin-induced thrombocytopenia in hemodialysis patients. Am J Kid Dis. 1996;28:82–5.
63. Amiral J, Peynaud-Debayle E, Wolf M, Bridey F, Vissac A, Meyer D. Generation of antibodies to heparin–PF4 complexes without thrombocytopenia in patients treated with unfractionated or low-molecular-weight heparin. Am J Haematol. 1996;52:90–5.

64. Luzzatto G, Cordiano I, Patrassi G, Fabris F. Heparin-induced thrombocytopenia: discrepancy between the presence of IgG cross-reacting in vitro with Fraxiparine and its successful clinical use. Thromb Haemost. 1995;74:1607–8.
65. Tomiyama Y, Kunicki TJ, Zipf TF, Ford SB, Aster RH. Response of human platelets to activating monoclonal antibodies: Importance of FcγRII (CD32) phenotype and level of expression. Blood. 1992;80:2261–8.
66. Rosenfeld SI, Ryan DH, Looney RJ, Anderson CL, Abraham GN, Leddy JP. Human Fcγ receptors: stable inter-donor variation in quantitative expression on platelets correlates with functional responses. J Immunol. 1987;138:2869–73.
67. McCrae KR, Shattil SJ, Cines DB. Platelet activation induces increased Fcγ receptor expression. J Immunol. 1990;144:3920–7.
68. Cassel DL, Keller MA, Surrey S et al. Differential expression of FcγRIIA, FcγRIIB and FcγRIIC in hematopoietic cells: analysis of transcripts. Molecular Immunol. 1993;30:451–60.
69. Warmerdam PAM, van de Winkel JGJ, Vlug A, Westerdaal NAC, Capel PJA. A single amino acid in the second Ig-like domain of the human Fcγ receptor II is critical for human IgG binding. J Immunol. 1991;147:1338–43.
70. Tate BJ, Witort E, McKenzie IF, Hogarth PM. Expression of the high responder/non-responder human FcγRII. Analysis by PCR and transfection into FcR-COS cells. Immunol Cell Biol. 1992;70:79–87.
71. Parren PW, Warmerdam PA, Boeije LC et al. On the interaction of IgG subclasses with the low affinity FcγRIIa (CD32) on human monocytes, neutrophils, and platelets. J Clin Invest. 1992;90:1537–46.
72. Bachelot C, Saffroy R, Gandrille S, Aiach M, Rendu F. Role of FcγRIIA gene polymorphism in human platelet activation by monoclonal antibodies. Thromb Haemostas. 1995;74:1557–63.
73. De Reys S, Blom C, Lepoudre B et al. Human platelet aggregation by murine monoclonal antiplatelet antibodies is subtype-dependent. Blood. 1993;81:1792–800.
74. Jefferis R, Kumarartne DS. Selective IgG subclass deficiency: quantification and clinical relevance. Clin Exp Immunol. 1990;81:357–67.
75. Brandt JT, Isenhart CE, Osborne JM, Ahmed A, Anderson CL. On the role of platelet FcγRIIa phenotype in heparin-induced thrombocytopenia. Thromb Haemost. 1995;74:1564–72.
76. Burgess JK, Lindeman R, Chesterman CN, Chong BH. Single amino acid mutation of Fcγ receptor is associated with the development of heparin-induced thrombocytopenia. Br J Haematol. 1995;91:761–6.
77. Arepally G, McKenzie SE, Jiang X, Poncz M, Cines DB. FcγRIIA H/R131 polymorphism, subclass-specific IgG anti-heparin/platelet factor 4 antibodies and clinical course in patients with heparin-induced thrombocytopenia and thrombosis. Blood. 1997;89:370–5.
78. Gachet C, Astier A, de la Salle H et al. Release of FcγRIIa2 by activated platelets and inhibition of anti-CD9-mediated platelet aggregation by recombinant FcγRIIa2. Blood. 1995;85:698–704.
79. Ierino FL, Powell Ms, McKenzie IF, Hogarth PM. Recombinant soluble human FcγRII: Production, characterisation and inhibition of the Arthus reaction. J Exp Med. 1994;178:1617–28.
80. Astier A, de la Salle H, Moncuit J, Freund M, Cazenave J, Fridman W et al. Detection and quantification of secreted soluble FcγRIIA in human sera by an enzyme-linked immunosorbent assay. J Immunol Meth. 1993;166:1–10.
81. Teillaud JL, Bouchard C, Astier A, Teillaud C, Tartour E, Michon J. Natural and recombinant soluble low-affinity FcγR: Detection, purification, and functional activities. Immunomethods. 1994;4:48–64.
82. Gavin AL, Wines BD, Powell MS, Hogarth PM. Recombinant soluble FcγRII inhibits immune complex precipitation. Clin Exp Immunol. 1995;102:1–6.

23
Fcγ receptor polymorphisms: clinical aspects

J. E. SALMON and R. P. KIMBERLY

INTRODUCTION

FcγR provide the link between cellular and humoral aspects of the immune cascade. Their structural diversity presents a rich framework within which one can understand how immunoglobulin complexes activate a broad range of cell programs relevant to autoimmunity, cancer and microbial host defence. A fundamental assumption is that different receptor structures have different biological functions. Such differences in structure among FcγR dictate differential signalling potentials and biologic functions and provide a diversity of FcγR within a given individual. Inherited or acquired differences in FcγR structure, expression, or function provide the basis for differences in FcγR function between individuals. Allelic variants of FcγR which confer distinct functional capacities to effector cells are a mechanism for inherited differences in disease susceptibility. Originally, allelic variants of FcγR were identified by phenotypic differences, such as binding to specific mAbs or IgG subclasses. More recently, allele-specific PCR, allele-specific oligomer hybridization of PCR products, SSCP, and direct automated sequencing have provided precise definition of each variant allowing characterization of the corresponding functional phenotype. Clinical significance of allelic polymorphisms of FcγR is the focus of this chapter.

Fcγ RECEPTOR IIA

FcγRIIa is the most widely distributed FcγR and is present on most haematopoietic cells. It is expressed on all phagocytes, antigen-presenting cells, and

J.G.J. van de Winkel and P.M. Hogarth (eds.), The Immunoglobulin Receptors and their Physiological and Pathological Roles in Immunity. 267–278.

microdermal endothelium. In a number of cell types, including platelets, it appears to be the only class of FcγR. There are two co-dominantly expressed alleles of FcγRIIa, R^{131} and H^{131}, which differ at amino acid position 131 in the second immunoglobulin-like proximal-membrane domain, arginine and histidine, respectively (Figure 23.1). FcγRIIa-R^{131} and FcγRIIa-H^{131} were previously known as HR (high responder) and LR (low responder), respectively, as a consequence of their relative capacity to bind $mIgG_1$ and provide a substrate for $mIgG_1$ anti-CD3 T cell proliferation assays ($R^{131} > H^{131}$) [1]. Recent studies of murine IgG_1 anti-CD3-mediated immunosuppression showing elevated TNF-α and lymphopenia only in FcγRIIa-R^{131} patients provide *in vivo* evidence for allele-specific FcγRIIa interactions with $mIgG_1$ [2]. A second polymorphic site at position 27 in the first extracellular domain has also been identified, but is not associated with differential binding characteristics [3,4].

Allelic forms of FcγRIIa, though initially identified on the basis of a functional polymorphism related to murine IgG_1 binding, differ substantially in capacity to bind of human IgG_2 [4–8]. *In vitro* studies of neutrophil-mediated internalization of erythrocytes coated with human IgG_2 or different bacterial strains opsonized by IgG_2 demonstrate efficient binding and phagocytosis by homozygous FcγRIIa-H^{131} PMN, whereas PMN from homozygous FcγRIIa-R^{131} individuals do not recognize this opsonin, independent of FcγRIIIb allotype [7–10]. We and others have obtained similar results in monocytes [7,8]. In fact, FcγRIIa-H^{131} is the only human FcγR which recognizes IgG_2 efficiently. From the perspective of the phagocyte, the efficiency of clearance of IgG_2-containing complexes depends upon FcγRIIa phenotype. Even with model immune complexes containing IgG_2 in combination with other IgG subclasses, there is differential handling in homozygous individuals related to host FcγRIIa genotype [7] (Figure 23.1).

To establish the functional phenotype of the heterozygous state in which half of the receptors on the phagocyte reflect the FcγRIIA-R^{131} gene product and half the FcγRIIA-H^{131} gene product, we directly measured phagocytosis of erythrocytes coated with $hIgG_2$ by normal donors homozygous or heterozygous

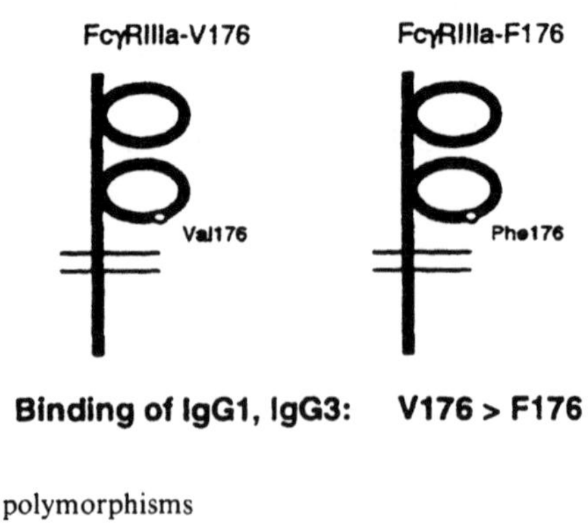

Figure 23.1 FcγRIIa polymorphisms

for R^{131} and H^{131} and found that heterozygotes have an intermediate capacity to recognize IgG_2 [11]. Thus, the FcγRIIA-R^{131} gene product is associated with deficient IgG_2 handling in both homozygotes and heterozygotes, whereas the FcγRIIA-H^{131} gene product provides optimal IgG_2 handling only in the homozygous state.

The allelic polymorphisms of FcγRIIa are of substantial clinical importance for host defense against infection, particularly infection with encapsulated bacteria known to elicit IgG_2 responses, such as *Neisseria meningitidis*, *Haemophilus influenzae*, and *Streptococcus pneumoniae*. Since IgG_2 is a poor activator of complement, optimal handling of IgG_2-opsonized bacteria is dependent upon FcγRIIA genotype. A role for FcγRIIa in determining susceptibility to these infectious diseases has been suggested by the observation that among otherwise healthy children suffering from recurrent respiratory tract infections or fulminant meningococcal sepsis there is an increased frequency of homozygosity for R^{131} [12,13]. We have recently found that in patients with systemic lupus erythematosus, FcγRIIa-R^{131} is a risk factor for invasive pneumococcal infection [14].

The distribution of FcγRIIa alleles differs according to ethnicity. Among Caucasians and African Americans the gene frequency is approximately 0.5 for each allele with about 25% of these populations homozygous for FcγRIIA-R^{131} and unable to efficiently bind IgG_2 sensitized targets [11,15–17]. In contrast, in Japanese and Chinese populations R^{131} homozygosity is uncommon ($<10\%$) and there is a very low incidence of *Haemophilus influenzae* infection [18]. These observations establish the precedent for clinical consequences of infection resulting from the interaction between the qualitative nature of the humoral response and defined FcγR genotypes.

That the qualitative nature of humoral autoimmune responses and their interactions with effector cells may dictate pathogenetic potential has been suggested in other pathological conditions. We have shown that autoimmune anticardiolipin antibodies (aCL) have an IgG_2 predominance and that IgG_2 aCL in particular are associated with arterial and/or venous thrombosis [19]. Only FcγRIIa-H^{131} can initiate monocyte, platelet, and endothelial cell activation in the context of an IgG_2 stimulus leading to a prothrombotic phenotype. In a cross-sectional study of 45 Caucasian high titer IgG aCL patients with thrombotic complications (42% arterial thrombosis and 67% venous thrombosis) and 103 Caucasian controls, we found that FcγRIIa-H^{131} is associated with thrombotic risk in aCL patients [19]. The FcγRIIa-H^{131} gene product provides optimal IgG_2 binding only in the homozygous state; non-H^{131} homozygotes (H^{131}/R^{131} and R^{131}/R^{131}) have decreased effector function when triggered by IgG_2. Therefore, we compared the frequency of H^{131}/H^{131} with that of R^{131}/H^{131} and R^{131}/R^{131} in patients and controls. 40% of the 45 patients with thrombosis were homozygous for FcγRIIa-H^{131}. In contrast, 24% of the 103 controls were homozygous for FcγRIIa-H^{131} ($p = 0.042$ Fisher's exact test). The presence of FcγRIIa-H^{131}, particularly in association with IgG_2 aCL, may be a useful clinical predictor of increased thrombotic risk in autoimmune IgG aCL patients. The role of FcγRIIa in heparin-induced thrombocytopenia, however, is less clear. While it has been proposed that thrombosis and thrombocytopenia are a consequence of platelet activation by cross-linking FcγRIIa, the relationship between IgG_2 anti-heparin/platelet factor 4 antibodies and FcγRIIa genotypes with risk for thrombosis or thrombocytopenia has not been confirmed [20–22].

Allelic variants of FcγR which alter IgG binding confer distinct phagocytic capacities and thus provide a mechanism for heritable susceptibility to immune complex diseases. Systemic lupus erythematosus (SLE), the prototype human immune complex disease, is characterized by tissue deposition of circulating antigen-antibody complexes leading to release of inflammatory mediators, influx of inflammatory cells, and clinically apparent disease, most prominently glomerulonephritis. The efficiency of the mononuclear phagocyte system in removing circulating immune complexes depends upon the FcγR and receptors for complement. From *in vivo* and *in vitro* studies with SLE patients, it is clear that inherited and acquired components of Fcγ receptor-dependent dysfunction may contribute to disease susceptibility and mechanisms of tissue injury [23–25].

To test the hypothesis that FcγRIIA genotypes are susceptibility factors for SLE, and particularly for renal involvement, we performed a two-stage cross-sectional study comparing the distribution of FcγRIIA alleles in African Americans with SLE to that in African American non-SLE controls [11]. African-Americans were selected because of the increased incidence and severity of SLE in this population. A pilot study of 43 SLE patients and 39 controls demonstrated a skewed distribution of FcγRIIA alleles, with only 9% of SLE patients homozygous for FcγRIIa-H^{131} compared with 30% of controls (odds ratio, 0.18; 95% CI, 0.05-0.69, $p = 0.009$). This was confirmed in a multicentre study of 214 SLE patients and 100 non-SLE controls. To determine whether FcγRIIA alleles relate specifically to nephritis, patients were stratified into two groups: those with and without ACR criteria for nephritis ($n = 100$ and 111, respectively). Patients with lupus nephritis have a profound deficit in immune complex removal [24]. Disease-induced dysfunction superimposed upon inherited FcγRIIA-R^{131} homozygosity – a potential accelerant of immune complex glomerular deposition – may increase the likelihood of nephritis, whereas FcγRIIA-H^{131} homozygosity may be protective in this situation. Indeed, the skewed distribution of FcγRIIA was most profound in the nephritis group compared to controls ($p = 0.003$). This subset of patients had the greatest increase in FcγRIIA-R^{131} allele frequency and reduction in FcγRIIA-H^{131}. The R^{131}/R^{131} genotype was particularly enriched in the nephritis group (nephritis vs control 42% vs 23%) while H^{131}/H^{131} was uncommon (12% versus 27%). Trend analysis of the genotype distribution showed a highly significant decrease in H^{131} as the likelihood for lupus nephritis increased ($p = 0.0004$), consistent with a protective effect of the H^{131} gene. The skewing in the distribution of FcγRIIA alleles identifies this gene as a risk factor for the SLE in African Americans. We have recently compared the distribution of FcγRIIA alleles in Mexican American SLE patients, another population with high prevalence of lupus renal disease, and found a marked decrease in homozygosity for FcγRIIa-H^{131} (8% vs 20%, $p<0.05$). The high prevalence of renal disease coupled with the high frequency of the FcγRIIa-R^{131} allele in these populations suggests that this gene may be an important determinant of susceptibility to lupus nephritis.

We and others have found that IgG_2 is a prominent component of anti-C1q antibodies, an autoantibody specifically correlated with severe renal disease in SLE [26–29]. These findings, taken together with a recent study showing that IgG_2 is a predominant IgG subclass found in the glomeruli of patients severe lupus nephritis [30], support the relationship between FcγRIIA genotype and predisposition to lupus nephritis. Though we found a dramatic skewing in FcγRIIA alleles in African Americans with nephritis, in our study of a heterogeneous Caucasian population ($n=262$ SLE patients) we were not able to demonstrate the association between FcγRIIA-R^{131} and nephritis described in 50 Dutch Caucasian lupus patients [31]. A recent study from the United Kingdom confirmed our findings [32]. Although the basis for the contrasting results in these distinct Caucasian populations is unclear they may relate to differences in disease phenotype, autoantibody profile, population ethnic homogeneity, or other genetic factors. Indeed, among Caucasian lupus patients with anti-C1q antibodies, we found that the presence of the FcγRIIA-R^{131} allele was highly associated with renal disease (any R^{131} vs H^{131}/H^{131}: 87% vs 55%, $p<0.03$) [29]. Thus, with precisely defined phenotypes, FcγRIIa variants are disease accelerants, increasing risk for nephritis in Caucasians.

In collaboration with Steve McKenzie, we have identified a novel functionally significant mutation in a healthy individual homozygous for FcγRIIa-R^{131}: a C to A substitution at codon 127, replacing glutamine (Q) with lysine (K) in one of the two FcγRIIA genes. It was predicted that mutations in the 14 amino acid stretch in the second extracellular domain (EC2) of FcγRIIa, which includes the polymorphic site at 131, would alter binding of IgG ligands. Indeed, the substitution of a charged basic residue for a neutral Q residue at the face of the molecule likely to contact the Fc end of IgG imparts to an FcγRIIa-R^{131} molecule the ability to more efficiently recognize $hIgG_2$, as shown by increased phagocytosis of E-$hIgG_2$ by blood monocytes and PMN [33]. In contrast, the mutant receptor retains the ability to bind $mIgG_1$, and thus has a unique phenotype. Identification of this variant establishes the precedent that novel point polymorphism which alter FcγRIIa function are detectable in the population and may provide a mechanisms for inherited differences in disease susceptibility.

Fcγ RECEPTOR IIIA

FcγRIIIa is expressed on natural killer cells, mononuclear phagocytes and renal mesangial cells [34], and recent observations demonstrate that it is polymorphic in both structure and function. In 1993, Vance *et al.* described a functional polymorphism in FcγRIIIa on natural killer cells in normal individuals [35]. These donors showed some differences in both in IgG binding and anti-CD16 reactivity consistent with variations in receptor expression. More recently, in 1996, de Haas *et al.* described a triallelic sequence polymorphism at nt 230 in FcγRIIIA [36]. This single nucleotide substitution, which is located in exon 3 encoding the first extracellular Ig-like domain (EC1), predicts an amino acid change from leucine (L) to arginine (R) or from leucine (L) to histidine (H). Some evidence suggests an association of this polymorphism with altered binding of human IgG and several anti-CD16 mAb [36,37]. Working from these observations, other investigators have speculated that structural variants of FcγRIIIa, recognized by altered patterns of anti-CD16 mAb binding, may be related to a clinical phenotype of repeated infections [38]. In order to test the hypothesis that a sequence polymorphism might explain the differences in natural killer cell FcγRIIIa and natural killer cell function, we have used direct sequencing of both genomic DNA and cell-type specific cDNA derived from purified natural killer cells and mononuclear phagocytes to characterize the FcγRIIIA nucleotide sequence of several of the normal donors studied by Guyre's group. Each of these donors was monomorphic at nt 230 and nt 248 with a leucine at amino acid position 66 and a serine at amino acid position 72. These residues, numbered such that the first amino acid of the signal sequence rather than the first amino acid of EC1 is designated amino acid 1 [39], correspond to residues 48 and 64 described by de Haas and colleagues. However, the donors were polymorphic at nt 559, a site noted by Ravetch and Perussia as potentially polymorphic [39]. This non-conservative T to G substitution predicts a change of phenylalanine (F) into valine (V) at position 176 in the EC2, the second immunoglobulin-like and membrane-proximal domain (Figure 23.2).

Since multiple studies suggest that the second immunoglobulin-like domain strongly influences ligand binding [40–44] (Chapter 2), we have characterized the biology of this 176F/V polymorphism by identifying normal donors homozygous for leucine at position 66 and serine at position 72, and homozygous for either phenylalanine (176F/F) or valine (176V/V) at position 176 [45]. Using multiparameter flow cytometry, analysis of CD56-positive natural

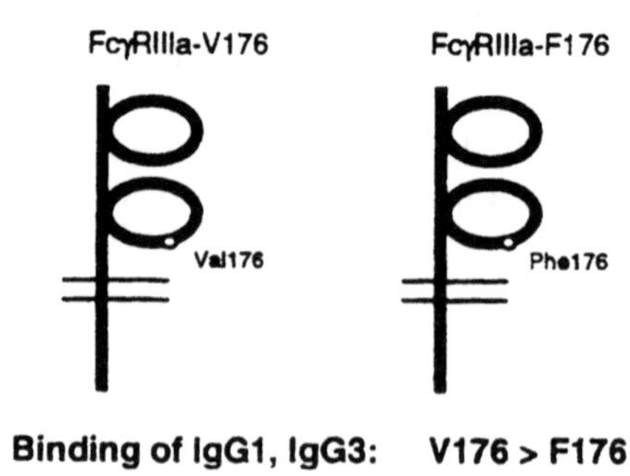

Figure 23.2 FcγRIIIa polymorphisms

killer cells has demonstrated identical reactivity of FcγRIIIa with the anti-CD16 mAb CLB-Gran1 for both 176F/F and 176V/V homozygotes. Other anti-CD16 mAbs (GRM1, Leu11a, 3G8, 30.2, 214.1, 135.9) showed similar reactivities, including B73.1 which is influenced by the residue 66 polymorphism. In contrast, mAbs ID3 and MEM154 showed markedly different reactivities with the two different residue 176 phenotypes. FcγRIIIa expressed in 176V/V homozygotes bound substantially more IgG_1 and IgG_3 relative to 176F/F homozygotes despite identical levels of receptor expression. Although identical in their function when bound by CLB-Gran1, in response to a standard IgG ligand 176V/V and F/F homozygotes differ in their ability to generate receptor-mediated early signals ($Ca^{2+}{}_i$), cell activation programs (IL-2 receptor expression) and cell death programs (apoptosis). Gene frequency is approximately 0.6 and 0.4 for the F and V alleles, respectively.

Since the data indicate that the polymorphism at residue 176 alters the affinity of FcγRIIIa on both natural killer cells and monocytes, this difference is not restricted to a single cell type and therefore not secondary to cell-type specific differences in glycosylation which can also influence receptor affinity [46]. We anticipate that these findings have profound implications for antibody-mediated immune surveillance, ADCC, antibody-mediated host defence against pathogens and autoimmune disease. For example, FcγRIIIa is the primary trigger molecule leading to destruction of *Toxoplasma gondii* by NK cells [47]. Similarly, FcγIIIa-expressing NK cells lyse HIV-infected target cells by ADCC, and ADCC activity correlates inversely with disease progression [48]. Thus, functionally significant polymorphisms of FcγRIIIa may alter the efficiency of antibody-dependent killing of infected cells expressing viral antigens. Considering the possibility that polymorphisms of FcγRIIIa on both mononuclear phagocytes and renal mesangial cells might be important in the predisposition to autoimmune immune complex disease, we have performed an initial analysis of 200 patients with SLE. This analysis indicates a strong association of the low binding 176F/F phenotype with SLE, especially SLE with nephritis, and a corresponding under-representation of the homozygous high binding 176V/V phenotype. This association has been confirmed in a large independent replication study of more than 450 SLE patients [49]. Thus, this polymorphism directly affects human biology, an effect which most probably extends beyond autoimmune disease with circulating immune complexes to many aspects of humoral immunity.

Fcγ RECEPTOR IIIB

Genetic polymorphisms of human FcγRIIIb have also been characterized which profoundly influence neutrophil function [7,39,50–56] (Figure 23.3). Two common allelic forms of neutrophil-specific FcγRIIIb were originally recognized by serological techniques as the NA1 and NA2 alloantigens [57]. The NA1-NA2 allotypic polymorphism has been known for several years because of its involvement in blood transfusion reactions and alloimmune neutropenias.

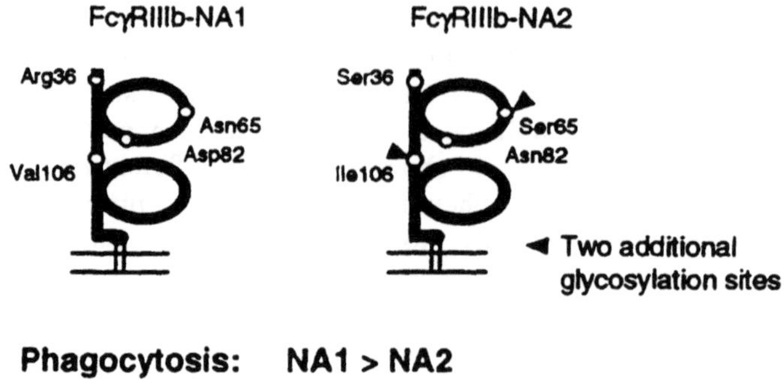

Figure 23.3 FcγRIIIb polymorphisms

The allelic frequencies of this polymorphism in Caucasians are approximately 37% for NA1 and 63% for NA2 [57,58]. The NA1 and NA2 alleles of FcγRIIIb are codominantly expressed. There is a clear gene dose effect, with NA1/NA2 heterozygotes expressing half a complement of each allele [54]. Available literature suggests that these two alleles probably differ by five nucleotides which results in four amino acid substitutions in EC1, but whether these five sites track as two alleles or whether they group combinatorially to comprise a larger number of alleles has been unclear. Systematic examination by gene-specific, direct sequencing indicates that the NA1 and NA2 serotypes reflect two predominant genotypes,each representing the predicted set of five polymorphic nucleotides and four polymorphic residues. We have identified a rare but previous unrecognized mutation at nucleotide position 266 where at C to A in codon 78 predicts an alanine to aspartic acid substitution [59]. This mutation has been confirmed as the 'SH' polymorphism which resides on the NA2 isoform and has an allele frequency of $<5\%$ in caucasians [60]. The basis for the serologically defined NB polymorphism is not known, but our studies of the sequence of FcγRIIIB in a large cohort make primary structure a highly unlikely explanation.

Although binding of IgG does not seem to be affected, the NA1 and NA2 allelic forms do have different levels of quantitative function [7,54–56]. By studying a genotyped panel of normal donors, we have shown that the NA1 isoform facilitates a more robust FcγR-mediated phagocytic response by neutrophils, independent of the allelic phenotype of the second type of Fcγ receptor on neutrophils. Homozygous NA2 individuals have a lower capacity to mediate phagocytosis compared to NA1, whether the FcγR is ligated by its classical IgG ligand binding site or by lectin-carbohydrate interactions, despite equivalent density of receptor [54,56]. This difference is apparent with human IgG_1 or IgG_3 as the opsonin and is enhanced upon blockade of FcγRIIa. The altered phagocytic capacity appears to be due, at least in part, to the ability of the NA1 allele to elicit a quantitatively larger oxidative burst and/or degranulation response as compared to the NA2 allele [55]. This oxidative burst and degranulation response in turns activates receptor specific internalization FcγRIIa which is also expressed on neutrophils. Although the molecular basis for the more robust activity of the NA1 allele has not been established, the NA1 and NA2 alleles do have different numbers of glycosylation sites and are differentially glycosylated *in vivo*. These post-translational modifications may mediate interactions with complement receptor 3 (CR3; Mac1; CD11b/18) in the neutrophil plasma membrane and alter net receptor function (see Chapter 13).

The functionally more active phenotype of the NA1 allele appears to be clinically significant. Homozygous NA1 individuals may be more resistant to bacterial infection, and conversely homozygous NA2 individuals may be more likely to have severe infection especially if FcγRIIa cannot be effectively engaged. This has been suggested by the findings of increased infections with *Neisseria meningiditis* among individuals with complement component 6 or 8 deficiency who are homozygous for FcγRIIIb-NA2 and FcγRIIa-R^{131}, than those with other FcγR alleles [61]. More surprising and perhaps even more robust has been the association of NA1 alleles with severe renal disease in Wegener's granulomatosis, a type of systemic vasculitis [59,62]. This observation, now established through a multinational collaboration, strongly suggests that neutrophils play an important role in the pathophysiology of at least some anti-neutrophil cytoplasmic antibody (ANCA)-positive vasculitides and that FcγRIIIb is an important activation molecule for neutrophils. Indeed, more efficient activation of neutrophils by ANCA appears to directly effect renal injury raising the possibility that receptor genotyping might identify the patients most at risk for severe disease and for disease exacerbations with changes in ANCA titre.

CONCLUSION

The physiology of FcγR alleles provide a new perspective within which to define the interplay between qualitative and quantitative humoral immune response and host genotype, and to identify heritable risk factors for disease susceptibility. The challenge for scientists and physicians interested in host defence and autoimmunity is to translate the diversity of FcγR structure into a framework providing insight into disease susceptibility, disease mechanism and therapeutic intervention.

References

1. Tax, WJM, Willems HW, Reekers PPM, Capel PJA, Koene RAP. Polymorphism in mitogenic effect of IgG_1 monoclonal antibodies against T3 antigen on human T cells. Nature. 1983;304: 445–7.

2. Tax WJ, Tamboer WP, Jacobs CW, Frenken LA, Koene RA. 1997. Role of polymorphic Fc receptor FcγRIIa in cytokine release and adverse effects of murine IgG_1 anti-CD3/T cell receptor antibody (WT31). Transplantation. 1997;63:106–12.
3. Warmerdam PAM, van de Winkel JGJ, Vlug A, Westerdal NAC, Capel PJAA single amino acid in the second Ig-like domain of the human Fcγ receptor II is critical in human IgG_2 binding. J Immunol 1991;147:1338–43.
4. Clark MR, Stuart SG, Kimberly RP, Ory PA, Goldstein IM. A single amino acid distinguishes the high-responder from low-responder form of Fc receptor II on human monocytes. Eur J Immunol. 1991;21:1911–16.
5. Warmerdam PAM, van de Winkel JGJ, Gosselin EJ, Capel PJA. Molecular basis for a polymorphism of human Fcγreceptor II (CD32). J Exp Med. 1990;172:19–25.
6. Tate BJ, Witort E, McKenzie IFC, Hogarth PM. Expression of the high responder/non-responder human FcγRII. Analysis by PCR and transfection into FcR^- COS cells. Immunol Cell Biol. 1992;70:79–87.
7. Salmon JE, Edberg JC, Brogle NL, Kimberly RP. Allelic polymorphisms of human Fcγ receptor IIA and Fcγ receptor IIIB. Independent mechanisms for differences in human phagocyte function. J Clin Invest. 1992;89:1274–8.
8. Parren PWHI, Warmerdam PAM, Boeije LCM et al. On the interaction of IgG subclasses with the low affinity FcγRIIA (CD32) on human monocytes, neutrophils, and platelets. Analysis of a functional polymorphism to human IgG_2. J Clin Invest. 1992;90:1537–46.
9. Sanders LA, Feldman RG, Voorhorst-Ogink MM et al. Human immunoglobulin G (IgG) Fc receptor IIa (CD32) polymorphism and IgG_2-mediated bacterial phagocytosis by neutrophils. Infect Immun. 1995;63:73–81.
10. Bredius RG, de Vries CE, Troelstra A et al. Phagocytosis of *Staphylococcus aureus*and *Haemophilus influenzae*type B opsonized with polyclonal human IgG_1 and IgG_2 antibodies. Functional hFcγ RIIa polymorphism to IgG_2. J Immunol. 1993;151:1463–72.
11. Salmon JE, Millard SS, Schachter LA et al. FcγRIIA alleles are heritable risk factors for lupus nepritis in African Americans. J Clin Invest. 1996;97:1348–54.
12. Sanders LA, van de Winkel JG, Rijkers GT et al. Fcγ receptor IIa (CD32) heterogeneity in patients with recurrent bacterial respiratory tract infections. J Infect Dis. 1994;170:854–61.
13. Bredius RG, Derkx BH, Fijen CA et al. Fcγ receptor IIa (CD32) polymorphism in fulminant meningococcal septic shock in children. J Infect Dis. 1994;170:848–53.
14. Yee AMF, Ng SC, Sobel RE, Salmon JE. FcγRIIA polymorphism as a risk factor for invasive pneumococcal infections in systemic lupus erythematosus. Arthritis Rheum. 1997; 40:1180–2.
15. Abo T, Tilden AB, Balch CM, Kumagai K, Troup GM, Cooper MD. Ethnic differences in the lymphocyte proliferative response induced by a murine IgG_1 antibody, Leu-4, to the T3 molecule. J Exp Med. 1984;160:303–9.
16. Reilly AF, Norris CF, Surrey S et al. Genetic diversity in human Fc receptor for immunoglobulin G: Fcγ receptor IIA ligand binding polymorphism. Clin Diag Lab Immunol. 1994;1:640–4.
17. Osborne JM, Chacko GW, Brandt JT, Anderson CL. Ethnic variation in frequency of an allelic polymorphism of human FcγRIIA determined with allele specific oligonucleotide probes. J Immunol Meth. 1994;173:207–17.
18. Musser JM, Kroll JS, Granoff DM et al. Global genetic structure and molecular epidemiology of encapsulated *Haemophilus influenzae*. Rev Infect Dis. 1990;12:75–111.
19. Sammaritano LR, Ng S, Sobel R et al. Anticardiolipin Immunoglobulin G subclasses: association of IgG_2 with arterial and/or venous thrombosis. Arthritis Rheum. 1997; in press.
20. Brandt JF, Osborne JM, Chacko G, Anderson CL. The role of FcγRIIa phenotype in heparin-induced thrombocytopenia. Thromb Haemost. 1995;74:1564.
21. Burgess JK, Lindeman R, Chesterman CN, Chong BH. Single amino acid mutation of Fcγ receptor is associated with the development of heparin-induced thrombocytopenia. Br J Haematol. 1995;91:761–6.
22. Arepally G, McKenzie SE, Jiang XM, Poncz M, Cines DB. FcγRIIA H/R 131 polymorphism, subclass-specific IgG anti-heparin/platelet factor 4 antibodies and clinical course in patients with heparin-induced thrombocytopenia and thrombosis. Blood. 1997;89:370–5.
23. Frank MM, Hamburger MI, Lawley TJ, Kimberly RP, Plotz, PH. Defective reticuloendothelial system Fc-receptor function in systemic lupus erythematosus. N Engl J Med. 1979;300: 518–23.

24. Parris TM, Kimberly RP, Inman RD, McDougal JS, Gibofsky A, Christian CL. Defective Fc receptor-mediated function of the mononuclear phagocyte system in lupus nephritis. Ann Int Med. 1982;97:526–32.
25. Salmon JE, Kimberly RP, Gibofsky A, Fotino M. Defective mononuclear phagocyte function in systemic lupus erythematosus: Dissociation of Fc receptor-ligand binding and internalization. J Immunol. 1984;133:2525–31.
26. Wisnieski JJ, Jones SM. Comparison of autoantibodies to the collagen-like region of C1q in hypocomplementemic urticarial vasculitis syndrome and systemic lupus erythematosus. J Immunol. 1992;148:1396–403.
27. Prada AE, Strife CF. IgG subclass restriction of autoantibody to solid-phase C1q in membrano-proliferative and lupus glomerulonephritis. Clin Immunol Immunopathol. 1992; 63:84–88.
28. Coremans IEM, Daha MR, van der Voort EAM, Sigertt CEH, Breedveld FC. Subclass distribution of IgA and IgG antibodies against C1q in patients with rheumatic diseases. Scand J Immunol. 1995;41:391–7.
29. Haseley LA, Wisnieski JJ, Denburg MR et al. Antibodies to C1q in systemic lupus erythematosus: characteristics and relation to FcγRIIA alleles. Kidney Int. 1997;52:1375–80.
30. Imai H, Hamai K, Komatsuda A, Ohtani H, Miura AB. IgG subclasses in patients with membranoproliferative glomerulonephritis, membranous nephropathy and lupus nephritis. Kidney Int. 1997;51:270–6.
31. Duits AJ, Bootsam H, Derksen RHW et al. Skewed distribution of IgG Fc receptor IIa (CD32) polymorphism is associated with renal disease in systemic lupus erythematosus patients. Arthritis Rheum. 1995;39:1832–6.
32. Botto M, Theodorisdis E, Thompson EM et al. FcγRIIa polymorphism in systemic lupus erythematosus: no association with disease. Clin Exp Immunol. 1996;104:264–8.
33. Norris CF, Pricop L, Millard SS et al. A naturally occurring mutation in FcγRIIA: a Q to K 127 change confers unique binding properties to the R^{131} allelic form of the receptor. Blood. 1998;91:656–62.
34. Radeke HH, Gessner JE, Uciechowski P, Magert HJ, Schmidt RE, Resch K. Intrinsic human glomerular mesangial cells can express receptors for IgG complexes (hFcγRIII-A) and the associated FcεRI γ-chain. J Immunol. 1994;153:1281–92.
35. Vance BA, Huizinga TWJ, Wardwell K, Guyre PM. Binding of monomeric human IgG defines an expression polymorphism of FcγRIII on large granular lymphocyte/natural killer cells. J Immunol. 1993;151:6429–39.
36. de Haas M, Koene HR, Kleijer M et al. A triallelic Fcγ receptor type IIIA polymorphism influences the binding of human IgG by NK cell FcγRIIIA. J Immunol. 1996;156:2948–55.
37. Koene HR, de Haas M, Roos D, von dem Borne AEGK. Soluble FcγRIII: biology and clinical implications. In: van de Winkel JGJ, Capel PJA, eds, Human IgG Fc Receptors. Austin: R.G. Landes Company; 1996:181–93.
38. Jawahar S, Moody C, Chan M, Finberg R, Geha R, Chatila T. Natural killer (NK) cell deficiency associated with an epitope-deficient Fcγ receptor type IIIA (CD16-II). Clin Exp Immunol. 1996;103:408–13.
39. Ravetch JV, Perussia B. Alternative membrane forms of FcγRIII (CD16) on human NK cells and neutrophils: cell-type specific expression of two genes which differ in single nucleotide substitutions. J Exp Med. 1989;170:481–97.
40. Hulett MD, Hogarth PM. Molecular basis of Fc receptor function. Adv Immunol. 1994;57: 1–124.
41. Hibbs ML, Tolvanen M, Carpen O. Membrane-proximal Ig-like domain of FcγRIII (CD16) contains residues critical for ligand binding. J Immunol. 1994;152:4466–74.
42. Hulett MD, Witort E, Brinkworth RI, McKenzie IFC, Hogarth PM. Multiple regions of human FcγRII (CD32) contribute to the binding of IgG. J Biol Chem. 1995;270:21188–94.
43. Tamm A, Kister A, Nolte KU, Gessner JU, Schmidt RE. The IgG binding site of human FcγRIIIB receptor involves CC′ and FG loops of the membrane-proximal domain. J Biol Chem. 1996;271:3659–66.
44. Tamm A, Schmidt RE. The binding epitopes of human CD16 (FcγRIII) monoclonal antibodies. Implications for ligand binding. J Immunol. 1996;157:1576–81.
45. Wu J, Edberg JC, Redecha PB, Bansal V et al. A novel polymorphism of FcγRIIIa (CD16) alters receptor function and predisposes to autoimmune disease. J Clin Invest. 1997;100:1059–70.

46. Edberg JC, Kimberly RP. Cell-specific glycoforms of FcγRIIIa: differential ligand binding. J Immunol. 1997;159:3849–57.
47. Erbe DV, Pffeferkorn ER, Fanger MW. Functions of the various IgG Fc receptors in mediating killing of Toxoplasma gondii. J Immunol. 1991;146:3145–51.
48. Tyler DS, Nastala CL, Stanley SD, Matthews TJ, Lyerly HK, Weinhold KJ. GP120 specific cellular cytotoxicity in HIV-1 seropositive individuals. Evidence for circulating CD16+ effector cells armed in vivo with cytophilic antibody. J Immunol. 1989;142:1177–82.
49. Wu J, Bansal V, Arnett F et al. The FcγRIIIA 176F/V polymorphism associates with the SLE phenotype in Caucasians and African Americans. Arthritis Rheum. 1997;40:S316.
50. Scallon BJ, Scigliano E., Freedman VH et al. A human immunoglobulin G receptor exists in both polypeptide-anchored and phosphatidylinositol-anchored forms. Proc Natl Acad Sci USA. 1989;86:5079–83.
51. Peltz GA, Grundy HO, Lebo RV, Yssel H, Barsh GS, Moore KW. Human FcγRIII: cloning, expression and identification of the chromosomal locus of the two Fcγ receptors for IgG. Proc Natl Acad Sci USA. 1989;86:1013–17.
52. Ory PA, Clark MR, Kwoh EE, Clarkson SB, Goldstein IM. Sequences of complementary DNAs that encode the NA1 and NA2 forms of Fcγ receptor III on human neutrophils. J Clin Invest. 1989;84:1688–92.
53. Huizinga TWJ, Kleijer M, Tetteroo PA, Roos D, von dem Borne AEGK. Biallelic neutrophil NA-antigen system is associated with a polymorphism on the phosphoinositol-linked Fcγ receptor III (CD16). Blood. 1990;75:213–17.
54. Salmon JE, Edberg JC, Kimberly RP. Fcγ receptor III on human neutrophils. Allelic variants have functionally distinct capacities. J Clin Invest. 1990;85:1287–95.
55. Salmon JE, Millard SS, Brogle NL, Kimberly RP. Fcγ receptor IIIb enhances Fcγ receptor IIa function in an oxidant dependent and allele-sensitive manner. J Clin Invest. 1995;95: 2877 85.
56. Bredius RGM, Fijen CAP, de Haas M et al. Role of neutrophil FcγRII (CD32) and FcγRIIIb (CD16) polymorphic forms in phagocytosis of human IgG_1- and IgG_3-opsonized bacteria and erythrocytes. Immunology. 1994;83:624 30.
57. Lalezari P. Granulocyte antigen systems. In: Engelfriet CP, Van Loghem JJ, Kr AEG, eds, Immunohaematology. Elsevier: Amsterdam; 1984:33.
58. Bux J, Stein EL, Santoso S et al. NA gene frequencies in the German population, determined by polymerase chain reaction with sequence-specific primers. Transfusion. 1995;35:54 7.
59. Wainstein E, Kocher M, Wu J et al. The neutrophil FcγRIIIB is associated with renal dysfunction in Wegeners granulomatosis (WG), 1997; Submitted for publication.
60. Bux J, Stein E-L, Bierling P et al. Characterization of a new alloantigen (SH) on the human neutrophil Fcγ receptor IIIb. Blood. 1998;89:1027–34.
61. Fijen CAP, Bredius RGM, Kuijper EJ. Polymorphism of IgG Fcγ receptors in meningococcal disease: risk marker in complement deficient patients. Ann Intern Med. 1993;119:636 41.
62. Wainstein E, Edberg E, Csernok E et al. FcγRIIIB alleles predict renal dysfunction in Wegeners granulomatosis (WG). Arthritis Rheum. 1996;39:S210

24
Soluble FcγR, a biological perspective

C. SAUTÈS, J. GALON, C. BOUCHARD, A. ASTIER, J.-L. TEILLAUD and W. H. FRIDMAN

INTRODUCTION

Soluble FcγR (sFcγR) were discovered in supernatants of cells of the immune system during the 1970s, and called immunoglobulin G-binding factors. Twenty years later, although the structure and mechanisms of generation of sFcγR have been largely clarified [see References 1–5 for reviews], questions remain regarding their biological role. In this chapter, we will review available evidence showing that these molecules can regulate immune reactions *in vitro*. The question of their *in vivo* role will be addressed on the basis of studies performed in various pathological situations and with transgenic animals.

SOLUBLE FcγR STRUCTURE AND GENERATION

Mouse

Soluble forms of low affinity FcγR II and/or III have been detected in supernatants of activated T and B lymphocytes, macrophage cell lines, and of epidermal Langerhans cells. They are produced by proteolytic cleavage of mFcγRII and/or III, and by alternative splicing of mFcγRII: a soluble form of mFcγRIIb3 corresponding to mFcγRIIb2 without the transmembrane sequence, has been detected in supernatants of macrophages and in Langerhans cells.

Human

Soluble forms of hFcγRII (CD32) are produced by proteolytic cleavage of the

J.G.J. van de Winkel and P.M. Hogarth (eds.), The Immunoglobulin Receptors and their Physiological and Pathological Roles in Immunity. 279–290.

membrane receptor and by alternative splicing. A soluble hFcγRIIa2, corresponding to hFcγRIIa1 without the transmembrane region, is produced in megakaryocyte, erythromyeloid and histiocytic cells as well as in epidermal Langerhans cells and platelets. Soluble hFcγRIII (CD16) are generated by enzymatic cleavage of hFcγRIIIa of NK cells and of hFcγRIIIb of neutrophils. Metalloproteases and serine proteases may be involved in this process.

INTERACTIONS WITH LIGANDS AND FUNCTIONAL CONSEQUENCES

Immunoglobulins

In vitro *studies*

Soluble FcγR produced by cleavage of low affinity FcγR or by alternative splicing contain the extracellular region of the receptors. The available evidence suggest that soluble FcγR have kept the overall structure of the external part of the membrane FcγR and also have a low affinity for IgG, with affinity constant (Ka) values for monomeric IgG ranging between 10^5 and 10^6 M^{-1}. In our laboratory, the method of total internal reflection (surface plasmon resonance) which monitors protein interactions in real time (using a Biacore instrument) was used to quantify the interactions of a recombinant shFcγRIIIb with human Ig. The Ka of shFcγRIIIb immobilized on the sensor chip for human IgG_1 and IgG_3 in solution at 22°C in HBS was $1.3 \pm 0.6 \times 10^6$ M^{-1} and $2.6 \pm 0.4 \times 10^6$ M^{-1} respectively (Galon *et al.* submitted). No detectable binding of monomeric human IgG_2, IgG_4 or IgA was observed. A Ka of 3.4×10^5 M^{-1} in PBS at 1°C was found by analytical centrifugation for hsFcγRIIIb and the Fc portion of human IgG_1 [6].

Binding of sFcγR to IgG-immune complexes should, by competition, inhibit Fc-dependent immune reactions: this been shown to be the case *in vitro*. Presentation of IgG immune complexes by APC occurs via low affinity FcγRs, which allow the internalization of immune complexes and a potentiated presentation of Ag to sensitized T cells. In mouse, preincubation of IgG-complexed Ag with smFcγRII or with mFcγRIIb3 led to a dose-dependent decrease in the Ag-presenting capacity of immune complexes by Langerhans cells [7]. Antibodies allow recognition of minute amounts of Ag that cannot be taken up efficiently by accessory cells. Thus, *in vivo*, by decreasing Ag-IgG presentation, sFcγR could locally decrease ongoing immune responses and participate to the termination of the response. A recombinant shFcγRII corresponding to the extracellular domain of hFcγRIIa was found to inhibit an anti-Leu4-induced T cell proliferation assay, showing that the soluble receptor has the capacity to block the binding of immune complexes to monocyte FcγRII [8].

Similarly, recombinant hFcγRIIa2, the TM-deleted counterpart of FcγRIIa1, was shown to efficiently compete with membrane-associted FcγR to prevent ADCC using PBMC as effector cells and IgG_1 coated target cells [3]. It also inhibited efficiently the binding of complexes to $Fc\gamma R^+$ cells [9]. In addition, hFcγRIIa2 was found to inhibit anti-CD9 antibody-induced platelet aggregation, suggesting that this molecule could play an important role in regulating the activation of platelets by immune complexes [10].

In vivo *relevance*

In all these studies, significant inhibition of IgG-dependent immune reactions required from 10 to 100 μg/ml of sFcγR, reflecting its low affinity binding to complexed IgG. This observation raises the question of the biological relevance of such phenomena. Serum smFcγR levels in healthy mice are far below these values (10–300 ng/ml, depending on the housing conditions). In man, serum levels of shFcγRII and shFcγRIII are around 10–100 ng/ml and 1.2 μg/ml respectively [3]. However, pathological conditions increase sFcγR levels especially in mouse up to 1 μg/ml [11], and the local concentration of sFcγR at inflammatory sites may be more important than in serum. Notably, human neutrophils which secrete high levels of shFcγRIIIb upon activation, may provide an abundant source of sFcγR at inflammatory sites.

The question of the *in vivo* relevance of blockade of IgG effector functions by sFcγR has been addressed in rats using a model of immune complex disease (reversed Arthus reaction) [8] . The i.d. injection of shFcγRII together with IgG anti-ovalbumin antibodies significantly inhibited the local inflammatory response in rats that received an i.v. injection of ovalbumin 2–6 h previously. A dose-dependent relationship between the amount of shFcγRII and the reduction in size and in severity of the inflammatory lesion was found. Although neutrophil chemotaxis was not inhibited, the perivascular infiltrate and red cell extravasation were less intense in the sFcγRII-injected group. The most likely mechanism which is responsible for the inhibition of reverse Arthus reaction in this model is that shFcγRII competes with the triggering of neutrophil membrane FcγRII by IgG-immune complexes.

Other ligands than Ig

The fact that one molecule can bind several ligands, an observation originally made in the immune system for immunoglobulins, is now a phenomenon extended to many biologically active cell surface and soluble molecules. The soluble form of the human low affinity type II IgE receptor, hFcεRII or CD23, binds not only IgE but also, in a sugar-dependent way, cell surface molecules such as CD21, CR3 and CR4 [12,13]. The complement receptors type 3 and 4 are β_2 integrin family members composed of two chains, CD11b and CD11c respectively, and of a common chain, CD18. The extracellular region of CD11b binds a large number of proteins such as iC3b, ICAM1, fibrinogen, factor X and of saccharides such as zymosan, and a series of glucans and mannans of variable sizes. Whereas the interaction with proteins involves the I domain, sugars bind to a lectin-like site, distinct from the I domain and located proximal to the transmembrane region. A series of experiments have shown that FcγR are also capable to interact with other ligands than IgG. For instance, neutrophil hFcγRIIIb is a GPI-linked glycoprotein with carbohydrates which confer its ability to bind lectins such as concanavalin A [14], and to interact with other cell surface glycoproteins such as CR3 [15]. Recent studies in our group using natural shFcγRIII purified from human serum and an eukaryotic recombinant form have indicated that shFcγRIII binds to the complement receptors CR3 and CR4 [16]. The contribution of the binding sites of CD11b in the binding was investigated by testing the effect of mAbs and of a series of mono and polysaccharides on the binding and by using non-glycosylated shFcγRIII made in *E. coli*. Sugars such as N-acetyl-D-glucosamine, α or β-methyl D-mannoside and zymosan were found to inhibit partly the binding, suggesting that the lectin-like site of CR3 is involved [16]. The procaryotic shFcγRIII also bound to CR3 but with a four-fold lower affinity than the glycosylated molecule [5]. Thus, the interactions between shFcγRIII and CR3 may involve not only the carbohydrates of the receptor but also the polypeptide backbone with a low affinity.

The interaction of sFcγR with non-Ig ligands may also trigger immune reactions. On neutrophils, the triggering of the phagocytosis of erythrocytes coated with iC3b (EiC3b) via CR3 requires the priming of CR3, in addition to the binding of iC3b to the I domain of CR3. The priming of CR3 leads to the formation of a new epitope and can be induced by stimulation of the lectin-like part of the molecules by sugars such as β-glucan [17]. Similarly, the lysis of EiC3b by NK cells requires β-glucan. The possible dual capacity of shFcγRIII to interact with the lectin like site of CR3 and to another part of the molecule such as the I domain may be particularly relevant in view of our recent results showing that endotoxin-free shFcγRIII induces the CR3-dependent release of proinflammatory cytokines such as IL-6 and IL-8 by monocytes [16]. Incubation of neutrophils with shFcγRIII increased their IL-8 production, 24–48 h after stimulus addition. Thus, shFcγRIII triggers activation signals and pro-inflammatory cytokine and chemokine release via its binding to CR3. These observations are reminiscent of those made with sCD23, that induces pro-inflammatory cytokine production by monocytes [13]. Epidermal dendritic cells are the most powerful antigen-presenting cells. Preliminary experiments have shown that the incubation of monocyte-derived dendritic cells with multivalent shFcγRIII leads to the CR3-dependent production of cytokines [18]. These findings open a new array of putative biological functions for sFcγR in immune defence.

ROLES OF sFcγR *IN VIVO*

Suppression of antibody production

In vitro *studies*

Several lines of evidence have suggested that human or mouse sFcγR isolated from supernatants of cells from the immune system such as lymphocytes, monocytes or neutrophils or of FcγR positive non-immune cells such as trophoblasts inhibit various immune responses including antibody production by normal or tumour B cells, and mitogen or alloantigen-driven lymphocyte proliferation [1,19, 20]. However, the use of recombinant purified sFcγR showed that this antibody production suppression is likely to be more complex process than previously thought. Notably, transforming growth factor β (TGF-β) is a potent immunosuppressive and anti-proliferative molecule produced by many cell types. An intriguing property of TGF-β family members is that, in addition to their cellular receptors, they bind several other proteins. We have recently shown that TGF-β bind immobilized-IgG and that the binding requires the Fc portion of IgG [21]. These data are in accordance with previous observations showing that immune complexes present in alloimmune human serum contain TGF-β-bound IgG which is responsible for suppression of cytotoxic T cell response by IgG-immune complexes [22]. These findings prompted us to investigate whether the suppressive activity of preparations containing sFcγR could be related to the presence of a TGF-β-like component. Mouse recombinant sFcγRII was used for these studies. Both TGF-β and semi-purified smFcγRII were found to inhibit the LPS- or anti-IgM-stimulated B cell reponses with similar kinetics.In addition, the antiproliferative activity of TGF-β and smFcγRII preparations was neutralized by antibodies against TGF-β. Finally, the suppressive activity eluted with TGF-β by gel exclusion chromatography [21]. Taken altogether, these results suggested that a TGF-β-like suppressive factor was present in recombinant smFcγRII preparations. Similar observations were made with smFcγRII isolated from supernatants of activated T cells by affinity chromatography on insolubilized IgG. Therefore, it is highly probable that the suppression of mouse antibody responses *in vitro* and *in vivo* by IgG-BF reflects the presence of TGF-β. Whether an immunosuppressive factor other than shFcγRIII is responsible for the inhibition of pokeweed mitogen-induced responses of human mononuclear cells by recombinant shFcγRIII [23] is still unclear.

Transgenic mice

In view of these results, and in order to investigate the contribution of soluble FcγR in immune responses *in vivo*, transgenic mice carrying a cDNA encoding a smFcγRII corresponding to the extracellular region of the receptor (174 amino acids) were generated [5]. The sFcγR transgene carried the mouse metallothionein I (MT-1) promoter and the human growth hormone polyadenylation signal. Three transgenic lines T1, T24 and T16 were derived after injection of the construct into pronuclei of fertilized F2 eggs obtained by mating C57BL/6 × DBA/2 F1 mice. Their DNA copy numbers were 1, 5 and 50 copies for the T1, T24 and T16 lines, respectively. The transgene mRNA was expressed in spleen, liver, kidney and brain of the T16 line and found in the liver, kidney, and brain of T1 and T24. The transgenic mice differed from control littermates in their serum smFcγRII/III levels. Whereas the T16 and T24 lines had only slight but significant increases (1.4- and 5-fold respectively) in sFcγR serum levels, the

T1 line had from 10- to 20-fold more smFcγRII/III than the control littermates (15±6 ng/ml). The respective contribution of endogeneous versus transgenic sFcγR in the transgenic mice is difficult to evaluate by ELISA. Nevertheless, the large and significant increase in sFcγR serum levels in T1 and T24 lines reflects most probably transgene expression.

To investigate the role of sFcγR in antibody production in vivo, mice from the trangenic lines received an i.p. injection of SRBC and were challenged 9 days later by another i.p. injection of the same dose of SRBC. Sera were collected at days 0, 6 (primary response) and/or 14 (secondary response). Mice were sacrificied at day 14. No significant differences were detected in IgM, IgG_1, IgG_{2a}, IgG_{2b} and IgG_3 levels between transgenic mice and control littermates before immunization. During the primary and secondary responses, the differences between Ig levels in transgenic and control mice remained unsignificant. Similarly, the anti-SRBC IgM and IgG antibody levels, as measured by ELISA, were not significantly different between transgenic mice and control littermates during the primary or secondary responses. The only exception was that of one group of T16 mice which exhibited higher IgG_1 anti-SRBC levels ($p<0.02$) than their controls. Finally, B cell numbers (60–78% B220 positive cells), T cell numbers (9.9–17.8% L3T4 and 5.1–11.0% Lyt2-positive cells) did not differ significantly in transgenics vs controls, with the exception of T16 mice which exhibited significantly higher L3T4 percentages than the control littermates [5]. This observation may be of interest in view of our recent findings in man showing that sFcγR induces cytokine production by dendritic cells [18]. Further experiments are in progress to study immune functions in these animals.

Soluble FcγR and disease

Mouse

Soluble forms of low affinity FcγR produced by proteolytic cleavage and by alternative splicing (mFcγRIIb3) have been described in mouse serum by biochemical techniques, immunodot assays and ELISA [3,5,11,24]. Because of the lack of mAb discriminating between FcγRII and III, no distinction has been made between smFcγRII and smFcγRIII produced by cleavage of their respective membrane receptors. Soluble mFcγRII/III levels are low in germ-free and in new born animals. In our pathogen free animal facility, levels range between 50 and 100 ng/ml in adult (6–8-week old) BALB/c mice. As summarized in Table 24.1, significant increases in serum smFcγRII/III have been described in a variety of pathological situations due to parasitic [25] or viral infections (Coutelier *et al*, in preparation), tumour growth [26] and autoimmune diseases [27]. Increases are generally modest (2- to 3-fold), with the exception of some tumours such as IgG_{2a} or IgG_{2b}-secreting hybridomas and lymphomas, which can increase sFcγR serum levels by up to 10–fold. Although sFcγR may come from FcγR-positive tumours, the fact that the FcγR-deficient IIA1.6 lymphoma, infections or autoimmune diseases increase sFcγR levels suggests

Table 24.1 Modification of total sFcγR levels in mice in various pathological situations

Disease	*Type*	*sFcγR increase (fold)*
Infection	*T. musculi*	⩽3
	T. cruzi	⩽2
	LDV	⩽2
Tumour	Hybridoma or lymphoma (IgG_{2a}, IgG_{2b})	4 to 12
	Hybridoma or lymphoma (IgG_1, IgA, non Ig)	⩽3
	Melanoma, thymoma	
Auto-immune	MRL, 1pr	⩽3
	NZW	⩽2
	NOD	1

that other factors are involved. The increase in sFcγR serum levels may reflect enhanced membrane FcγR expression. Increased frequencies of FcγR-positive cells have been described in patients bearing IgG-secreting myeloma and in mice bearing IgG-secreting plasmocytomas [26]. Although the mechanisms underlying the IgG-induced up-regulation of FcγR described *in vitro* on T cells [28], such as protection from cell surface degradation such as in the case of hFcεRII, or increase in synthesis, need further investigation, IgG may also participate in the regulation of sFcγR levels. The injection of high doses of mouse IgG_{2a} has been shown to increase slightly sFcγR levels *in vivo*. Increased immunoglobulin production, either restricted to one isotype such as IgG_{2a} during LDV infection, or concerning all IgG isotypes, is observed in most of these cases. Other factors than IgG may participate to the regulation of sFcγR levels *in vivo*. Interferon γ was shown to up regulate FcγR expression *in vitro* [29]. Proteases produced or activated during infections may also play an important role in FcγR release *in vivo*.

The mFcγRIIb3 product is also present in serum. It can be specifically detected from cleaved sFcγR by SDS-PAGE, because of its distinct size from cleaved smFcγRII or III, or by ELISA using anti-intracytoplasmic antibodies. The nonobese diabetic (NOD) mouse, which represents a useful model of human insulin-dependent diabetes melitus (IDDM), displays an elevation of IgG serum levels and production of autoantibodies. Recent studies have shown that NOD mice have lower mFcγRIIb3 serum levels than controls, reflecting a specific defect of mFcγRII gene expression in macrophages [30]. Studies of congenic mice showed that this defect was closely linked to *fcgr2* locus and with IgG_1 and IgG_{2b} up-regulation. Although it cannot be excluded that serum IgG concentration and mFcγRII expression are controlled by distinct but closely linked genes, it is tempting to speculate that macrophage low affinity FcγR play a role in the regulation of IgG levels. An FcγRIIb2-dependent diminished catabolism of immune complexes and/or an FcγRIIb3-dependent decrease of inhibition of Ag presentation by IgG-immune complexes may occur in NOD mice. Of note, serum smFcγRII/III total levels were unchanged in NOD mice, contrasting with the mFcγRIIb3 decrease, suggesting that the production of cleaved smFcγRII/III may be increased in these animals, like in other autoimmune strains of mice. The use of smFcγRII and mFcγRIIb3 transgenic animals will help in understanding the respective roles of these soluble FcγR *in vivo*.

Human

Soluble forms of hFcγRII and hFcγRIII circulate in serum, the most abundant form, shFcγRIIIb, being produced by cleavage of neutrophil FcγRIIIb. In serum from healthy individuals total shFcγRIII levels are 1.25 ± 0.28 μg/ml (30 donors tested), as demonstrated by ELISA [3]. As shown in Table 24.2, significant decreases in serum shFcγRIII levels have been detected in a variety of pathological situations such as multiple myeloma [3,31,32], acute myeloid leukemia (AML) [33] and AIDS [34] (Galon *et al.*, in preparation]. These may reflect a blockade of shFcγRIII in the detection assays, a decrease of neutrophil numbers or of hFcγRIII expression, or a decrease in the enzymatic process leading to hFcγRIII cleavage. Acute myeloid leukemia is generally characterized by granulopenia. However whether the defect in granulopoiesis is responsible for the lower shFcγRIII levels found in AML patients is unclear. A correlation was found with the absolute number of $CD16^+/CD3^+$ cells. These results should be interpreted with caution since a FcγRIII-negative neutrophil population (25% of neutrohils) which may be related to bacterial infections has been detected in patients with AIDS [35].

In a recent study, plasma shFcγRIII levels were compared in 35 patients with multiple myeloma, 54 individuals with monoclonal gammopathy of unknown

Table 24.2 Modifications of sFcγR serum levels in human disease

Disease	*sCD 32 (FcγRIIa$_2$)*	*sCD16*
Tumour		
Multiple myeloma	Constant	Stage-dependent decrease
MGUS	nd	Decrease in rapid progressors
AML	nd	Decrease
BCLL	Increase (stage C)	Unmodified
Infection		
HIV	nd	Decrease (AIDS)
Autoimmune		
SLE	Constant	Increase
Primary Sjögren's syndrome	nd	Increase

significance (MGUS) and 29 healthy individuals. A low level of shFcγRIII (values below 1.3 μg/ml) was found to discriminate between MM patients and controls, confirming previous observations [32]. Moreover, a low shFcγRIII level was also found to identify among MGUS patients a subgroup of patients who rapidly progressed towards multiple myeloma. Of note, the stage-dependent shFcγRIII decrease found in multiple myeloma patients was not correlated with the Ig isotype, suggesting that it was not due to blockade of shFcγRIII by IgG. Recent studies performed in our group showed that although the number of hFcγRIII sites per neutrophil is not significantly different in myeloma patients and healthy individuals, patients with multiple myeloma have significantly lower absolute numbers of polymorphonuclear cells than controls (Galon *et al.*, in preparation). Indeed it should be kept in mind that the vast majority of neutrophils is present in tissues.

Soluble FcγRIIa2 levels were found to be around 10 ng/ml when polyclonal antibodies directed against the intra-cytoplasmic region were used as revealing antibodies [36]. When ELISA was used to detect both hFcγRIIa2 and putative cleaved shFcγRIIa1, higher concentrations (ranging from 30–150 ng/ml) were found (P. Ghillani, manuscript in preparation). Of note, in patients with multiple myeloma, no modification of hFcγRIIa2 serum levels was found. In contrast, in BCLL patients in which no modification of shFcγRIII serum levels were detected [37], hFcγRIIa2 concentration was found to increase in stage C patients. Whether this reflects the production of hFcγRIIa2 by tumour B cells or an increase expression in other cell types remains unclear.

Higher amounts of shFcγRIII have been detected in serum from some patients with autoimmune diseases [38,39]. However, these results need further confirmation, in view of the low number of individuals and of the low sensitivity of the assay used. Soluble hFcγRIIa (cleaved shFcγRIIa$_1$ + shFcγRIIa$_2$) serum levels have been examined in our laboratory in a group of SLE patients and of patients with Behçet's disease. No significant difference was observed with a group of healthy donors. However, 22.2% SLE patients exhibited shFcγRIIa serum levels above 90 ng/ml while only 6.8% of normal donors had shFcγRIIa values above this level. Conversely, 21.8% of patients with Behçet's disease had shFcγRIIa serum levels lower than 20 ng/ml, compared with only 14.5% of healthy donors. No clinical or biological correlations with known parameters could be defined in this study.

Serum levels of shFcγRIIa were also studied in patients with chronic hepatitis C and compared to healthy donors (P. Ghillani *et al.*, manuscript in preparation). A subgroup of HCV-infected patients exhibited significantly higher shFcγRIIa serum levels. No correlation was found between shFcγRIIa serum levels and patient age, duration of liver desease or transaminase serum levels. Soluble hFcγRIIa serum levels weakly correlated with the increased IgG serum levels. There was no significant difference in patients with or without cryoglobulinemia. We also investigated the serum levels of shFcγRIII. As observed with shFcγRIIa, a significant increase of shFcγRIII serum levels was found in HCV-infected patients when compared to healthy donors. No correlation with patient age, duration of liver disease, transaminase and IgG levels, or with cryoglobulinemia was found. No correlation was observed between shFcγRIII and shFcγRIIa serum levels in HCV patients as well as in healthy donors. Similarly, when examining HCV patients with elevated shFcγRIIa or shFcγRIII, no correlation was found.

As discussed above, variations of sFcγR levels could be due to various factors, including modifications of cytokine production since cytokines such as TGF-β, or IFN-γ increase FcγRIII expression [40,41] and whereas IL-4 and IL-10 have an opposite effect on all FcγR types [42,43].

CONCLUSIONS AND PERSPECTIVES

Soluble FcγR are useful tools for studying the binding specificity of membrane FcγR and are markers of the number and activation of the cells which synthesize them. Moreover, because of their direct interaction with cells and with immune complexes, sFcγR may exert regulatory functions at various stages of immune responses.

Although the mechanisms of generation of sFcγR have been elucidated, the precise nature of the enzymes involved in the cleavage from membrane FcγR has not yet been determined. The active field of research dealing with the characterization of the enzymes which generate soluble cytokines such as IL-1 or TNF-α and soluble cytokine receptors should allow to resolve this question for sFcγR.

The identification of ligands other than Ig for sFcγRIII and sCD23, opens a new array of interactions with cells of the immune system. Indeed, the β_2 integrins are molecules involved in adhesion, as well as in signalling, and their interaction with FcγR-bearing cells, or with sFcγR, may modify the activities of the cells which express them. In addition, it is also possible that the activity of the FcγR-bearing cells are also modified through such interactions.

Finally, the role of the modifications of plasmatic levels of sFcγR in pathological situations is difficult to interpret. It is not clear if it is only a marker or if it can be functionnally associated to disease evolution. The production of transgenic mice and the study of well-defined animal models should help to resolve these questions, an answer to which may help defining novel therapeutic approaches.

Acknowledgements

We wish to thank the technical assitance of A. Galinha, J. Moncuit, and P. Paulet. Portions of this work were performed in collaboration with H.J. Garchon (INSERM U25 Hopital Necker, Paris), J. Jami (INSERM U257, Faculté de médecine Cochin), D. Hanau (CJF INSERM, ETS Strasbourg), N. Mazières and R. Spagnoli (Roussel Uclaf, Romainville). This work was supported by INSERM, Institut Curie, and grants from Roussel Uclaf, ARC and LNFCC.

References

1. Fridman WH, Rabourdin-Combe C, Néauport-Sautès C, Gisler RH. Characterization and function of T cell Fcγ receptor. Immunol Rev. 1981;56:51–88.
2. Fridman WH, Bonnerot C, Daëron M, Amigorena S, Teillaud JL, Sautès C. Structural bases of Fcγ receptor functions. Immunol Rev. 1992;125:49–76.
3. Teillaud J-L, Bouchard C, Astier A et al. Natural and recombinant soluble low-affinity FcγR: detection, purification, and functional activities. Immuno Meth. 1994;4:48–64.
4. Sautès C. Soluble Fc receptors. In: Fridman WH and Sautès C, eds, Cell Mediated Effects of Immunoglobulins. Austin: R.G. Landes Co; 1997:139–64
5. Galon J, Paulet P, Galinha A et al. Soluble Fcγ receptors: interaction with ligands and biological consequences. Int Rev Immunol. 1997;16:87–111.
6. Ghirlando R, Keown MB, Mackay GA, Lewis MS, Unkeless JC, Gould HJ. Stoichiometry and thermodynamics of the interaction between the Fc fragment of human IgG_1 and its low affinity receptor Fc gamma RIII. Biochemistry. 1995;34:13320–7.
7. Esposito-Farèse ME, Sautès C, de la Salle H. Membrane and soluble FcγRII/III modulate the antigen presenting capacity of murine dendritic epidermal Langerhans cells for IgG-complexed antigens. J Immunol. 1995;155:1725–36.
8. Ierino FL, Powell MS, McKenzie IF, Hogarth PM. Recombinant soluble human Fc gamma RII: production, characterization, and inhibition of the Arthus reaction. J Exp Med. 1993; 178:1617–28.
9. Astier A, de la Salle H, de la Salle C et al. Human epidermal Langerhans cells secrete a soluble receptor for IgG (FcγRII/CD32) that inhibits the binding of immune-complexes to FcγR+ cells. J Immunol. 1994;152:201–12.
10. Gachet C, Astier A, de la Salle H et al. Release of FcγRIIa2 by activated platelets and inhibition of anti-CD9-mediated platelet aggregation by recombinant FcγRIIa2. Blood. 1995;85:698–704.
11. Lynch A, Tartour E, Teillaud JL, Asselain B, Fridman WH, Sautès C. Increased levels of soluble low affinity Fcγ receptor (IgG-binding factor) in the sera of tumour-bearing mice. Clin Exp Immunol. 1992;87:208–14.
12. Aubry J-P, Pochon S, Graber P, Jansen KU, Bonnefoy J-Y. CD21 is a ligand for CD23 and regulates IgE production. Nature. 1992;358:505–7.
13. Lecoanet-Henchoz S, Gauchat J, Aubry J et al. CD23 regulates monocyte activation through a novel interaction with the adhesion molecules CD11b-CD18 and CD11c-CD18. Immunity. 1995;3:119–25.
14. Kimberly RP, Tappe NJ, Merriam LT et al. Carbohydrates on human Fc gamma receptors. Interdependence of the classical IgG and nonclassical lectin-binding sites on human Fc gamma RIII expressed on neutrophils. J Immunol. 1989;142:3923–30.
15. Petty HR, Todd RF III. Receptor-receptor interactions of complement receptor type 3 in neutrophil membranes. J Leuk Biol. 1993;54:492–4.
16. Galon J, Gauchat JF, Mazières N et al. Soluble Fcγ Receptor type III (FcγRIII, CD16) triggers cell activation through interaction with complement receptors. J Immunol. 1996;157: 1184–92.
17. Ross GD, Vetvicka V. CR3 (CD11b, CD18): a phagocyte and NK cell membrane receptor with multiple ligand specificities and functions. Clin Exp Immunol. 1993;92:181–4.
18. de la Salle H, Galon J, Bausinger H et al. Binding of recombinant soluble FcγRIII to human dendritic cells induces phenotype changes, soluble FcγRIII internalization and cytokine secretion. In: Ricciardi-Castagnol P, ed., Dendritic Cells in Fundamental and Clinical Immunology. Plenum Press. 1997;3:345–52.
19. Brunati S, Moncuit J, Fridman WH, Teillaud JL. Regulation of IgG production by suppressor FcγRII+ T hybridomas. Eur J Immunol. 1990;20:55–61.
20. Fridman WH. Fc Receptors and Immunoglobulin Binding Factors. FASEB J. 1991;5:2684–90.
21. Bouchard C, Galinha A, Tartour E, Fridman WH, Sautes C. A transforming growth factor β-like immunosuppressive factor in immunoglobulin G-binding factor. J Exp Med. 1995;182: 1717–26.
22. Stach RM, Rowley A. A first or dominant immunization. II Induced immunoglobulin carries transforming growth factor b and suppresses cytolytic T Cell responses to unrelated alloantigens. J Exp Med. 1993;178:841–52.

23. Teillaud C, Galon J, Zilber M-T et al. Soluble CD16 binds peripheral blood mononuclear cells and inhibits pokeweed-mitogen (PWM)-induced responses. Blood. 1993;82:3081–90.
24. Tartour E, de la Salle H, de la Salle C et al. Identification, in mouse macrophages and in serum, of a soluble receptor for the Fc portion of IgG (FcγR) encoded by an alternatively spliced transcript of the FcγRII gene. Intern Immunol. 1993;5:859–68.
25. Araujo-Jorge T, El Bouhdidi A, Rivera M-T, Daëron M, Carlier Y. Trypanosoma cruzi infection in mice enhances the membrane expression of low-affinity Fc receptors for IgG and the release of their soluble forms. Parasite Immunol. 1993;15:539–46.
26. Mathur A, Lynch RG. Increased Tg and Tm cells in BALB/c mice with IgG and IgM plasmacytomas and hybridomas. J Immunol. 1986;136:521–5.
27. Pure E, Durie CJ, Summerill CK, Unkeless JC. Identification of soluble Fc receptors in mouse serum and the conditioned medium of stimulated B cells. J Exp Med. 1984;160:1836–49.
28. Löwy I, Brézin C, Néauport-Sautès C, Thèze J, Fridman WH. Isotype regulation of antibody production: T cell hybrids can be selectively induced to produce subclass specific suppressive immunoglobulin-binding factors. Proc Natl Acad Sci USA. 1983;80:2323–7.
29. Weinshank RL, Luster AD, Ravetch JV. Function and regulation of a murine macrophage-specific IgG Fc receptor, FcγR-a. J Exp Med. 1988;167:1909–25.
30. Luan JJ, Monteiro RC, Sautès C et al. Defective FcγRII gene expression in macrophages of NOD mice. J Immunol. 1996;157:4707–16.
31. Mathiot C, Teillaud JL, Elmalek M et al. Correlation between serum soluble CD16 (sCD16) levels and disease stage in patients with multiple myeloma. J Clin Immunol. 1993;13:41–8.
32. Mathiot C, Mary JP, Tartour E et al. Soluble CD16 (sCD16), a marker of malignancy in individulas with monoclonal gammapathy of undetermined significance (MGUS). Br J Haematol. 1996;95:660–5.
33. Fleit HB, Kobasiuk CD, Daly C, Furie R, Levy PC, Webster RO. A soluble form of Fc gRIII is present in human serum and other body fluids and is elevated at sites of inflammation. Blood. 1992;79:2721–8.
34. Khayat D, Soubrane C, Andrieu JM et al. Changes of soluble CD16 levels in serum of HIV-infected patients:correlation with clinical and biologic prognostic factors. J Infect Dis. 1990; 161:430–5.
35. Boros P, Gardos E, Bekesi GJ, Unkeless JC. Change in expression of Fc gamma RIII (CD16) on neutrophils from human immunodeficiency virus-infected individuals. Clin Immunol Immunopathol. 1990;54:281–9.
36. Astier A, de la Salle H, Moncuit J et al. Detection and quantification of secreted soluble FcγRIIA in human sera by an enzyme-liked immunoadsorbent assay. J Immunol Meth. 1993;166:1–10.
37. Fridman WH, Mathiot C, Moncuit J, Teillaud JL. Fc receptors, immunoglobulin-binding factors and B chronic lymphocytic leukemia. Nouvelle Revue Française d'Hématologie. 1988;30:311–15.
38. Lamour A, Soubrane C, Ichen M, Pennec YL, Khayat D, Youinou P. Fc-gamma receptor III shedding by polymorphonuclear cells in primary Sjögren's syndrome. Eur J Clin Invest. 1993;23:97–101.
39. Hutin P, Lamour A, Pennec YL et al. Cell-free Fc-gamma receptor III in sera from patients with systemic lupus erythematosus:correlation with clinical and biological features. Int Arch Allerg Immunol. 1994;103:23–7.
40. Welch GR, Wong HL, Wahl SM. Selective induction of FcγRIII on human monocytes by transforming-growth factor-b. J Immunol. 1990;1990:3444–8.
41. Hartnell A, Barry Kay A, Wardlaw AJ. IFN-g induces expression of FcγRIII (CD16) on human eosinophils. J Immunol. 1992;148:1471–8.
42. Te Velde AA, Huibens RJF, de Vries JE, Figdor CG. IL-4 decreases FcγR membrane expression and FcγR-mediated cytotoxic activity of human monocytes. J Immunol. 1990; 144:3046–51.
43. Te Velde AA, de Wall R, Huibens RJF, de Vries JE, Figdor CG. IL-10 stimulates monocyte surface FcγR expression and cytotoxic activity. J Immunol. 1992;149:4048–52.

25
FcγR-directed immunotherapies

C. A. GUYRE, P. K. WALLACE and M. W. FANGER

INTRODUCTION

Fc receptors for IgG (FcγR) mediate a variety of effects including antibody-dependent cellular cytotoxicity (ADCC), clearance of immune complexes, enhancement of antigen presentation, secretion of reactive oxygen intermediates, and phagocytosis of antibody-coated pathogens [1]. Additionally, these receptors appear to be the only molecules on human myeloid cells capable of mediating ADCC of tumour cells; they can, therefore, play a key role in antibody therapy for cancer. Preclinical and clinical studies show that bispecific antibody (BsAb) therapies which target tumour cells to FcγR-bearing effector cells lead to arming of effector cells, increased killing of tumour cells, and vaccine-like effects.

FcγR

Three different classes of human FcγR have been defined, all of which are structurally related but functionally distinct members of the Ig supergene family [1]. FcγRI (CD64) is a 70 kDa glycoprotein that binds with high affinity to monomeric human IgG_1 and IgG_3, mouse IgG_{2a} and IgG_3, as well as IgG immune complexes [1–4]. The features of FcγRI are listed in Table 25.1.

FcγRII (CD32) is a 40 kDa glycoprotein that binds IgG with low affinity. It is expressed on a broad range of cell types, including PMNs, platelets, NK cells, and B lymphocytes, as well as monocytes and macrophages [1–4]. FcγRII does not bind monomeric immunoglobulin, but reacts with IgG immune complexes and opsonized particles [5]. It is a trigger molecule on monocytes, macrophages, PMNs, and eosinophils that mediates lysis and/or phagocytosis of tumour cell targets. Triggering of FcγRII has been shown to require that targets be coated

J.G.J. van de Winkel and P.M. Hogarth (eds.), The Immunoglobulin Receptors and their Physiological and Pathological Roles in Immunity. 291–305.

Table 25.1 Properties and functions of FcγRI

1. Expressed only on monocytes, macrophages, dendritic cells, and IFN-γ or G-CSF activated PMNs.
2. High affinity Fc receptor; occupied by IgG *in vivo*.
3. Expression and function enhanced by IFN-γ or G-CSF.
4. Mediates antibody-dependent cellular cytotoxicity, phagocytosis, cytokine release, and superoxide generation.
5. Mediates enhanced antigen presentation.
6. Killing of opsonized targets requires less antibody than other Fc receptors.

with 10 times as much antibody as for triggering of FcγRI [6], suggesting that FcγRII may be a less efficient trigger molecule than FcγRI.

FcγRIIIa (CD16) is a 50–80 kDa transmembrane glycoprotein with moderate affinity for IgG . It is found predominantly on NK cells, macrophages and a small subset (5–15%) of monocytes, and is a cytotoxic and/or phagocytic trigger molecule on these cells [1–4]. FcγRIIIb is a glycosyl phosphatidylinositol linked glycoprotein with low affinity for IgG that is selectively expressed on granulocytes. Cross-linking FcγRIII triggers phagocytosis on monocytes, but not on granulocytes where instead it may participate in the capture of immune complexes which can then be phagocytosed through other receptors [7].

MDX-33, A mAb TO FcγRI

Self-reactive antibodies are a significant cause of morbidity and mortality and contribute to both cytotoxic and immune complex-triggered inflammatory disorders, typified by rheumatic diseases, autoimmune haemolytic anaemia, and thrombocytopenia. A role for Fc receptors in these diseases has been proposed, although their specific role and the contributions of the individual FcγR classes to autoimmune injury remains unclear. Patients with immune thrombocytopenia purpura (ITP) who do not respond to conventional therapy may receive high dose IVIgG or anti-Rh globulin [8–10] to raise the platelet count. These treatments are aimed at blocking the ability of phagocytic cells to destroy IgG-coated platelets. Their effectiveness has been attributed in part to 'FcγR blockade', a mechanism in which the transfused IgG competes with platelet bound IgG for FcγR on phagocytic cells [9,11]. Support for this mechanism comes from the finding that IVIgG slows the clearance of radiolabelled RBC opsonized with anti-D in patients with ITP [9,12] and by the success of anti-RhD in raising platelet counts in Rh^+ ITP patients [10,13]. Further support for FcγR blockade comes from studies demonstrating that an intact Fcγ region, but not the Fab domain, is required for IVIgG to achieve a clinical response [14].

Recently, Ericson *et al.* [15] treated a 44-year-old women with refractory ITP with mAb 197. This murine IgG_{2a} binds to two distinct epitopes on FcγRI, through its Fc domain to the ligand binding site on FcγRI and through its Fab domain to an epitope of FcγRI outside the ligand binding site [16,17]. Binding of this mAb results in extensive cross-linking of FcγRI. Such cross-linking not only blocks the ligand binding site but also significantly down-modulates surface expression of FcγRI on myeloid cell lines and monocytes both *in vitro* and *in vivo* [15,16]. Following infusion of mAb 197 there was down-modulation of FcγRI on all circulating monocytes and a marked clinical improvement was reported, with resolution of oral ecchymoses and epistaxis. While no appreciable effect on platelet count was observed during the 5 day course of mAb treatment, mAb 197 appeared to dramatically increase responsiveness to subsequent IVIgG therapy.

These characteristics would make mAb 197 useful for determining the role of FcγRI in ITP. However, repeated administration of a murine antibody may induce the development of antibodies to the foreign Ig and result in immune complexes which could have adverse effects. Recently, another FcγRI-specific mAb (mAb 22) was humanized by grafting its complementary determining regions onto human IgG_1 constant domains [17,18]. Humanized mAb 22 (MDX-33) also recognizes an epitope on FcγRI outside the ligand binding domain. Because the humanized form is an IgG_1, it also binds through its Fc domain to the FcγRI ligand binding site [18]. Binding of MDX-33 results in extensive cross-linking of FcγRI and modulation of surface expression of FcγRI. Recent studies also demonstrate that MDX-33 blocks phagocytosis of opsonized RBC much more effectively than an irrelevant IgG [19]. These characteristics should make MDX-33 useful for determining the role of FcγRI in ITP, and, potentially, as a therapy for antibody-mediated autoimmune disease.

Clinical Trial. A human clinical trial has recently been initiated using MDX-33. The initial focus of this trial is to determine the effect of MDX-33 on monocyte FcγR from normal volunteers. In subsequent trials MDX-33 will be used in patients with ITP.

BISPECIFIC ANTIBODIES (BsAb)

BsAb are chimeric antibodies with specificity for two different antigens and can be designed to link viral or tumour-associated antigens with effector cells. For BsAb to mediate cytolytic activity, one of their specificities must be to a cytolytic trigger molecule (e.g. FcγR) on these cells. Such BsAb are able to direct or redirect the cytolytic ability of the effector cell to lyse appropriate targets, including tumour cells. This section will focus on BsAb which target antigens to two such trigger molecules, FcγRI and FcγRIIIa. BsAb have been produced using various methods, including:

(1) Fusion of two hybridoma lines to obtain a quadroma which makes and secretes antibodies composed of all combinations of L and H chains of the parent antibodies, only some of which are true BsAbs;

(2) Chemically cross-linking intact, $F(ab')_2$, Fab′, or Fab antibody species of two different antibodies; and

(3) Genetically linking the genes of two different single chain antibodies (scFvs), composed of fused v regions, to produce a bispecific fusion molecule [20].

BsAbs TARGETING THROUGH FcγRI (CD64).

The development of antibodies which bind FcγRI outside the ligand binding domain, yet trigger FcγRI function, has allowed for targeting of antigen to effector cells, even *in vivo* where FcγRI is occupied by IgG. One such antibody, mAb 22, has been humanized (mAb H22) and used in several BsAbs, including MDX-210, MDX-447 and MDX-240 (Table 25.3).

MDX-210

HER-2/neu, or c-erbB-2, is a human proto-oncogene homologous to the rat neu oncogene and related in sequence to the epidermal growth factor receptor [21–23]. Some normal tissues express low levels of HER-2/neu; however, over-expression of HER-2/neu is observed in a variety of human cancers and may correlate with poor prognosis [22,24]. Although it is still unclear whether HER-2/neu will have significant prognostic implications, it is one of the more promising target antigens for BsAb therapy. MDX-210, a BsAb constructed by chemically linking Fab′ fragments of mAb H22 with Fab′ fragments of a mAb to HER-2/neu, has been developed for the treatment of tumours which over-express HER-2/neu including breast, ovarian, and prostate cancers [17,25].

MDX-210 demonstrates specific, dose-dependent and saturable binding to FcγRI-expressing cells, as well as to HER-2/neu-expressing cells, as determined by indirect immunofluorescence and flow cytometry [26]. Moreover, it binds to both FcγRI and to HER-2/neu simultaneously, as determined using a solid phase ‘bispecific’ ELISA as well as ‘bispecific’ immunofluorescent flow cytometric analyses. The binding affinity of the BsAb appeared consistent with that of its respective Fab′ fragments.

MDX-210 mediates monocyte dependent lysis of HER-2/neu over-expressing cell lines from breast, ovarian, and lung carcinomas. This cytolytic activity is completely inhibited by $F(ab')_2$ fragments of anti-FcγRI mAb 22, but is not blocked by human serum or IgG, indicating that MDX-210 could be used *in vivo* where the ligand binding site of FcγRI is normally occupied [26,27]. IFN-γ treatment of monocytes enhances BsAb-mediated ADCC whereas IFN-γ and/

or G-CSF are required for PMN-mediated ADCC. Repp *et al.* [28] have demonstrated that treatment of patients with G-CSF induces FcγRI expression on their PMNs, which could then be targeted with MDX-210 to lyse HER-2/neu-positive tumours. Similar results have been obtained by us with PMN from patients treated with IFN-γ. Keler *et al.* [26] observed cytotoxic activity by IFN-γ-activated, MDX-210 targeted monocytes against several cell lines over-expressing HER-2/neu but not against the A431 target which expresses low levels of HER-2/neu. Because over-expression of the proto-oncogene receptor was required for significant BsAb-mediated lysis, normal cells expressing low levels of HER-2/neu may be less likely targets for MDX-210 dependent lysis.

MDX-210 also elicits the release of TNF-α from monocytes when cultured in the presence of HER-2/neu-positive cell lines but not when cultured with HER-2/neu-negative cell lines [26]. As expected, the BsAb alone does not elicit TNF-α release, since MDX-210 is monovalent and signaling through FcγRI requires receptor cross-linking. While secretion of TNF-α and other soluble mediators has been demonstrated upon cross-linking FcγR and is associated with monocyte anti-tumour activity, it is not known whether the induction of TNF-α release is involved in MDX-210-mediated cytotoxicity.

Clinical Trials. Based on these preclinical observations, MDX-210 seemed an excellent candidate for efficiently directing the cytolytic activity of FcγRI-expressing effector cells *in vivo* to HER-2/neu-expressing tumours. MDX-210 clinical trial data are summarized in Table 25.2. Two phase Ia/Ib clinical trials

Table 25.2 Summary of data on patients treated with MDX-210 (all patients had failed standard therapy, had progressive disease and were HER-2/neu positive)

1)	Effector cells	
		Transient monocyte depletion – return to normal within 24 h
		Little effect on lymphocyte or PMN populations
		All monocytes armed with BsAb at doses of 3.5 mg/m^2 or higher
		Many monocytes remain armed for more than 48 h
2)	Cytokine response	
		Increased TNF-α, IL-6, G-CSF
3)	Toxicity	
		With few exceptions – grade 1, 2 – at all doses
4)	Tumour	
		BsAb found in tumour biopsies
		Monocyte infiltration into tumour biopsies
5)	Clinical responses (in some patients)	
		Evidence of substantial shrinkage of tumour
		Stable disease for periods of 2–5 months
		Increased anti-tumour antibodies after therapy (vaccine-like effect)
		Reduced circulating HER-2/neu levels months after therapy

Table 25.3 Clinical trials of BsAbs targeting CD64

BsAb	*Dose*	*Cytokine*	*Target*	*Indication*	*Phase*	*Location*	*Principal investigator*
MDX-33	M	–	CD64	ITP	I	Philadelphia	Zielonka
MDX-210*	S	–	HER2/neu	B, O	I	Dartmouth	Valone
MDX-210*	M	–	HER2/neu	B, O	I/II	Dartmouth	Valone
MDX-210	M	–	HER2/neu	Mostly B	I/II	NCI	Cowan
MDX-210*	M	–	HER2/neu	P	I/II	Dartmouth	Ernstoff
MDX-210	S	G-CSF	HER2/neu	B	I/II	Europe	Repp
MDX-210	M	G-CSF	HER2/neu	B	I/II	USC	Weber
MDX-210	M	GM-CSF	HER2/neu	P, R	II	Europe	James
MDX-210	M	GM-CSF	HER2/neu	B, C, L, O, R	I/II	Georgetown	Hawkins
MDX-210	M	GM-CSF	HER2/neu	C	II	Europe	Mellstedt
MDX-210	M	IFN-γ	HER2/neu	B, C, P	I/II	Dartmouth	Kaufman
MDX-240	M	–	gp41	AIDS	I/II	Europe	Pasquali
MDX-447	M	–	EGF-R	H, N, P, R, Sq	I/II	Sloan-Kett.	Pfister

*Completed

Abbreviations: M, multiple; S, single; B, breast; C, colon; H, head; L, lung; N, neck; O, ovarian; P, prostate; R, renal; Sq, squamous; ITP, immune thrombocytopenia purpura

using MDX-210 for treatment of patients with stage IV breast cancer or stage III or IV ovarian cancer over-expressing HER-2/neu have been completed [24,29]. In the first, a single dose escalation trial, 15 patients were treated at doses of 0.35–10 mg/m^2 and in the second trial 10 patients received 3 alternate day or weekly infusions of MDX-210 over a similar dose range. Non-haematological toxicities in both trials were mild, with most patients experiencing only grade 1–2 flu-like side-effects. Of interest, one patient with extensive pulmonary metastases developed grade 4 pulmonary toxicity after BsAb infusion, suggesting that MDX-210 was clinically active at this site. All toxicity resolved fully within 4–18 h of completing the infusion. In both trials, no significant haematological toxicity was noted, although a monocytopenia consistently occurred 1–2 h after each infusion of MDX-210 which resolves within 24 h.

For doses >3.5 mg/m^2, essentially all monocytes bound MDX-210 at the end of the infusion, and greater than 80% continued to exhibit MDX-210 binding 48 h after infusion at the highest doses. MDX-210 saturated available monocyte FcγRI in a dose-related manner, maximal saturation of 80–90% being achieved at doses >3.5 mg/m^2 dose level. At the highest doses, 20–40% of monocyte receptors for FcγRI remained occupied by MDX-210 48 h after antibody infusion.

Plasma levels of MDX-210 >1.0 μg/ml were achieved at doses >3.5 mg/m^2. MDX-210 concentrations of 0.1 μg/ml effectively target monocyte and PMN cytotoxicity *in vitro*. Furthermore, plasma concentration–time profiles of MDX-210 after infusion exhibited a single exponential decay with an elimination $t_{\frac{1}{2}}$ which ranged from 8.2 to 16.6 h over the dose range 1–10 mg/m^2.

Treatment with MDX-210 induced cytokine release in a complex manner. Increased plasma concentrations of TNF-α, IL-6, and G-CSF were observed by h 1–6. There appeared to be a threshold effect for stimulation of IL-6 and G-CSF in which low doses of MDX-210 did not stimulate release of these cytokines, but once an active dose was reached cytokines were released in a non-dose-dependent manner. In the multi-dose trial, increased plasma levels of TNF-α, IL-6 and G-CSF were noted after the first infusion, but subsequent infusions of MDX-210 resulted in much lower plasma cytokine levels.

Overall, treatment with MDX-210 was well tolerated, and associated with remarkable evidence of immunological activity, as demonstrated by transient monocytopenia and cytokine release. Localization of MDX-210 and a mononuclear leukocyte infiltrate in tumour specimens from two patients who underwent biopsies of skin metastases after treatment was further evidence for the immunological efficacy of MDX-210. Encouraging evidence for clinical efficacy was also seen. One patient with breast cancer had a >50% decrease in the size of chest wall and nodal metastases. Another patient with ovarian cancer had >50% decrease in cervical adenopathy without a significant change in bulky abdominal masses. Of particular interest, MDX-210 induced active anti-tumour immunity to tumour cell antigens in several patients. Induction of specific anti-tumour immunity has been noted with another BsAb targeting FcγRIII effector cells [30], suggesting that FcγR-directed BsAbs may be important in the induction of vaccine-like effects.

Based on studies in which *in vitro* assays have shown that FcγRI-positive PMNs constitute a major effector cell population during G-CSF therapy, two on-going Phase I trials of MDX-210 in combination with G-CSF for patients with stage IV breast cancer have been initiated. Side effects in both trials were similar to the MDX-210 alone trials, consisting mainly of fever and short periods of chills, which were related in time to elevated plasma cytokine levels. In one study [31], patients received G-CSF for eight consecutive days, followed by a single dose of MDX-210. During the time of G-CSF administration, isolated PMNs were highly cytotoxic in the presence of MDX-210 *in vitro*. Elevated plasma levels of soluble HER-2/neu increased after MDX-210, and fell below baseline at the end of the study in most patients, suggesting tumour cell lysis *in vivo*. In a second trial with G-CSF, Weber *et al.* [32] reported on 10 breast cancer patients receiving multiple doses of MDX-210 in combination with G-CSF. With repetitive dosing, there were no differences between peak BsAb values. Haematological effects included significant mean WBC increases 24 h after infusion of BsAb and a cycle of G-CSF. Soluble HER-2/neu levels also decreased in the 7 patients tested so far during therapy.

MDX-447

Biochemical and histological studies of tumour biopsies and human cell lines have demonstrated over-expression of the normal and mutated forms of EGF-R in almost all head and neck tumours and in approximately one-third of breast, ovarian, brain, prostate, bladder, lung, pancreatic and gastrointestinal tumours [33]. A mAb against the extracellular domain of the EGF-R, mAb 425, has been generated and developed into a therapeutic compound, EMD 55900 [34]. Eighty-six patients with malignant gliomas were treated with this mAb in Phase I and I/II trials.

Although no major tumour shrinkage was seen after treatment with EMD 55900, it seemed reasonable to suppose that a BsAb which efficiently targeted FcγR bearing effector cells to the tumour might be more effective. A BsAb, designated MDX-447, was therefore constructed from the Fab fragments of EMD 55900 and mAb H22 to target and trigger FcγRI bearing leukocytes. MDX-447 demonstrated specific, dose-dependent, and saturable binding to FcγRI expressing cells and to EGF-R-positive cells. The ability of MDX-447 to direct monocyte-derived macrophage killing of EGF-R overexpressing cell lines was measured by a standard ^{51}Cr release assay that detects extracellular killing and a novel two-colour flow cytometric method that assays phagocytosis [35]. MDX-447 was shown to significantly enhance macrophage phagocytosis of these cells even without activation of macrophages, but BsAb-mediated ADCC required preincubation of macrophages with IFN-γ.

Using a slightly different approach to target the EGF-R, Goldstein *et al.* [36] constructed a novel bispecific fusion protein consisting of EGF, the natural ligand for EGF-R, and the Fab fragment of mAb H22. The H22-EGF construct binds simultaneously to soluble FcγRI and EGF-R bearing cells and mediates ADCC by IFN-γ activated monocytes of A-431 cells, an EGF-R over-expressing cell line. This fusion protein also mediates phagocytosis of A-431 cells comparable to that mediated by MDX-447. In further experiments, Goldstein *et al.* found that both EGF and the H22-EGF fusion protein significantly inhibit cell growth in a dose dependent fashion. Thus, H22-EGF exhibits both cytotoxic functions associated with FcγRI targeting and cytostatic functions associated with the EGF ligand, suggesting that this fusion protein may have therapeutic utility for EGF-R over-expressing malignancies.

Clinical Trial. A phase I/II clinical trial of MDX-447 in patients with tumours over-expressing EGF-R is currently ongoing at Sloan-Kettering in New York.

MDX-240

Broadly reactive neutralizing antibodies to the HIV-1 gp120 envelope glycoprotein appear early in infected individuals and interfere with the binding of gp120 to CD4 [37], although the presence of such antibodies does not interfere with the development of acquired immune deficiency syndrome (AIDS). This may be related to the highly mutagenic nature of HIV-1 and the resulting in changes in the gp120 glycoprotein which affect antibody binding [38]. After development of an anti-HIV-1 antibody response, it seems likely that antibodies mediate clearance of HIV-1 through interaction with FcγR on myeloid effector cells. Although antibody-dependent enhancement (ADE) of HIV-1 infection of FcγR-bearing monocytes has been demonstrated in the presence of subneutralizing concentrations of HIV-1 antibody-positive sera [39,40], several studies have demonstrated that ADE is dependent upon gp120-CD4 binding [41–43]. Thus, FcγR-mediated binding and internalization of HIV-1 complexed with antibody does not mediate infectivity.

BsAbs have now been used to directly evaluate the role of FcγR in mediating ADE of HIV-1 infection of monocytes. BsAbs prepared using Fab′ fragments of a murine neutralizing mAb that binds gp120 at a site distinct from the CD4 binding site, coupled to Fab′ fragments of mAbs recognizing either FcγRI, FcγRII, or FcγRIII significantly inhibited virus production in monocytes compared to monocytes infected in the absence of BsAb [44]. Infection of monocytes was not inhibited when HIV-1 was targeted to either HLA class I or CD33. These studies also suggested that the degree of antibody opsonization may be a critical factor in determining the outcome of HIV–FcγR interaction. Moreover, they indicate a requirement for FcγR cross-linking for effective endocytosis and degradation of antibody–virus complexes. Thus, insufficient opsonization may result in little or no FcγR cross-linking on the monocyte, and little or no virus inactivation, but could stabilize virus at the cell membrane permitting infection to proceed through interaction with CD4 [45].

A BsAb constructed from Fab fragments of a human anti-gp41 mAb and mAb H22 (MDX-240) reduced infection of monocyte-derived macrophages, although the intact human anti-gp41 mAb alone did not confer protection [46]. Again, interaction with CD4 was critical to infectivity even under conditions of antibody-mediated binding of HIV-1 to the cell surface. In addition, PMNs, which express FcγR and are not infected by HIV-1, mediate killing of HIV-1. IFN-γ-activated PMN significantly inhibited HIV-1 infection of the human T-cell lymphoma cell line H9 [47]. With the addition of another BsAb (anti-FcγRI × anti-gp120), IFN-γ-activated PMN completely suppressed production of HIV-1 by H9 cells. Overall, it appears that the effective interaction of antibody-opsonized HIV-1 with FcγR expressed on human monocytes and PMNs reduces viral infectivity through FcγR-mediated cytotoxic mechanisms.

Clinical Trial. MDX-240 has been administered to 12 HIV infected patients with CD4+ T cell counts <400 cells/mm^2 and whose viral isolates were shown to bind MDX-240 [48]. Three patients each received six doses of 0.3, 1.0, 3.0, or 10.0 mg/m^2 MDX-240 over a 2-week period. Peak plasma levels (>2 μg/ml) of MDX-240 were observed in patients receiving doses >3.0 mg/m^2, and pharmacokinetic studies showed a $t_{\frac{1}{2}}$ of 3–6 h. Up to 90% of monocytes were armed with MDX-240 by 1 h post-infusion in groups receiving 3.0 and 10.0 mg/m^2, and BsAb remained bound up to 48 h. Acute monocytopenia was observed in these patients within 3 h post-infusion. Viral levels in both plasma and PBMCs remained stable for one month following initial treatment at all doses. Most significantly, CD4+ T cell levels increased 2 to 3-fold in four patients during the course of treatment and remained stable in the remaining patients. No change in cytokine levels were observed, with the exception of a minor increase in IL-6 in some patients. By day 28 post-infusion, no anti-MDX-240 antibody responses were detected in any of the patients. These results indicate that MDX-240 treatment demonstrates haematological activity, does not mediate enhancement of infection, and is well tolerated. This therapy may be useful in increasing CD4+ T cell levels in HIV infected patients.

Anti-CD64 × anti-CD19 and anti-CD64 × anti-CD37

BsAbs have been constructed from mAb 22 and mAbs to the B cell differentiation antigens CD19 and CD37. The widespread expression of these antigens on B-cell lymphomas and their absence from haemopoietic stem cells makes them especially suitable targets for therapy. CD19, a member of the Ig supergene family, is a B cell-specific phosphoglycoprotein involved in signal transduction [49]. It is present on more than 90% of all B cell tumours [50,51]. CD37 is a pan B cell marker found on mature B cells and a high percentage of tumours of B cell origin [51,52]. *In vitro*, the anti-CD19 × mAb 22 and the anti-CD37 × mAb 22 BsAbs have similar ability to mediate phagocytosis by monocyte-derived macrophages of both Burkitt's lymphoma cell lines and patient tumour cells [35]. Moreover, BsAb-mediated phagocytosis is not inhibited by human IgG, whereas phagocytosis mediated by both parental anti-B cell mAbs (murine IgG_{2a} mAbs) is completely inhibited in the presence of human IgG, indicating that BsAbs constructed from mAb 22 circumvent blocking by human IgG of FcγRI-dependent function *in vivo*. BsAb-mediated phagocytosis is enhanced in the presence of IFN-γ and M-CSF suggesting that using BsAb with haematopoietic growth factors to influence the functional activity of myeloid cells may be an extremely effective strategy for the treatment of minimal residual disease.

Anti-CD64 × anti-CD15

Another BsAb (anti-CD64 × anti-CD15), was prepared by chemically cross-linking Fab fragments of an anti-FcγRI mAb (mAb 32.2) to an IgM anti-CD15 mAb, PM-81. This BsAb was able to lyse $CD15^+$ tumour cells by two different mechanisms, complement- and cell-mediated lysis, even in the presence of human IgG [53].

Clinical Trial. This anti-CD15 × anti-CD64 BsAb was used in a phase I clinical trial involving four patients with $CD15^+$ tumours: one with acute myelogenous leukemia, one with SCCL, one with breast cancer, and one with pancreatic islet cell carcinoma [53]. In patients with solid tumours, it was not possible to monitor the effects of this BsAb on the tumour population, although its activity was indicated by the fact that the circulating neutrophils decreased dramatically after each infusion. In the patient with AML, there was a transient 30–60% reduction in circulating leukaemic blast cells during each of six infusions over a 2-week period, with no symptoms of toxicity. Moreover, a reduction in circulating cell counts were observed when peak serum concentrations were as low as 50 ng/ml.

2B1, a BsAb targeting through FcγRIII (CD16)

Because FcγRIIIa has been shown to mediate cytotoxicity by NK cells, BsAbs have been developed which target through this receptor. BsAb 2B1, which targets HER-2/neu-expressing tumour cells to FcγRIII, has been shown to mediate lysis of tumour cell targets by NK cells [54]. Maximal saturation of both human peripheral blood leukocytes and HER-2/neu-expressing SK-OV-3 (ovarian carcinoma) cells occurs at 1 μg/ml of 2B1, while 1000-fold less antibody is required for significant lysis of targets. Although 2B1 binds to PMNs, it does not mediate cytotoxicity of tumour targets by these cells [55]. *In vivo* studies using SK-OV-3 xenografted mice demonstrated a significantly longer survival time in mice treated with 2B1, IL-2, and human peripheral blood lymphocytes or lymphokine-activated killer cells, than in untreated mice. In fact, 70% of treated mice appeared to be tumour-free at day 150 [56].

Clinical Trial. In a Phase I clinical trial, 15 patients with HER-2/neu over-expressing tumours were treated with either 1.0, 2.5, or 5.0 mg/m^2 i.v. injections of 2B1 on days 1, 4, 5, 6, 7 and 8. Non-dose-limiting toxicities included fever, rigors, nausea, vomiting, and leukopenia. Thrombocytopenia was dose-limiting in two patients treated at the 5.0 mg/m^2 dose level who had received extensive myelosuppressive chaemotherapy prior to the study. Murine BsAb, which was detected in serum following 2B1 administration, maintained bispecific binding characteristics and bound to all neutrophils and to a proportion of monocytes and lymphocytes. The pharmacokinetics were variable with an average $t_{\frac{1}{2}}$ of 20 h, as described by nonlinear kinetics. Following the initial 2B1 treatment, more than 100-fold increases in circulating levels of TNF-α, IL-6, and IL-8 and lesser rises in GM-CSF and IFN-γ were observed. Human anti-mouse antibody responses were observed in 14 of the 15 patients. Several minor clinical responses were noted, including resolutions of pleural effusions, ascites, and a liver metastasis, respectively, in three patients with metastatic colon cancer, and a reduction in thickness of chest wall disease in one breast cancer patient. Phase II trials are in progress at a daily dose of 2.5 mg/m^2, the maximum tolerated dose (MTD) for patients with extensive prior myelosuppressive chemotherapy. Dose escalation studies will be performed to identify the MTD for patients who have been less heavily pretreated [57].

THE FUTURE OF BsAb TECHNOLOGY TARGETING FcγR

Because FcγR naturally mediate cytolytic functions, they serve as particularly effective targets for BsAb therapy. By bridging tumour cells with effector cells, BsAbs targeting FcγRI are able to mediate killing of tumour cells, even when the ligand binding site is occupied with irrelevant IgG. Encouraging clinical results are beginning to accumulate from trials using BsAbs to target leukocytes. In HIV-1 infected patients, BsAb therapy can increase the number of $CD4^+$ T cells. In patients with solid tumours, BsAbs can reduce tumour burden and have vaccine-like effects. Many solid tumours have a significant macrophage component which *in vitro* studies have shown to be non-tumouricidal. BsAbs seem able to permeate the tumour, and should activate and trigger these resident macrophages. Moreover, the production of cytokines should recruit additional leukocytes. Future trials with BsAbs will focus on minimal residual disease where the impact of BsAbs should be most beneficial. Other trials will combine BsAbs with cytokine therapy to further enhance the immune response.

BsAb technology continues to evolve with the development of single chain antibodies (scFvs). These genetic constructs allow for the production of bispecific fusion molecules, proteins which can be produced quickly and purified more easily than traditional BsAbs. Such fusion proteins are smaller, bi-functional molecules capable of bridging effector cells with target cells, while reducing the potential for human anti-BsAb antibodies. In addition, scFv–ligand fusion molecules are being developed to target tumour cells over-expressing receptors such as EGF-R. An scFv specific to FcγRI is also being used in the production of targeted peptide vaccines, protein vaccines, and DNA vaccines [36,58]. These vaccine approaches take advantage of the expression of FcγRI exclusively on antigen-presenting cells (APC). Because APC are potent T cell stimulators, targeting tumour antigens, antigenic peptides, or DNA expression vectors encoding tumour antigens specifically to APC should lead to anti-tumour T cell responses (see Chapter 17).

References

1. Fanger MW, Shen L, Graziano RF, Guyre PM. Cytotoxicity mediated by human Fc receptors for IgG. Immunol Today. 1989;10:92–9.
2. Wallace PK, Howell AL, Fanger MW. Role of Fc gamma receptors in cancer and infectious disease. J Leukoc Biol. 1994;55:816–26.
3. van de Winkel JGJ, Anderson CL. Biology of human immunoglobulin G Fc receptors. J Leukoc Biol. 1991;49:511–24.
4. van de Winkel JGJ, Capel PJ. Human IgG Fc receptor heterogeneity: molecular aspects and clinical implications. Immunol Today. 1993;14:215–21.
5. Looney RJ, Abraham GN, Anderson CL. Human monocytes and U937 cells bear two distinct Fc receptors for IgG. J Immunol. 1986;136:1641–7.
6. van de Winkel JGJ, Boonen GJ, Janssen PL, Vlug A, Hogg NT, Tax WJ. Activity of two types of Fc receptors, Fc gamma RI, and Fc gamma RII, in human monocyte cytotoxicity to sensitized erythrocytes. Scand J Immunol. 1989;29:23–31.
7. Shen L, Graziano RF, Fanger MW. The functional properties of Fc gamma RI, II and III on human myeloid cells: a comparative study of killing of erythrocytes and tumour cells mediated through the different Fc receptors. Mol Immunol. 1989;26:959–69.

8. Bussel JB, Szatrowski TP. Uses of intravenous gammaglobulin in immune hematologic disease. Immunol Invest. 1995;24:451–6.
9. Newland AC, Macey MG. Immune thrombocytopenia and Fc receptor-mediated phagocyte function. Ann Hematol. 1994;69:61–7.
10. Bussel JB, Graziano JN, Kimberly RP, Pahwa S, Aledort LM. Intravenous anti-D treatment of immune thrombocytopenic purpura: analysis of efficacy, toxicity, and mechanism of effect. Blood. 1991;77:1884–93.
11. Bussel JB. Modulation of Fc receptor clearance and antiplatelet antibodies as a consequence of intravenous immune globulin infusion in patients with immune thrombocytopenic purpura. J Allergy Clin Immunol. 1989;84:566–78.
12. Bussel JB, Kimberly RP, Inman RD, et al. Intravenous gammaglobulin treatment of chronic idiopathic thrombocytopenic purpura. Blood 1983;62:480–6.
13. Salama A, Kiefel V, Mueller-Eckhardt C. Effect of IgG anti-Rho(D) in adult patients with chronic autoimmune thrombocytopenia. Am J Hematol 1986;22:241–50.
14. Debré M, Bonnet MC, Fridman WH et al. Infusion of Fc gamma fragments for treatment of children with acute immune thrombocytopenic purpura. Lancet. 1993;342:945–9.
15. Ericson SG, Coleman KD, Wardwell K et al. Monoclonal antibody 197 (anti-FcγRI) infusion in a patient with immune thrombocytopenia purpura (ITP) results in down-modulation of FcγRI on circulating monocytes. Br J Hematol. 1996;92:718–24.
16. Pfefferkorn LC, Fanger MW. Transient activation of the NADPH oxidase through Fc gamma RI. Oxidase deactivation precedes internalization of cross-linked receptors. J Immunol. 1989;143:2640–9.
17. Guyre PM, Graziano RF, Vance BA, Morganelli PM, Fanger MW. Monoclonal antibodies that bind to distinct epitopes on Fc gamma RI are able to trigger receptor function. J Immunol. 1989;143:1650–5.
18. Graziano RF, Tempest PR, White P et al. Construction and characterization of a humanized anti-gamma-immunoglobulin receptor type I (Fc gamma RI) monoclonal antibody. J Immunol. 1995;155:4996–5002.
19. Wallace PK, Keler T, Coleman K et al. Humanized mAb H22 binds the human high affinity Fc receptor for IgG (FcγRI), blocks phagocytosis and modulates receptor expression. J Leukoc Biol. 1997;62:469–79.
20. Graziano RF, Somasundaram C, Goldstein J. The production of bispecific antibodies. In: Fanger MW, ed., Bispecific Antibodies. Heidelberg: R.G. Landes Co; 1995:1–26.
21. Maguire HCJ, Greene MI. The neu (c-erbB-2) oncogene. Semin Oncol. 1989;16:148–55.
22. Slamon DJ, Godolphin W, Jones LA, et al. Studies of the HER-2/neu proto-oncogene in human breast and ovarian cancer. Science (Washington DC). 1989;244:707–12.
23. Levine MN, Andrulis I. The HER-2/neu oncogene in breast cancer: so what is new? J Clin Oncol. 1992;10:1034–6.
24. Kaufman PA, Guyre PM, Lewis LD,et al. Her-2/*neu* targeted immunotherapy: a pilot study of multi-dose MDX-210 in patients with breast or ovarian cancers that overexpress HER-2/*neu* and a report of an increased incidence of HER-2/*neu* overexpression in metastatic breast cancer. Tumor Targeting. 1996;2:17–28.
25. Ring DB, Clark R, Saxena A. Identity of BCA200 and c-erbB-2 indicated by reactivity of monoclonal antibodies with recombinant c-erbB-2. Mol Immunol. 1991;28:915–17.
26. Keler T, Graziano RF, Mandal A et al. Bispecific antibody-dependent cellular cytotoxicity of HER2/neu-overexpressing tumor cells by Fcγ receptor type I-expressing effector cells. Cancer Res. 1997;57:4008–14.
27. Stockmeyer B, Valerius T, Repp R et al. Preclinical studies with FcγR bispecific antibodies and granulocyte colony-stimulating factor-primed neutrophils as effector cells against HER-2/neu overexpressing breast cancer. Cancer Res. 1997;57:696–701.
28. Repp R, Valerius T, Sendler A et al. Neutrophils express the high affinity receptor for IgG (Fc gamma RI, CD64) after in vivo application of recombinant human granulocyte colony-stimulating factor. Blood. 1991;78:885–9.
29. Valone FH, Kaufman PA, Guyre PM et al. Phase Ia/Ib trial of bispecific antibody MDX210 (anti-HER-2/*neu* x anti-Fc gamma RI) in patients with advanced breast or ovarian cancer that over-expresses the proto-oncogene HER-2/*neu*. J Clin Oncol. 1995;13:2281–92.
30. Gralow J, Weiner LM, Ring D et al. Her-2/*neu* specific immunity can be induced by therapy with 2B1, a bispecific monoclonal antibody binding to Her-2/*neu* and CD16. Proc Am Soc Clin Oncol. 1997;14:1807A.

31. Valerius T, Repp R, Wieland G et al. Bispecific antibody MDX210 (FcgammaRI × HER-2/NEU) in combination with G-CSF: results of a Phase I trial in patients with metastatic breast cancer (Meeting abstract). Proc Am Soc Clin Oncol. 1996;15:A97.
32. Weber JS, Spears LA, Marty VS, Deo Y, Link JL. Biologic effects of a HER2/neu bispecific antibody with G-CSF in patients with metastatic breast cancer (Meeting abstract). Proc Am Assoc Cancer Res. 1996;37:A1162.
33. King CR, Kraus MH, DiFiore PP, Paik S, Kasprzyk PG. Implications of erbB-2 over-expression for basic science and clinical medicine. Semin Cancer Biol. 1990;1:329–37.
34. Westphal M, Hamel W, Zirkel D et al. Epidermal growth factor receptor expression in human malignant glioma: in vitro and in vivo effects of application of monoclonal antibodies to the epidermal growth factor receptor. In: Weistler O, Schlegel U, Schramm J, eds, Molecular Neuro-oncology and its Impact in the Clinical Management of Brain Tumors. Berlin: Springer-Verlag; 1994:171–84.
35. Ely P, Wallace PK, Givan AL, Graziano RF, Guyre PM, Fanger MW. Bispecific-armed, IFNγ-primed macrophage-mediated phagocytosis of malignant non-Hodgkin's lymphoma. Blood. 1996;87:3813–21.
36. Goldstein J, Graziano RF, Sundarapandiyan K, Somasundaram C, Deo YM. Cytolytic and cytostatic properties of an anti-human Fc gammaRI (CD64) × epidermal growth factor bispecific fusion protein. J Immunol. 1997;158:872–9.
37. Thali M, Moore JP, Furman C et al. Characterization of conserved human immunodeficiency virus type 1 gp120 neutralization epitopes exposed upon gp120-CD4 binding. J Virol. 1993;67:3978–88.
38. Haynes BF. Immune response to HIV infection. In: DeVita VT Jr., Hellman S, Rosenberg SA, eds, AIDS: Etiology, Diagnosis, Treatment, and Prevention, 3rd edn. Philadelphia: J.B. Lippincott Co; 1992:77–86.
39. Takeda A, Tuazon CU, Ennis FA. Antibody-enhanced infection by HIV-1 via Fc receptor-mediated entry. Science (Washington DC). 1988;242:580–3.
40. Homsy J, Meyer M, Tateno M, Clarkson SB, Levy JA. The Fc and not CD4 receptor mediates antibody enhancement of HIV infection in human cells. Science (Washington DC). 1989;244:1357–60.
41. Jouault T, Chapuis F, Olivier R, Parrvicini C, Bahraoui E, Gluckman JC. HIV infection of monocytic cells: role of antibody-mediated virus binding to Fc-gamma receptors. AIDS. 1989;3:125–33.
42. Zeira M, Byrn RA, Groopman JE. Inhibition of serum-enhanced HIV-1 infection of U937 monocytoid cells by recombinant soluble CD4 and anti-CD4 monoclonal antibody. AIDS Res Hum Retroviruses. 1990;6:629–39.
43. Perno CF, Baseler MW, Broder S, Yarchoan R. Infection of monocytes by human immunodeficiency virus type 1, blocked by inhibitors of CD4-gp120 binding, even in the presence of enhancing antibodies. J Exp Med. 1990;171:1043–56.
44. Connor RI, Dinces NB, Howell AL, Romet-Lemonne JL, Pasquali JL, Fanger MW. Fc receptors for IgG (Fc gamma R) on human monocytes and macrophages are not infectivity receptors for human immunodeficiency virus type 1 (HIV-1): studies using bispecific antibodies to target HIV-1 to various myeloid cell surface molecules, including the Fc gamma R. Proc Natl Acad Sci USA. 1991;88:9593–7.
45. Bolognesi DP. AIDS. Do antibodies enhance the infection of cells by HIV? Nature (London). 1989;340:431.
46. Mabondzo A, Aussage P, Bartholeyns J et al. Bispecific antibody targeting of human immunodeficiency virus type 1 (HIV-1) glycoprotein 41 to human macrophages through the Fc IgG receptor I mediates neutralizing effects in HIV-1 infection. J Infect Dis. 1992;166:93–9.
47. Howell AL, Guyre PM, You K-S, Fanger MW. Targeting HIV-1 to Fc gamma R on human phagocytes via bispecific antibodies reduces infectivity of HIV-1 to T cells. J Leukoc Biol. 1994;55:385–91.
48. Pasquali JL, Vandercam B, Romet-Lemonne JL, Dormont D, Deo YM. Phase I clinical study of an anti-FcγRI × anti-gp41 bispecific antibody (MDX-240) in HIV infected patients. (Meeting abstract). Frontiers of HIV Therapy, Palm Springs, CA; 1996.
49. Kozmik Z, Wang S, Dorfler P, Adams B, Busslinger M. The promoter of the CD19 gene is a target for the B-cell- specific transcription factor BSAP. Mol Cell Biol. 1992;12:2662–72.

50. Campana D, Janossy G, Bofill M et al. Human B cell development. I. Phenotypic differences of B lymphocytes in the bone marrow and peripheral lymphoid tissue. J Immunol. 1985;134: 1524–30.
51. Dorken B, Moller P, Pezzutto A, Schwartz-Albiez R, Moldenhauer G. B-cell antigens: section report. In: Knapp W, Dorken B, Gilks WR et al, eds. Leukocyte Typing IV: White Cell Differentiation Antigens. Oxford: Oxford University Press; 1989:15–224.
52. Moore K, Cooper SA, Jones DB. Use of the monoclonal antibody WR17, identifying the CD37 gp40-45 Kd antigen complex, in the diagnosis of B-lymphoid malignancy. J Pathol. 1987;152:13–21.
53. Ball ED, Guyre PM, Mills L, Fisher J, Dinces NB, Fanger MW. Initial trial of bispecific antibody-mediated immunotherapy of CD15-bearing tumors: cytotoxicity of human tumor cells using a bispecific antibody comprised of anti-CD15 (MoAb PM81) and anti-CD64/FcγR1 (MoAb 32). J Hematother. 1992;1:85–94.
54. Weiner LM, Holmes M, Richeson A, et al. Binding and cytotoxicity characteristics of the bispecific murine monoclonal antibody 2B1. J Immunol. 1993;151:2877–86.
55. Weiner LM, Alpaugh RK, Amoroso AR, Adams GP, Ring DB, Barth MW. Human neutrophil interactions of a bispecific monoclonal antibody targeting tumor and human Fc gamma RIII. Cancer Immunol Immunother. 1996;42:141–50.
56. Weiner LM, Holmes M, Adams GP, LaCreta F, Watts P, Garcia de Palazzo I. A human tumor xenograft model of therapy with a bispecific monoclonal antibody targeting c-erbB-2 and CD16. Cancer Res. 1993;53:94–100.
57. Weiner LM, Clark JI, Davey M et al. Phase I trial of 2B1, a bispecific monoclonal antibody targeting c-erbB-2 and Fc gamma RIII. Cancer Res. 1995;55:4586–93.
58. Liu C, Goldstein J, Graziano RF et al. F(c)gammaRI-targeted fusion proteins result in efficient presentation by human monocytes of antigenic and antagonist T cell epitopes. J Clin Invest. 1996;98:2001–7.

26
The rewards of conversion to 'FcRism'

W. BOYLE

During the span of what can be termed a scientific generation, the knowledge of Fc receptors has expanded vigorously. For those involved directly in FcR research during that time, the response of other immunologists to their findings must often have seemed less than generous: yet the reasons are not hard to find. This generation of mainstream immunologists was intellectually obsessed with unravelling the specificity of lymphocyte recognition of antigen and lymphocyte-based regulation of immunity. Immunological induction was 'where the action is' and effector molecules and mechanisms were second class concerns. For me, as for many, the incident on the road to Damascus which incited conversion was the demonstration that at least some FcRs displayed the same cytoplasmic-tail ITAMs as the antigen-specific receptor complexes on T and B lymphocytes.

The implications were profound. Did we have to accept that these 'upstart molecules' might do more than capture immune complexes? Did we have to really learn about all those isoforms and polymorphic variants and their differing functions? Could FcRs actually be involved in regulating immunity and, if so, could they add a whole new dimension to the understanding and possible control of a range of immune-based disease states? The pleasures of reading the contributions in this section have been not only been in obtaining exciting new knowledge in a palatable form, but to see so many of these questions being answered simply by presentation of how each is already having an impact on understanding of human disease states.

Are all the isoforms and polymorphic variations and their apparent differences in function of any significance? In most articles in this chapter this form of variation of FcγR and FcεR are mentioned in various contexts, attesting to their importance, but they also form the focus of Chapter 23. They review poly-

J.G.J. van de Winkel and P.M. Hogarth (eds.), The Immunoglobulin Receptors and their Physiological and Pathological Roles in Immunity. 307–309.

morphism in FcγRIIA and FcγRIIIB and relate both to incidence of infection and propensity to immune-mediate disease. Both are good examples of the 'swings and roundabouts' aspect of risk associated with specific alleles. Increased incidence of recurrent respiratory tract infection in children is associated with the FcγRIIA-R^{131} homozygotes while homozygosity for the alternative allele H^{131} confers an apparent thrombotic risk in patients with anti-cardiolipin antibodies. Similarly the NA1 allele of FcγRIIIB may confer increased resistance to bacterial infections but is associated with increased incidence of severe renal disease in those with Wegener's granulomatosis. It appears likely that further investigations of polymorphisim could provide further valuable parameters to assess potential risk and modify prognostic decisions for an increasing range of diseases.

The production of soluble forms of FcγRII and FcγIII either by alternative splicing or enzymatic cleavage introduces almost a new set of players into the field. This is raised by Baker and Dale (Chapter 22) as a potential modulating factor early in HIT and Sautès and colleagues (Chapter 24) extend the documentation of altered sFcγRII and sFcγRIII levels in various human tumours states, in autoimmune diseases as well as in HIV-1 infection. Their cautious appraisal of their role is fully justified by the rest of their chapter which clearly identifies our need to expand our vision of the role of FcRs in view of their demonstrable capacity to interact with non-Ig ligands such as CD21, β_2 integrins and cytokines, notably TGF-β. It seems likely that further analysis of these 'upstart molecules' in their networking with these different biological cascades will provide greater insights to various disease states. Will other knowledge of FcRs increase further our capacity to control specific disease states? The major prerequisite to control of any disease is precise definition of the disease stages and improvement of diagnostic and prognostic evaluation. Baker and Dale provide an excellent example in their detailed analysis of the Ig–Heparin–PT4–FcγIIR interplay and the derivation of alternative tests which may improve the laboratory diagnosis of HIT. FcR variation either in altered levels of cell-bound or soluble forms, of changed functional activity, (independent of or associated with isoform expression) in a range of disease states are extensively reviewed by Repp and van de Winkel (Chapter 21) and raise the thought that the results in HIT may be a paradigm soon to be extended to many others diseases.

While most chapters are written with the underlying perception that this field offers many potential strategies for intervention in human disease, two chapters directly address the issue. Guyre, Wallace and Fanger (Chapter 25) review the use of bispecific antibodies (BSA) which are already in clinical trial in ITP, AIDS or cancer patients. Many of these use BSA with one arm specific for FcγRI and the other arm directed against the HER2/neu epitope commonly expressed at distinctive levels in a range of cancers. The range of trials with or without concurrent cytokine therapy involving these BSA and other combinations is an impressive testimony to the commitments of researchers in this area to rapidly apply their findings to patients. The vitality of this field is further established by comparing the Guyre *et al.* contribution with that from Powell and Hogarth (Chapter 20). Whereas the former article specifically utilizes BSA to recruit specific effector function via FcγRI, Powell and Hogarth recount a series of studies designed to block, compete with or reverse FcR (and CR) involvement in a range of antibody-mediated inflammatory conditions. Their particular brief is to review the extensive use of soluble recombinant proteins (FcγRII, anti-FcγRII, CTLA4, CRI etc.) To shed light on the pathology of a number of disease conditions where damage is perceived to be mediated by Type II or III hypersensitivity mechanisms. Although this range of products with putative application may at first seem to be an embarrassment of riches, this breadth of attack should be maintained since different strategies are likely to be necessary to prevent onset, arrest disease progression or reverse recurrent flare-ups.

Two final points which encourage enthusiasm for the future arise from attempting to link the contribution of Repp and van de Winkel with that from Powell and Hogarth. The former provides the perception that the pathology of many conditions involving FcR binding of the appropriate ligands may derive from both outside-in and inside-out induced signalling. Powell and Hogarth cite, primarily as an answer to putative glycosylation or folding deficiencies of recombinant proteins, that the search for alternative ligands for FcRs will in the future employ combinatorial libraries of peptides or peptoids. Considering how single – amino acid substitutions in peptides have already led to the dissociation of signals for cytokine or proliferation via MHC-peptide and TCR interaction, the future of such manipulations via the FcR signalling offers immense potential. Perhaps past neglect of this area by many can now be forgiven, since the enthusiasm of new converts will provide many in fully-supportive-mode for the future!

27
Fc receptors: historical perspectives and a look to the future

P. M. HOGARTH and J. G. J. van de WINKEL

INTRODUCTION

It is over 20 years since Pareskevas *et al.* identified specific receptors for immunoglobulins on the surface of cells. These observations were to point the way for the generation of an entire field based on the cellular effector mechanisms induced by humoral immune responses and mediated through the Fc portion of immunoglobulin. The nature of specific receptors was to take some time to characterize biochemically and as biochemical technology improved, and reagents developed, the pace of discovery was rapid. The ensuing decade or so saw a wide range of investigators take an interest in these specific receptors and make observations that indicated the importance of Fc receptors in the uptake of immune complexes and perhaps even the regulation of the immune system. Even soluble forms of Fc receptors, also known as 'immunoglobulin-binding factors', were identified. During this time, it was also obvious that a multitude of these receptors existed – at least one receptor for each immunoglobulin class. Over time, subsets of receptors were observed but specific roles difficult to identify with the possible exception of the IgE receptor on mast cells which was to ultimately become the most biochemically studied of all Fc receptors because of its potent pharmacological role in a life-threatening disease process.

As often happens in science, new technologies reveal surprising new properties of the molecules under investigation. The development of monoclonal antibodies, combined with new and sensitive emerging biochemical technologies, indicated that there was indeed a great diversity of Fc receptors. In some cases there was clearly more than one class of receptor for the same

J.G.J. van de Winkel and P.M. Hogarth (eds.), The Immunoglobulin Receptors and their Physiological and Pathological Roles in Immunity. 311–313.

immunoglobulin subclass. Antibodies against the mouse Fc receptor and against human Fc receptors broadened the data base on Fc receptor identification and provided key reagents for the future.

The next revolution was to come with the gene cloning technologies developed in the early to mid-1980s and the pace of change was frenetic. Purification of protein with monoclonal antibodies in the mouse and the subsequent use of this information to clone human receptors, in addition to application of new technologies revolutionized the identification of individual Fc receptor classes and confirmed the diversity within receptor families. No-one could have predicted the multitude of biological effects would be mediated by so many different receptor isoforms. As things move forward we do a great deal to further understand the biological functions of Fc receptors as they participate in adaptive and innate immunity. We also apply FcR technologies to the treatment of disease. Monoclonal antibodies have been used in animal models of disease, and engineered as bispecific antibodies for targeting effector cells to tumours. These are now in the clinic. Soluble Fc receptors are now being viewed as possible antagonists of immune complex-induced tissue damage, and for screening combinatorial libraries for the more traditional receptor antagonists.

The study of Fc receptors has always been 'leukocentric': Fc receptor studies have been dominated by those of us who work on leukocyte receptors. It would be remiss not to consider those Fc receptors which do not appear on leukocytes and have completely different biological roles. Indeed, outside the leukocytes receptors for immunoglobulins are less about cell activation and more about transport. Indeed we now know that FcRn is responsible, at least in part, for maintaining the prodigious half-life of immunoglobulins in the circulation. Such a receptor was postulated by Professor Brambell in the early 1960s and may also play a major role in maternal Ig uptake. Similarly, secretory component long identified as an essential Fc binding protein is now known as the receptor for polymeric immunoglobulins with a crucial role in transepithelial Ig transport. These latter two receptors are clearly major 'workhorses' that subsequently enable the immune system to undertake its functions.

Outside mammals, Fc receptors have been well studied. In fact, it could well be argued that the most useful Fc receptor defined is, of course, bacterial in origin (i.e. protein A). Indeed the evolution of pathogens has included structures purpose built for the avoidance of the host immune system and included amongst these are the bacterial and viral Fc-binding proteins.

THE WAY TO THE FUTURE

Effective therapies are currently developed to harness the biological properties of FcR, e.g. bispecific anti-FcR and anti-cancer antibodies. Intervention in FcR:Ig interactions in inflammation is also a major future task. Beyond the application of Fc receptor technology to disease, other scientific questions remain to be answered. How and under what circumstances can the 'b' form of FcγRII negatively regulate immunity *in vivo*? A great deal of speculation based on some elegant experiments *in vitro* points the way, but the molecular mechanisms *in vivo* are still not understood. The use of homologous recombination to inactivate genes, 'knock-in' mutants and the use of transgenic technologies to establish receptor expression in unique places are all technologies which are now being effectively applied. What of the three-dimensional structure of these receptors? A great deal of speculation and solid, pioneering science has been performed on the nature of the receptor structure but the genuine structure still eludes us.

It is indeed interesting that over 20 years ago, identification of immune complex binding to cells set in train a series of events which has taken us from the observation of an interaction between complexes on the cell surface up to the molecular identification of the structures; understanding how that interaction takes place and the affinity of that interaction; into the cloning of the receptors and on to the nature of signal transduction, and now events in the nucleus. On the biological side of things, the Fc receptor has come from being merely an entity on the surface able to bind immunoglobulin to having a variety of sophisticated roles including the regulation of immunity and induction of potent responses that aid the humoral immune response, to being key players in the inflammatory process in general.

On the pathogenic side of immunity Fc receptors are major players in the induction of inflammation more so than previously recognized. Our future holds key directions. What of the relationship between complement receptors and Fc receptors in inducing inflammation? What is the precise role of these receptors *in vivo*; what of the IgA receptor and its definitive role in mucosal and pathological immunity; the biochemical definition of receptors for IgM which may be key players in immunity; their existence is still an issue.

The future of the non-leukocyte receptors will be an interesting one indeed, although we tend to dwell on the role of leukocyte receptors in autoimmune diseases, what of FcRn and the poly-Ig receptor? They are 'slaves' to the immune system because of their transporting role and in essence will be sources of major and no doubt novel science in the years to come.

For the pathogen-related Fc receptors a great deal is to be done to understand the precise mechanisms by which these assist the survival of the pathogen.

The search for new ligands for Fc receptors will continue and it is noteworthy that a number have already been described including other cell surface molecules, CY-5 a chromophore and measles virus nucleocapsid protein. Whilst the identification of new ligands is interesting in itself, the biological implications of these will be of paramount importance as they broaden the spectrum of activities of FcR and imply a wider role in immune processes than previously considered.

Index

J.G.J. van de Winkel and P.M. Hogarth (eds.), The Immunoglobulin Receptors and their Physiological and Pathological Roles in Immunity. 315–323.

Immunology and Medicine Series

1. A.M. McGregor (ed.). *Immunology of Endocrine Diseases.* 1986 ISBN: 0-85200-963-1
2. L. Ivanyi (ed.). *Immunological Aspects of Oral Diseases.* 1986 ISBN: 0-85200-961-5
3. M.A.H. French (ed.). *Immunoglobulins in Health and Disease.* 1986 ISBN: 0-85200-962-3
4. K. Whaley (ed.). *Complement in Health and Disease.* 1987 ISBN: 0-85200-954-2
5. G.R.D. Catto (ed.). *Clinical Transplantation: Current Practice and Future Prospects.* 1987 ISBN: 0-85200-960-7
6. V.S. Byers and R.W. Baldwin (eds.). *Immunology of Malignant Diseases.* 1987 ISBN: 0-85200-964-X
7. S.T. Holgate (ed.). *Mast Cells, Mediators and Disease.* 1988 ISBN: 0-85200-968-2
8. D.J.M. Wright (ed.). *Immunology of Sexually Transmitted Diseases.* 1988 ISBN: 0-74620-087-0
9. A.D.B. Webster (ed.). *Immunodeficiency and Disease.* 1988 ISBN: 0-85200-688-8
10. C. Stern (ed.). *Immunology of Pregnancy and its Disorders.* 1989 ISBN: 0-7462-0065-X
11. M.S. Klempner, B. Styrt and J. Ho (eds.). *Phagocytes and Disease.* 1989 ISBN: 0-85200-842-2
12. A.J. Zuckerman (ed.). *Recent Developments in Prophylactic Immunization.* 1989 ISBN: 0-7923-8910-7
13. S. Lightman (ed.). *Immunology of Eye Disease.* 1989 ISBN: 0-7923-8908-5
14. T.J. Hamblin (ed.). *Immunotherapy of Disease.* 1990 ISBN: 0-7462-0045-5
15. D.B. Jones and D.H. Wright (eds.). *Lymphoproliferative Diseases.* 1990 ISBN: 0-85200-965-8
16. C.D. Pusey (ed.). *Immunology of Renal Diseases.* 1991 ISBN: 0-7923-8964-6
17. A.G. Bird (ed.). *Immunology of HIV Infection.* 1991 ISBN: 0-7923-8962-X
18. J.T. Whicher and S.W. Evans (eds.). *Biochemistry of Inflammation.* 1992 ISBN: 0-7923-8985-9
19. T.T. MacDonald (ed.). *Immunology of Gastrointestinal Diseases.* 1992 ISBN: 0-7923-8961-1
20. K. Whaley, M. Loos and J.M. Weiler (eds.). *Complement in Health and Disease, 2nd Edn.* 1993 ISBN: 0-7923-8823-2
21. H.C. Thomas and J. Waters (eds.). *Immunology of Liver Disease.* 1994 ISBN: 0-7923-8975-1
22. G.S. Panayi (ed.). *Immunology of Connective Tissue Diseases.* 1994 ISBN: 0-7923-8988-3
23. G. Scadding (ed.). *Immunology of ENT Disorders.* 1994 ISBN: 0-7923-8914-X
24. R. Hohlfeld (ed.). *Immunology of Neuromuscular Disease.* 1994 ISBN: 0-7923-8844-5
25. J.G.P. Sissons, L.K. Borysiewicz and J. Cohen (eds.). *Immunology of Infection.* 1994 ISBN: 0-7923-8968-9
26. J.G.J. van de Winkel and P.M. Hogarth (eds.). *The Immunoglobulin Receptors and their Physiological and Pathological Roles in Immunity.* 1998 ISBN: 0-7923-5021-9
27. A.P. Weetman (ed.). *Endocrine Autoimmunity and Associated Conditions.* 1998 ISBN: 0-7923-5042-1

Kluwer Academic Publishers – Dordrecht / Boston / London

FSC
www.fsc.org
MIX
Papier aus verantwortungsvollen Quellen
Paper from responsible sources
FSC® C105338

If you have any concerns about our products,
you can contact us on
ProductSafety@springernature.com

In case Publisher is established outside the EU,
the EU authorized representative is:
Springer Nature Customer Service Center GmbH
Europaplatz 3, 69115 Heidelberg, Germany

Printed by Libri Plureos GmbH
in Hamburg, Germany